Spectrophotometry and Spectrofluorimetry

A Practical Approach

Edited by

Michael. G. Gore

Division of Biochemistry and Molecular Biology,
School of Biological Sciences, University of
Southampton.

OXFORD
UNIVERSITY PRESS

OXFORD
UNIVERSITY PRESS

Great Clarendon Street, Oxford OX2 6DP

Oxford University Press is a department of the University of Oxford.
It furthers the University's objective of excellence in research,
scholarship, and education by publishing worldwide in

Oxford New York

Athens Auckland Bangkok Bogotá Buenos Aires Calcutta Cape Town
Chennai Dar es Salaam Delhi Florence Hong Kong Istanbul Karachi
Kuala Lumpur Madrid Melbourne Mexico City Mumbai Nairobi Paris
São Paulo Singapore Taipei Tokyo Toronto Warsaw

with associated companies in Berlin Ibadan

Oxford is a registered trade mark of Oxford University Press in the UK
and in certain other countries

Published in the United States by Oxford University Press Inc., New York

A catalogue record for this book is available from the British Library

Library of Congress Cataloguing in Publication Data

Spectrophotometry and spectrofluorimetry : a practical approach /
edited by Michael G. Gore.
    (The practical approach series ; 225)
    1. Spectrophotometry. 2. Fluorimetry. 3. Biochemistry–Technique.
  I. Gore, Michael G. II. Series.
    QP519.9.S58 S643 2000   572'.36–dc21   99-057576

ISBN 0 19 963813 6 (Hbk)
ISBN 0 19 963812 8 (Pbk.)

1 3 5 7 9 10 8 6 4 2

Typeset in Swift by Footnote Graphics, Warminster, Wilts
Printed in Great Britain on acid-free paper
by The Bath Press, Avon

Preface

Optical spectroscopy underpins the day to day operations of most laboratories in the chemical, biological and medical sciences, and this book is the second edition of a text, first published in 1987, intended to help the reader understand the background concepts to spectrophotometry and spectrofluorimetry. The intervening years have seen a steady development in the design and quality of laboratory instrumentation in contrast to the dramatic increase in the use of microprocessors and computers to assist data handling and presentation. It is therefore only too easy for an inexperienced researcher to obtain processed data without perhaps appreciating the limitations of the instrumentation or technique, or the assumptions made in the final data output. An understanding of the underlying physical principles greatly assists its successful use and helps avoid experiments which are too demanding for the technique or instrumentation available.

This edition is intended to build upon and extend concepts established in the first edition of the series. It therefore contains chapters addressing the principles of spectrophotometry, spectrophotometric assays, spectrofluorimetry, time resolved fluorescence and phosphorescence studies, circular dichroism and pre-equilibrium spectroscopic techniques. In all of the chapters the emphasis is placed upon practical aspects, with protocols to guide readers through test experiments, with sufficient theory and references to provide interested readers with a starting point for more in-depth studies. Other chapters are included to introduce subjects that have traditionally depended upon spectroscopy, such as basic enzyme kinetics, ligand binding, data handling and the more recently established interest in the study of protein and DNA stability. Finally, the concept of 'global analysis' is introduced to provide the reader with an insight to this method of utilizing the vast arrays of experimental data provided by current methodologies.

It is assumed throughout, that readers will have available to them commercially available instrumentation. In these chapters, authors have sometimes described named instruments used in their studies and laboratories. Readers should note that these are only given as examples, not as recommendations, comparable instruments are available from other manufacturers.

Finally, I would like thank the authors for their participation and contributions to the text.

It was with deep regret that I learned of the demise of Professor Jorge E. Churchich shortly before the completion of this book. He was a dedicated scientist who inspired all who worked with him.

Southampton   M.G.G.
1999

Contents

CONTENTS

8 Stopped-flow spectroscopy *209*

M. T. Wilson and J. Torres

CONTENTS

xiv

Protocol list

Abbreviations

A	absorbance, absorption
c, C	concentration
D	attenuance (A)
DSC	differential scanning calorimetry
ε	molar absorbance coefficient
εATP	1, N^6–ethenoadenosine diphosphate
FAD(H)	Flavin adenine dinucleotide (reduced form)
Gdn-HCl	guanidine hydrochloride
HEPES	N-2-hydroxyethylpiperazine-N'-2-ethanesulphonic acid
I	intensity of transmitted light
I_o	intensity of incident light
ITC	isothermal titration calorimetry
l	pathlength
MES	2-(N-morpholino)ethanesulphonic acid
MOPS	3-(N-morpholino)propanesulphonic acid
NAD(H)	Nicotinamide adenine dinucleotide (reduced form)
OD	optical density
PIPES	piperazine-N,N'-bis(2-ethanesulphonic acid)
PMT	Photomultiplier tube
Φ_F	Quantum yield of fluorescence
s	second
T	transmittance, transmission or temperature
TNP-ATP	2'(3')-o-(2,4,6-trinitrophenyl)adenosine 5'-triphosphate
Tris	tris(hydroxymethyl)aminomethane
TRP	Tryptophan
TYR	Tyrosine
UV	ultraviolet

Chapter 1

Introduction to light absorption: visible and ultraviolet spectra

Robert K. Poole* and Uldis Kalnenieks[†]

* Krebs Institute for Biomolecular Research, Dept. of Molecular Biology and Biotechnology, The University of Sheffield, Firth Court, Western Bank, Sheffield S10 2TN

[†] Institute of Microbiology and Biotechnology, University of Latvia, Kronvalda Boulevard 4, Riga, LV-1586, Latvia

1 Introduction

1.1 Radiation and light

Light is a form of electromagnetic radiation, usually a mixture of waves having different wavelengths. The wavelength of light, expressed by the symbol λ, is defined as the distance between two crests (or troughs) of a wave, measured in the direction of its progression. The unit used is the nanometre (nm, 10^{-9} m). Light that the human eye can sense is called visible light. Each colour that we perceive corresponds to a certain wavelength band in the 400–700 nm region (*Figure 1*). Spectrophotometry in its biochemical applications is generally concerned with the ultraviolet (UV, 185–400 nm), visible (400–700 nm) and infrared (700–15 000 nm) regions of the electromagnetic radiation spectrum, the former two being most common in laboratory practice.

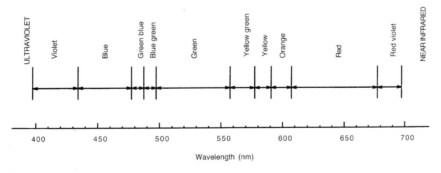

Figure 1 The visible spectrum. Colours shown are of the light beam at the wavelengths on the scale so the colour the eye perceives of a reflectant object in 'white' light is that due to wavelengths that are not absorbed. Thus, the 'green oxidase' (cytochrome *d*) looks green but absorbs light in the blue and red regions of the spectrum.

1.2 UV and visible spectra

The wavelength of light is inversely related to its energy (E), according to the equation:

$$E = ch/\lambda \qquad\qquad 1$$

where c denotes the speed of light, and h is Planck's constant. UV radiation, therefore, has greater energy than the visible, and visible radiation has greater energy than the infrared. Light of certain wavelengths can be selectively absorbed by a substance according to its molecular structure. Absorption of light energy occurs when the incident photon carries energy equal to the difference in energy between two allowed states of the valency electrons, the photon promoting the transition of an electron from the lower to the higher energy state. Thus biochemical spectrophotometry may be referred to as electronic absorption spectroscopy. The excited electrons afterwards lose energy by the process of heat radiation, and return to the initial ground state. An absorption spectrum is obtained by successively changing the wavelength of mono-chromatic light falling on the substance, and recording the change of light absorption. Spectra are presented by plotting the wavelengths (generally nm or μm) on the abscissa and the degree of absorption (transmittance or absorbance) on the ordinate. For more information on the theory of light absorption, see Brown (1) and Chapters 2, 3 and 4.

2 Spectrophotometry

2.1 The Beer–Lambert law

The most widespread use of UV and visible spectroscopy in biochemistry is in the quantitative determination of absorbing species (chromophores), known as spectrophotometry. All spectrophotometric methods that measure absorption, including various enzyme assays, detection of proteins, nucleic acids and dif-ferent metabolites, reside upon two basic rules, which combined are known as the Beer–Lambert law. Lambert's law states that *the fraction of light absorbed by a transparent medium is independent of the incident light intensity, and each successive layer of the medium absorbs an equal fraction of the light passing through it.* This leads to an exponential decay of the light intensity along the light path in the sample, which can be expressed mathematically, as follows:

$$\log_{10} (I_0/I) = kl \qquad\qquad 2$$

where I_0 is the intensity of the incident light, I is the intensity of transmitted light, l is the length of the light-path in the spectrophotometer cuvette, and k is a constant for the medium, which is deciphered by Beer's law. Beer's law claims that *the amount of light absorbed is proportional to the number of molecules of the chromo-phore through which the light passes.* In other words, the constant k is proportional to the concentration (c) of the chromophore: $k = \varepsilon c$, where ε is the molar absorption coefficient, a property of the chromophore itself. ε is numerically equal to the absorption of a molar (1 mol litre^{-1}) solution in a 1 cm light-path. The units are litre mol^{-1} cm^{-1} or M^{-1} cm^{-1}, and *not* cm^2 mol^{-1}.

Hence, the expression of the combined Beer–Lambert law is:

$$\log_{10}(I_o/I) = \varepsilon cl \qquad\qquad 3$$

The term $\log_{10}(I_o/I)$ is called the absorbance (*A*), synonymous with attenuance (*D*). Extinction (*E*) has been used in the past but is now discouraged. Pathlength (usually in cm) and wavelength (in nm, without units) are sometimes given as subscripts, e.g. $A_{1\ cm,\ 550}$. Note that, in difference spectroscopy, a difference molar absorption coefficient (*ε*) may be defined (2) by the equation:

Δ=difference

$$\Delta A = \Delta \varepsilon cl \qquad\qquad 4$$

where ΔA is the difference in absorption for the chromophore in the two environments.

Sometimes the passage of light through the cuvette is described in terms of transmittance, or transmission (*T*): $T = I/I_o$ and is generally expressed as a percentage. It is important to note, however, that only absorbance, not transmittance, is linearly proportional to the chromophore concentration. In quantitative analysis, where it is required to obtain the concentration of substance, therefore, absorbance is more commonly used. The relation between these two parameters is given by the following:

$$A = \log_{10}(1/T) \qquad\qquad 5$$

Thus, when $A = 2$ only 1% of the incident light is transmitted, while at $A = 3$, only 0.1% is transmitted.

2.2 Deviations from the Beer–Lambert law

According to the Beer–Lambert law, absorbance should be linearly proportional to the concentration of the chromophore. However, there are systems that apparently do not obey this rule. The examples that follow are amplified in (2).

2.2.1 High absorbance

limitation

At high absorbances, deviation from linearity is caused by stray light. This is defined as light received at the detector that is not anticipated in the spectral band isolated at the monochromator. Stray light is usually 'white', i.e. having the same composition as the source or ambient illumination, and arises from (i) the monochromator and/or (ii) light leaks in the sample and detector region. The first ought to be eliminated by good monochromator design and filters, the second by good design and construction or masking. When the selected wavelength is absorbed by the sample, the proportional contribution of the stray light to the transmitted intensity increases because it is of a wavelength that is not absorbed. For practical purposes, it is necessary to keep light of the chosen wavelength (I_c) about 10 times greater than the stray light (I_s). It is prudent to check the stray light characteristics of any spectrophotometer by measuring A for a range of standards. Severe deviations from linearity may occur at A values as low as 2 or as high as 6. A more detailed treatment is given in (3).

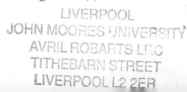

2.2.2 Concentration variations

Apparent deviations from the Beer–Lambert law arise when concentration variation causes changes in the distribution of several chromophore species in the solution. In these cases, if the absorption contribution from each species is considered, the Beer–Lambert law is obeyed by each species, but it is the variation in species concentrations which produces the apparent deviation. For example, dimerization of the chromophore molecules at higher concentrations may take place, providing a dimer with an ε value other than that of the monomeric form, causing a non-linear dependence of absorbance upon the solute concentration.

2.2.3 Coupled reactions

In colorimetric assays, where the compound under study has first to react with a colour reagent in order to be determined, non-linearity at higher concentrations can arise when insufficient colour reagent is present. Likewise, non-linearity at low concentrations may be caused if insufficient time is allowed for completion of the colour reaction, since the rate of reaction at low concentrations is lowest.

2.3 Absorbance or light scattering?

Spectrophotometers are designed for measurement of light *absorption*, but in laboratory practice it is common for the sample under investigation to scatter light as well as absorb it. As we shall see (Sections 4.2 to 4.4), instruments have been designed to allow maximum sensitivity of absorbance measurements even in the face of a very high degree of light scattering. It is quite possible to measure absorbance changes of $0.005A$ in a suspension of cells that has an apparent absorbance of 3 (see later, *Figure 15*). However, in some applications, it is scattering that we wish to measure, not absorbance. One of the most common of such applications occurs in microbiology where turbidity is routinely used as a measure of biomass concentration.

It has been argued before (4) that measurements on bacterial cultures, for example, where scattering is considerable should not be referred to as 'absorbance' or 'attenuance' but as 'apparent absorbance' or 'optical density'. The spectrophotometer output (chart recorder, analogue or digital readout, or computer monitor) may say 'absorbance' but in fact most of the signal will be due to light scattering. When describing measurements of 'apparent absorbance' or 'optical density', the pathlength (or type of tube etc.) and wavelength (in nm) of the incident light should be given (e.g. OD_{600}) and, very importantly, the type of spectrophotometer (maker and model) should be cited (because optical design greatly influences readings; see below) and the extent of any dilution used should be cited. It is prudent to prepare a calibration curve using a range of cell concentrations and determine the 'apparent absorbance' or 'optical density' at which the response of the spectrophotometer becomes non-linear, which may be as low as 0.6–0.7. The use of instruments specifically designed to measure turbid samples, such as nephelometers or Klett meters, should be reported in appropriate units. It is sometimes said that Klett units are meaningless (because they cannot be defined as A can) but, in fact, they are more useful: the variation between Klett meters is likely to be much smaller than the variation obtained when measuring a turbid sample in different 'biochemical' spectrophotometers.

Arthur Koch has written eloquently (5, 6) of the factors that influence light scattering by particles in suspension and of our ability to detect sensitively such light scattering. In the most practically oriented of these papers (6), Koch compares experimentally various laboratory spectrophotometers and wavelengths for their suitability as 'turbidimeters'; variation is due largely to what portion of the scattered light is viewed by the detector. Most of the light scattered by a microbial suspension deviates by only a few degrees (7°) from the incident beam (*Figure 2*). Thus, using an instrument in which the detector is placed as far as possible from the cuvette will greatly enhance sensitivity, since a low concentration of scattering particles will be sufficient to scatter some of the light from the incident beam. It can be very instructive to measure optical density of a single sample (perhaps fixed with formaldehyde to prevent further growth and

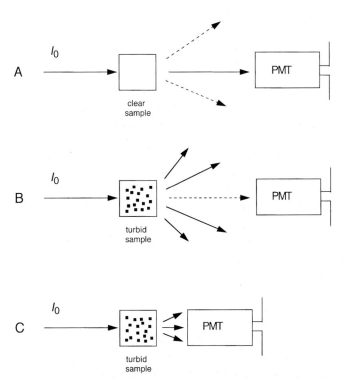

Figure 2 Properties of ideal turbidimeter and spectrophotometers. For a clear sample, the position of the detector (PMT, photomultiplier tube) is unimportant (A). If a turbid sample is examined in a spectrophotometer where the PMT is far removed from the cuvette (B), the instrument will behave as a sensitive turbidimeter because it will detect little of the scattered light. If a turbid sample is examined in a spectrophotometer where the PMT is close to the cuvette (C), the instrument will detect absorbance even though much of the light is scattered.

changes in cell size) in as many spectrophotometers as possible, as Koch has done for some now-obsolete instruments. As a rule, choosing a lower wavelength will offer greater sensitivity of the optical density measurement, but a longer wavelength will give greater linearity over the same range of cell suspensions. Although an ideal turbidimeter would receive none of the scattered light (*Figure 2*), many biochemical spectrophotometers are designed to view as much of this as possible generally by locating the detector as close to the sample as possible. Beware of using spectrophotometers that offer a choice of cuvette positions in terms of distance from the detector; frighteningly large discrepancies between OD readings at such positions are likely.

3 Spectra of some important naturally occurring chromophores

A very few examples of biochemical applications follow, betraying our personal interests in respiratory metabolism.

3.1 Amino acids and proteins

The spectra of amino acids are determined by the nature of their side chains. None of the amino acids has an absorption extending into the visible region, and, in the absence of additional chromophores (haem groups, Cu ions, flavin, etc.), proteins are therefore colourless. Amino acids with aromatic side chains, in particular tryptophan and tyrosine, account for most of the absorption seen in proteins below 300 nm. The spectral properties of amino acids absorbing in the UV region are given in Section 6.3 of Chapter 2. Due to the contribution of tyrosine and tryptophan side chains, protein absorption is maximal at about 280 nm, and can be used for approximate estimation of protein concentration in purified samples or those not containing other species which absorb in the same wavelength region. Absorption at 280 nm is a convenient protein assay in following the elution of a sample from a chromatographic column, and the A_{260}/A_{280} ratio is a common method of assessing samples of nucleic acid for freedom from protein (7). Note that the tyrosine absorption spectrum is pH-dependent, due to its hydroxyl group, which dissociates at high–medium pH.

3.2 Nucleic acids

The spectra of a free base and the corresponding nucleoside are quite similar, since only the purine or pyrimidine base component is a chromophore, giving rise to absorption in the region between about 250 and 270 nm at neutral pH. Because of protonation of the base nitrogen atoms, spectra of these compounds are also highly pH-dependent. Nucleic acids have an absorption maximum close to 260 nm. The variation in absorption coefficient of DNA and RNA per nucleotide residue is small; hence the absorption at 260 nm can be used as a measure of total nucleic acid concentration (7). However, the absorption of a native nucleic acid molecule cannot be expressed as a sum of absorbances of all the component nucleotides. This is known as the hypochromism of nucleic acids, and arises from the weak interactions of the neighbouring residues with one another. Nucleic acids can be denatured by exposure to heat or extreme pH. During the denaturation process, the double helix of the native DNA unwinds, the two strands separate as random coils, and weak interactions between neighbouring nucleotides are diminished. As a result, denaturation is accompanied by an absorbance increase at 260 nm, the so-called hyperchromic effect. Measurement of the extent of the hyperchromic effect allows the progress of denaturation to be followed, for example, as a function of temperature, and the melting point of a DNA sample to be determined (see Chapter 13 for more details).

3.3 NAD(P)H

Upon reduction of both nicotinamide adenine dinucleotide (NAD) and nicotinamide adenine dinucleotide phosphate (NADP) to NADH and NADPH, respectively, a strong absorption peak appears at 340 nm. The millimolar absorption coefficient at this wavelength for both reduced coenzymes is 6.22 mM^{-1} cm^{-1}. As these compounds function as coenzymes of many dehydrogenases, a large number of enzyme reactions can be followed (either directly, or using coupled assays) by monitoring the change of absorbance at 340 nm (8).

3.4 Carotenoids

Carotenoids are members of a large class of red, yellow and orange plant pigments, also found in the membranes of some bacteria. Carotenoids contain long carbon chains with many conjugated double bonds. Delocalization of electrons is responsible for their strong absorption in the visible region. In general, the more conjugated double bonds a molecule contains, the stronger and more redshifted its light absorption is. Carotenoids are used as intrinsic membrane probes of transmembrane electric potential in photosynthetic energy-transducing membranes (9) due to their very rapid electrochromic shift of absorption maximum in response to transmembrane electric field.

3.5 Haem proteins

The haem group consists of a porphyrin ring with a ferrous or ferric iron co-ordinated centrally. It serves as a prosthetic group of haemoglobins, myoglobin, hydroperoxidases, and a vast number of cytochromes. The conjugated double bond system of the porphyrin ring causes a strong absorption in haemoproteins, termed the α, β and γ bands. Typically, α bands occur at longest wavelengths (550 to 650 nm), γ bands at the shortest wavelengths (also called Soret bands, after the Swiss scientist who first examined the near UV region of cytochromes), and β bands lie between. The characteristic absorbances for reduced (ferrous) haemoproteins are generally studied in reduced *minus* oxidized difference spectra (see *Figure 3* for absolute and difference spectra of cytochrome *c*). The positions of these absorbances are determined by the nature of the

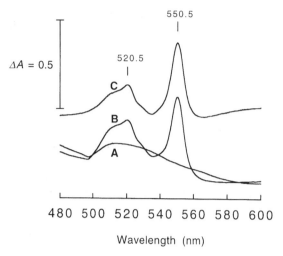

Figure 3 Absolute and difference spectra of cytochrome *c*. A is the absolute spectrum of the oxidized (Fe^{III}) form of purified horse heart cytochrome *c*, recorded with reference to a sample of buffer, with 500 nm as the reference wavelength in a dual wavelength scanning spectrophotometer. B is the absolute spectrum of the reduced (Fe^{II}) form of the same sample after treatment with a few grains of sodium dithionite. Isosbestic points occur wherever the spectra of the oxidized and reference forms cross. C (shifted up) is the difference spectrum obtained by subtraction of A from B. Spectra were recorded in a Johnson Foundation/Current Designs Inc. SDB4 Dual-Wavelength Spectrophotometer and manipulated using *Soft* SDB software. Conditions were: room temperature (about 20 °C); 10 mm pathlength; scan speed, 4.25 nm s^{-1}; spectral bandwidth, 2 nm. The positions of the β (520.5 nm) and α (550.5 nm) bands are marked.

substituent groups attached to the porphyrin ring and thus the particular type of haem, its redox state, the spin state of the iron and the presence of bound ligands.

The haem iron of cytochromes, globins, oxygenases and hydroperoxidases frequently bind ligands resulting in characteristic absorbance changes. For example, carbon monoxide (CO) binds to the reduced form of certain haem proteins, acting as a competitive inhibitor with respect to oxygen in respiration. The most common laboratory application is in CO difference spectroscopy, i.e. the identification of CO-binding haem proteins by recording the difference between a reduced sample and another that has been treated with CO. For example, a terminal oxidase common in bacteria, cytochrome *bo′*, binds CO at the oxygen-reactive haem O to give an adduct with an absorption maximum at about 415 nm. Since the reduced but unligated haem O has an absorption maxi-mum at about 430 nm, a difference spectrum (reduced + CO *minus* reduced) will show a peak near 415 nm and a trough near 430 nm. Also, the CO–Fe bond is photodissociable so that for certain haem proteins a second type of difference spectrum, the photodissociation spectrum, can be recorded. This is the differ-ence between the CO-ligated sample after photolysis (i.e. reduced, unligated) *minus* the same sample before photolysis (reduced, CO-ligated). For an excellent account of the use of cytochrome spectroscopy as applied to bacterial cytochromes, see Wood (10).

4 Spectrophotometer configurations

4.1 Single beam spectrophotometers

Single beam instruments are widely used for routine laboratory measurements at a single wavelength. As shown in *Figure 4*, only a single light beam from the monochromator passes through the sample compartment. The absorbance zero or 100 per cent transmittance is adjusted with a cuvette containing buffer or solvent. Then the absorbance (or transmittance) value of a sample cuvette containing sample solution is measured. It is necessary to put the sample and the reference in the optical beam alternately, by manual operation, typically with an interval of a few seconds required to change the cuvettes. However, if full spectra of the sample and reference are required, the interval between the two measurements is likely to be several minutes, with the consequent risk that lamp drift or other sources of instability can lead to errors.

It is impossible to give more detailed information or protocols for using the wide range of single-beam instruments on the market. The advice given here is that which must apply throughout this chapter: consult the manufacturer's manual.

4.2 Split beam or 'double beam' spectrophotometers

Confusion of nomenclature is rife here. Certain manufacturers and the authors prefer 'split beam' since it emphasizes that a single beam of monochromatic light is split spatially. 'Double beam' is better reserved for two beams of different wavelengths (see Section 4.3).

In the split beam mode, corrections for variations in light source intensity are made automatically because the incident beam is divided between reference and sample materials, as shown in *Figure 5*. The monochromatic light from the monochromator is split (chopped) into reference and sample beams by a chopper mirror. The two transmitted light beams of identical wavelength are now separated in time and space. one beam passes through the sample and the other through a standard or reference. The beams are sequentially detected as

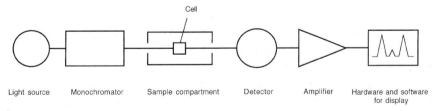

Light source   Monochromator   Sample compartment   Detector   Amplifier   Hardware and software for display

Figure 4 Components and their arrangement in a single beam spectrophotometer. For details, see text.

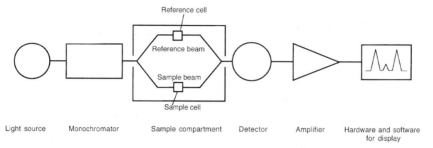

Light source   Monochromator   Sample compartment   Detector   Amplifier   Hardware and software for display

Figure 5 Components and their arrangement in a split beam ('double beam') spectrophotometer. For details, see text.

the reference signal I_o and the sample signal I, and the sample absorbance is obtained as the logarithm of the ratio of the two signals, which, according to the Beer–Lambert law, is independent of the incident light intensity. In comparing a sample and reference at each wavelength, the split beam spectrophotometer produces a 'difference' spectrum (instead of two separately acquired 'absolute' spectra, one for the reference and one for the sample, in a single-beam device). In the difference spectrum, those features common to sample and reference, like buffer absorption, moderate turbidity, or optical imperfections, are cancelled out.

The split beam mode is used to measure either absolute spectra (i.e. the difference between a sample in one beam and the solvent, or buffer, or even air in the other beam) or a difference spectrum. For the latter, the intensities of the two beams after passing through the cuvettes are again compared but, instead of comparing a sample to a non-absorbing reference, two nearly identical samples are compared. The difference between the two may be amplified because any interfering or background absorbance or light scattering is common to both samples. It should be noted that in difference spectra both peaks (positive differences) and troughs (negative differences) are anticipated according to whether the 'sample' absorbs more or less light, respectively, than the reference.

The use of a chopper, usually vibrating or rotating with a frequency of about 50 Hz, limits the kinetic resolution of the split-beam instruments. Spectral changes occurring on a faster timescale cannot be monitored. Recent advances in component design with improved stability have to some extent limited the advantages of split-beam methods and there has been a resurgence of interest in and availability of simpler single-beam designs.

4.3 Dual-wavelength spectrophotometers

The most distinctive feature of the dual-wavelength spectrophotometer is the presence of *two* monochromators, the outputs from which are directed to a single sample, again traditionally via a chopper mirror. Consider first the operation of a dual-wavelength non-scanning spectrophotometer. Rather than comparing the absorbance of a sample and reference at the *same* wavelength (as

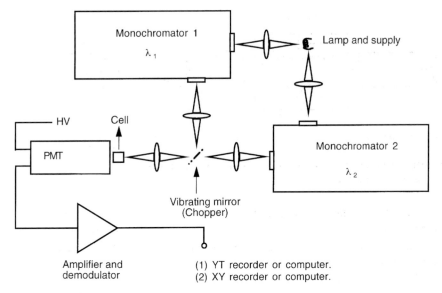

Figure 6 Components and their arrangement in a dual-wavelength spectrophotometer. In its basic mode of operation, both λ_1 and λ_2 are fixed and kinetic data may be obtained ($\Delta A_{\lambda_1 - \lambda_2}$ versus time). In the dual-wavelength scanning mode, λ_2 is set at a reference wavelength and λ_1 is scanned to generate a spectrum referenced to λ_2. HV, high voltage; PMT, photomultiplier tube. For details, see text.

in the split beam mode; Section 4.2), the absorbance of a single sample is compared at *two* wavelengths—'sample' (λ_1 in *Figure* 6) and 'reference' (λ_2 in *Figure* 6). The sample λ_1 is selected to correspond to the absorbance maximum of the signal under study (e.g. 550 nm for cytochrome *c*) and the reference λ_2 is chosen to be, ideally, at an isosbestic point (see Section 6.3) or other wavelength close to λ_1 where significant absorbance changes are *not* anticipated. Wavelength pairs in the fixed dual λ mode should be chosen to optimize:

(1) sensitivity, i.e. high signal:noise (choose λ for high absorbance coefficient);

(2) specificity (select for freedom from other components); and

(3) freedom from light scattering changes (select pairs as close as possible, ideally 15–40 nm apart).

In this way, λ_2 serves as a reference measurement for the observed absorbance change to be studied at λ_1. Thus, if there is an undesired change in the absolute spectrum of the sample during measurement, e.g. due to particles settling out, the effects on $\lambda_1 - \lambda_2$ will be minimized since the undesired change will be recorded at *both* wavelengths. Results are presented on a YT recorder, computer etc.

The dual-wavelength scanning spectrophotometer was developed by Britton Chance in the Johnson Foundation in the early 1950s and remains immensely useful and in production (*Figure 7*). A single light source is used for two independent monochromators; one is set to a reference wavelength while the

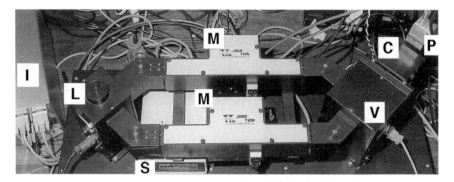

Figure 7 Photograph of a custom-built dual-wavelength spectrophotometer designed and constructed by the University of Pennsylvania School of Medicine Research Instrumentation Shop and Current Designs Inc. The instrument has the optical components in a classical dual-wavelength configuration, except that use of monochromators with entrance and exit slits 'in-line' (not at 90°) to each other requires 'elbow'-type mirror lens adapters to bring the two beams to the chopper. Light from a 45 W tungsten halogen source (L) is focused on the entrance slits of two Jobin–Yvon H20 monochromators (M), having aberration-corrected holographic gratings with focal length 200 mm, f/4.2. One of the monochromators is driven by a J-Y TTL Stepper Interface (S) over the desired spectral range while the second is set at the reference wavelength, selected to lie at an isosbestic point or a region of the spectrum adjacent to regions of interest. The output of the monochromators is modulated by a resonant tuning fork vibrating mirror (V) and focused on the cuvette (C). Transmitted light is measured by a Hamamatsu R928 side window photomultiplier tube (P) positioned about 50 mm from the nearest edge of the cuvette. The log of the photocurrent is taken prior to digitization of the data. A Scanner Interface Unit (I) incorporates the high voltage supply and servo circuit, log amp, chopper driver and demodulator 18 bit ADC, and interface to a Macintosh computer. Spectral data are analysed using *Soft*SDB (Current Designs Inc.) and subsequently by CA-Cricket Graph III, Microsoft Excel or other desired software.

other scans the desired spectral range. The two monochromator exit beams intercept at right angles at a vibrating chopper mirror which transmits only one of the beams at a time through the cell (the other beam being reflected). The two beams arrive at the detector, generally a photomultiplier tube, separated in time; the signals are separated electronically, compared and recorded. Results are presented on an XY recorder, computer etc.

In the non-scanning mode, dual-wavelength instruments are generally used to study kinetic changes at a pair of wavelengths (sample–reference) allowing the observation of small absorbance changes despite a very high initial absorbance or in the face of light scattering changes arising from particles settling, for example. In the dual-wavelength scanning mode, spectra are obtained with reference to an isosbestic point or other suitable reference wavelength. Note that since only one cuvette is scanned at a time, difference spectra are obtained only by computing the difference between two scans. In either mode, the dual-wavelength instrument is valuable for optically isolating components in mixtures using wavelength(s) in which the desired component changes absorption and other components do not.

4.4 Multi-wavelength spectrophotometers

A novel design was introduced by Chance and colleagues in 1975 (11). It allows four- or eight-channel measurements to be made and can serve as fluorometer, reflectometer or spectrophotometer. In essence, four or eight interference filters are mounted in a disc, rotating about an axis parallel to the light beam from a halogen source. The speed of the disc (driven as a turbine from a filtered, compressed air supply or by an electric motor) provides a variable chopping frequency, typically of 100 Hz. Full details are given in reference (11).

4.5 Diode array spectrophotometers

Diode array spectrophotometers are particularly suited to the single-beam mode since spectra are acquired very quickly (see Section 4.1). In conventional spectro-photometers, it has been customary to use as the detector a photomultiplier or photodiode which is sensitive to a wide range of wavelengths (wavelength specificity being provided by a monochromator between the lamp emitting polychromatic light and the sample). The development of diode array spectro-photometry and the introduction in 1979 of commercially available instruments has led to a wide range of applications, most of which exploit the ability of the diode array to detect light at a large number of discrete but closely spaced wavelengths simultaneously.

A diode array consist of a series of photodiode detectors arranged side-by-side on a silicon crystal. The diode array typically has a photosensitive area of a few square millimetres. Each diode is dedicated to measuring a finite but narrow band of the spectrum and all diodes are connected to a common output line. By measuring the variation in light intensity over the entire wavelength range, the absorption spectrum is measured.

Figure 8 shows the arrangement of the basic components. Polychromatic light from an appropriate source is passed through the sample area and focused on the entrance slit of a grating or polychromator, which disperses light onto the array. Note that the relative positions of the sample and dispersive element are reversed relative to a conventional spectrophotometer ('reverse optics'). The bandwidth of light detected by each diode is related to the size of the poly-chromator entrance slit and the physical size of the diode. A shutter is used to cut off light from the source until a measurement is to be made. This prevents continuous illumination of the sample with high intensity radiation, minim-izing undesirable photochemical effects (see later).

In conventional scanning spectrophotometers, the instrumental spectral bandwidth (see later, Section 5.2) is primarily a function of the entrance and exit slit widths of the monochromator and the dispersion generated by the grating; resolutions of 0.5 to 2 nm are common. In a diode array instrument, the spectral resolution is dependent on the number of photodiodes dedicated to measurement in the specified wavelength range (see *Figure 9*). The spectrum can be considered to be digitized, with the distance between successive diodes the sampling interval. Higher resolution requires more (and smaller) diodes.

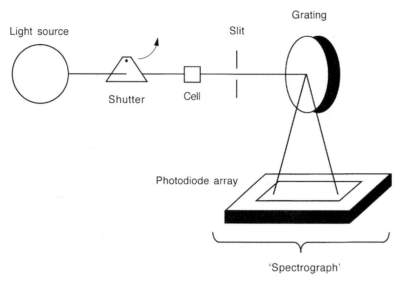

Figure 8 Components and their arrangement in a diode array spectrophotometer.

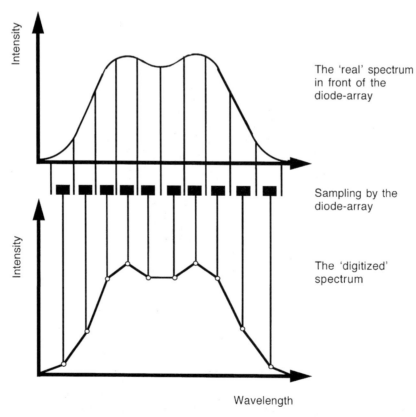

Figure 9 The discontinuous spectrum produced by a diode array spectrophotometer.

The main advantages of diode array spectrophotometry are:

- it is very fast because of parallel data acquisition and electronic scanning
- excellent wavelength reproducibility
- reliability, due to the absence of moving parts in a monochromator, and solid-state technology

The most common applications currently are in rapid reaction kinetic studies where a diode array detector can be mounted to a stopped-flow spectrophotometer (several commercial instruments of this type are available, for example, from Hi-Tech Scientific Instruments and Applied Photophysics) and in monitoring fractions during chromatography. *Figure 10* shows an application in the study of a flavohaemoglobin (12) and illustrates the ability to record spectra over a wide wavelength range, revealing simultaneous changes in three chromophores over an observation period in which temporal resolution from milliseconds to hundreds of seconds was achieved. Again exploiting the ability to obtain a spectrum virtually instantaneously, another recent application involves obtaining absorbance spectra during transient steady states in a growing bacterial culture (13).

Potential problems include the use of very high light intensities with possible

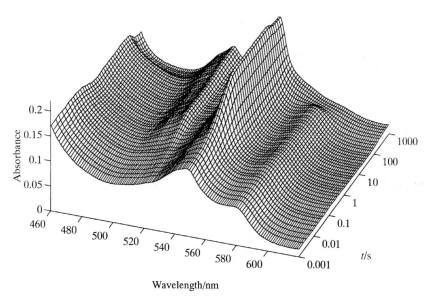

Figure 10 An example of a diode array stopped-flow spectrophotometric experiment showing the ability to display complete spectral changes over time. The reduction of 5 μM cytochrome *c* (550 and 520 nm signals) by *Escherichia coli* flavohaemoglobin (Hmp, 4 μM) in the presence of oxygen and NADH is shown. The FAD of Hmp is largely responsible for the trough centred near 480 nm and the oxygenated haem absorbs at 540 and 580 nm. The spectra were recorded by Dr Yutaka Orii (Kyoto University) on a rapid-scan stopped-flow spectrophotometer constructed in his laboratory. For further details see (12). This figure is reproduced from Poole, R. K., Rogers, N. J., D'Mello, R. A. M., Hughes, M. N. and Orii, Y. (1997). *Microbiology* **143**, 1557–1565 with permission from the Society for Microbiology.

15

implications for photosensitive materials (like flavins), cost (although direct comparisons with conventional instruments are difficult), and difficulties in obtaining high resolution spectra with extremely turbid light-scattering samples.

4.6 Microwell plate-reading spectrophotometers

Microplate readers have been available for over 10 years, but those available now offer a range of useful operating modes that can find numerous applications in screening multiple samples in 96-well microwells of 'microtitre' plates. Applications include analysis of microbial growth turbidimetrically, colorimetric assays (e.g. NAD(P)H-linked or dye reduction), immunoassays, platelet aggregation assays and measurements of nucleic acids and proteins (see Chapter 7). In one current instrument, Molecular Devices Spectramax 340PC, wavelengths between 340 and 85 nm can be scanned in 1 nm increments. The depth of the liquid in each well is determined and values normalized to 1 cm. Sample volumes between 30 and 100 μl can be measured. The dual-scanning monochromator versions of such readers now available for fluorescence work are outside the scope of this chapter.

4.7 Reflectance methods

For opaque samples, such as microbial cell cultures, animal tissues and cell films, it is possible to obtain a spectrum by measuring not the (very small) amount of transmitted light but rather the reflectance of light from the sample surface. An example of the use of fibre optic light guides to record diffuse reflectance spectra from brain *in vivo* was given by Bashford (14). Essentially, a bifurcated light guide was used to transmit light from the source to the brain surface and the reflected light was transmitted to the photomultiplier detector by the second limb of the light guide. A further example using a custom-built apparatus can be found in (15). For reflectance measurements from skin, a Dermaspectrometer (Cortex Technology, Hadsund, Denmark) has diodes for emission centred at 568 nm (for haemoglobin) and 655 nm (for melanin) (16). Note that the method devised by Brown *et al.* (17) for screening yeast colonies on agar plates is actually a transmittance method but a fibre optic light guide is used to direct light from the monochromator to the colony *in situ* on the Petri dish; a photodiode is placed beneath the colony.

4.8 Novel double monochromator methods

A unique design has been implemented by Olis Inc. in the Olis RSM-1000 incorporating a patented subtractive double-grating monochromator, capable of 1000 scans/s. In the middle plane of the monochromator is a spinning disc containing a number of slits which acquires millisecond 'snap-shots' of a rapid reaction. The RSM 1000 Spectrophotometer offers an alternative to diode array systems for millisecond spectral scanning and avoids the high light intensities required for diode array measurements. The instrument can be used for absorbance, fluorescence and circular dichroism.

4.9 Computing and spectrophotometry

It is now so common for spectrophotometers of any of the above designs to be equipped with computer control, that it is easily forgotten that optically excellent instruments abound that predate the Macintosh and PC era. For instance, dual wavelength instruments designed and constructed in the Johnson Research Foundation laboratory of Britton Chance in the 1950s were fitted with on-board micro-processing for spectrum acquisition, subtraction and other manipulations a quarter of century ago. These were fitted with superb optics but, with the ability to 'store' only four spectra, now appear rather primitive. Fortunately a few designers and manufacturers offer packages for computerization of an existing spectrophotometer. One such example, for Macintosh-based spectroscopy, is Current Designs, Inc. The digital signal processor (DSP)-based 'Scanner Interface Unit' is powerful and flexible, so that it can be readily used as the hardware basis for this task. The DSP system can perform signal averaging, wavelength scan, control, chopped beam demodulation, and almost any other sort of processing, with the specifics handled in software. Second, since it is easy to add other channels of analogue or digital inputs, oxygen electrode readings, for instance, can be acquired and displayed in real time along with absorbance. Furthermore, the Current Designs Macintosh software, for example, has been put in the public domain, allowing researchers and other developers to build an ever-enlarging library of enhancements, analysis functions and hardware control packages.

5 Choice of spectrophotometer operating conditions

An increasing number of commercial spectrophotometers offer programs designed for certain tasks—kinetics, wavelength scan, etc. However, unless the reason why such a program may specify a particular scan rate or slit width, for example, is understood, serious mistakes can be made.

5.1 Wavelength range and light source

The wavelength settings or wavelength range may be set manually by turning the monochromator(s) to the desired λ or inserting filters, or by instructing a microprocessor or computer which controls the apparatus to select the wavelength. Decisions will generally be based on published optical properties or prior experience. References cited in (18) provide useful lists of absorbance maxima.

5.2 Spectral versus natural bandwidth

This is perhaps the least understood of the commonly set variables. The ability of a spectrophotometer to (a) resolve peaks, that is to distinguish between two absorption bands close together, and (b) to measure their intensity accurately, is dependent on spectral bandwidth and natural bandwidth. The former, but not the latter, can be set by the operator.

To understand spectral bandwidth, it must be appreciated that 'monochromatic' light from a monochromator never consists only of radiation of precisely one wavelength (as does the light from a laser). Instead, it has a certain bandwidth, including smaller amounts of light with shorter and with longer wavelengths than the particular chosen wavelength. The spectral bandwidth is defined as the band of wavelengths contained in the central half of the entire band passed by the exit slit of the monochromator (*Figure 11*). It is the exit slit (generally a few mm across) whose physical width can be set by the investigator. To calculate spectral bandwidth from this slit width, we need to know the reciprocal dispersion of the monochromator, which can be found in manufacturers' literature. Typical values of $d\lambda/dx$, where x is the slit width, are 2 or 4 nm mm^{-1}. Thus, in the latter case, when the exit slit is opened to 0.5 mm, say, the spectral bandwidth is 2 nm.

In contrast, the natural bandwidth is an intrinsic property of the sample, independent of the instrument bandwidth, and is defined as the width (in nm) at half the height of the sample absorption peak, as shown in *Figure 11*. For example, the value for the natural bandwidth of the 340 nm peak of NADH is 58 nm, whereas for most cytochromes at room temperature the natural bandwidths in the α-region are of the order of 10 nm. It is easy to conceive that having too broad a spectral bandwidth would result in an apparent decrease of sample absorption. This is because the incident light would contain a large fraction of radiation with wavelengths poorly absorbed by the sample.

A rarely seen graph, allowing errors in peak height measurements to be predicted, is shown in *Figure 12*. This shows the dependence of the absorbance peak magnitude on the ratio of the spectral bandwidth to the natural band-

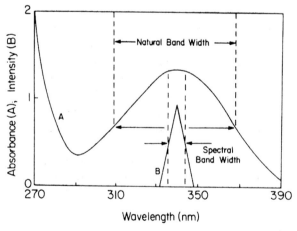

Figure 11 The relationship between natural and spectral bandwidths. The natural bandwidth shown is for NADH (58 nm, curve A). Curve B is a schematic representation of the spectral bandwidth of the monochromator exit beam. Taken from (18) where the original source is cited.

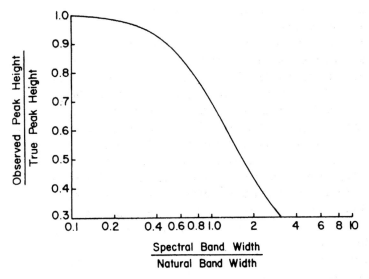

Figure 12 Graph allowing calculation of the error in measured peak height for a given monochromator spectral bandwidth and natural bandwidth of an absorption band. Spectral bandwidth can be obtained from manufacturers' information or by knowing the physical slit width of the monochromator and the reciprocal dispersion (nm/mm). Since natural bandwidths of, for example, cytochrome absorption bands are about 10 nm at room temperature a spectral bandwidth of 2.5 nm (ratio on the abscissa of 0.25) will introduce no more than a 3% error in measured peak height. However, for low temperature spectra, spectral bandwidths of about 0.5 nm are required.

width. Obviously, the best spectral bandwidth to be selected on a spectrophotometer is 1/8 to 1/10 as wide as the natural bandwidth of the sample absorption peak. However, if too narrow a spectral bandwidth is selected, too small an amount of light energy will reach the sample and, as a result, electronic noise (defined as any undesired recorder pen or display excursion) will be higher, decreasing the accuracy of the measurement. On the other hand, for example, use of a spectral bandwidth of 8 nm would give rise to an error of about 22% if attempting to measure the intensity of a cytochrome absorption band with a natural bandwidth of 10 nm, but in a measurement of NADH the error would be less than 0.5%. For quantitative work with compounds having narrow natural bandwidths, a compromise must be reached. Probably the best solution is to run the spectrum trying different spectral bandwidths remembering that because of sample light scattering and variations in detector response, one slit width may not be ideal for all wavelengths. Examples of such a test were shown by Poole and Bashford (18) in their Figure 7.

5.3 Spectral resolution

Spectral resolution is a measure of the capability of a spectrophotometer to distinguish between two closely adjacent wavelengths. The wavelengths are considered to be resolved if the trough between the two peaks in the spectrum

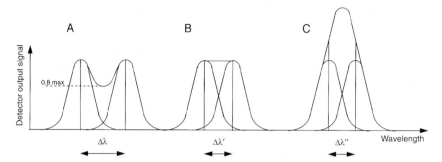

Figure 13 Resolution of two adjacent bands. In A, the separation between the bands (Δλ) is sufficient for them to be resolved using the criterion that the trough between the adjacent peaks is no higher than 0.8 of the peak height. In B, the separation between the bands (Δλ′) is just sufficient for the presence of two bands to be seen. In C, the separation between the bands (Δλ′′) is insufficient for resolution and the bands cannot be distinguished.

obtained is lower than 80% of the peak height. This is sometimes called the Rayleigh criterion (see *Figure 13*). Although spectral resolution is a parameter that greatly concerns the practical spectroscopist, it is inextricably linked to spectral bandwidth (see above).

5.4 Scan speed and instrument response time

These variables are closely inter-related and again their selection is a matter of compromise. Scanning a spectrum too fast for the instrument's response time can lead to serious skewing of the observed peak, with loss of peak intensity and blurring of fine detail in the signal (see the experiment shown in Figure 8 of (18)). On the other hand, scanning too slowly can allow the sample to change chemically or physically (e.g. by cell or membrane particles settling out of suspension).

5.5 Temperature

The temperature-controlled sample holders available in most commercial spectrophotometers, when coupled to an external circulating heater or cooler, can control temperature in the approximate range 0 to 40 °C. Specialized cuvettes are not necessary and the device is useful for kinetic measurements or observation of a labile sample. However, there are many advantages in kinetics and wavelength scanning of being able to operate at much lower temperatures.

5.5.1 77 K

Spectroscopy at 77 K (liquid nitrogen temperatures) has been widely used to detect differences in the absorption spectra of closely related haemoproteins, to trap unstable intermediates and steady-states of oxidation and reduction, to slow down rapid rates of reaction, and to detect and measure very low concentrations of haemoproteins. The main effects are a sharpening and enhancement of the absorption bands and a blue shift of 1–4 nm. The absorbance changes are due to:

(1) a true temperature dependence of the absorbance of the chromophore, resulting in narrowing, sharpening and shifting to shorter wavelengths of the bands;

(2) light-scattering changes in the medium, which result in an effective increase in pathlength by multiple internal reflections from ice crystals. [It is worth noting that enhancement effects of related origin may be observed in highly scattering suspensions (e.g. of intact cells) at room temperature.]

Reference (19) is a classic account of the practical uses of low temperature spectroscopy. The enhancement effects are strongly dependent on the suspending medium and the method used to freeze it (20). For example, a devitrified (i.e. polycrystalline) 1.4 M sucrose solution can give a 25-fold intensification at −190 °C, compared with 20 °C. The presence of an organic solvent also makes the enhancement factors more reproducible than in dilute buffers. Glycerol has the added advantage of suppressing the pH changes that occur on freezing, as does a careful choice of buffer. The pH of Tris buffer changes dramatically (increasing) on lowering the temperature, while phosphate and acetate buffers, for example, are relatively temperature-independent. For a detailed account of the choice of buffers in low temperature work, Douzou (21) should be consulted.

Accessories for recording spectra at 77 K are available for many but not all spectrophotometers. The attachments are generally expensive and many workers have designed and constructed simple devices for their own instruments. References to examples are given by Jones and Poole (22).

5.5.2 Other sub-zero temperatures

In media, such as those containing mannitol or sucrose, which do not exhibit phase transitions in the experimental range of temperatures, the absorption peak heights and peak areas are nearly linear functions of temperature between −40 °C and liquid nitrogen temperatures. Furthermore, if the sample is warmed within this temperature range, and cooled again to the original temperature, the resultant spectrum is indistinguishable from that recorded originally (19). At temperatures higher than −40 °C, the absorbance intensity decreases abruptly and re-freezing does not fully restore the enhancement.

For kinetic studies at sub-zero temperatures, or where it is required to study the effect of temperatures on an absorption spectrum (e.g. in investigating haemoprotein spin-states), it is sometimes necessary to use temperatures other than 77 K. Temperatures between about 0 and −40 °C can be maintained simply by circulating cooling water with ethylene glycol (as antifreeze) around a cuvette containing the ethylene glycol-supplemented sample. For temperatures between about −40 °C and −140 °C, as required for kinetic studies of ligand binding to terminal oxidases and of subsequent electron transfer, the procedure developed by Chance and co-workers (23) for their 'triple trapping' studies of mitochondrial cytochrome c oxidases is appropriate. Here, the temperature is maintained by a steady flow of nitrogen gas, cooled by passing through a copper coil immersed in liquid nitrogen and re-heated by a thermostatically controlled resistor. Measurements of the sample temperature with a thermocouple next to the cuvette show that temperatures are maintained to better than 0.5 °C by the thermostat. This system is simple, reliable and relatively inexpensive, but not available commercially. The techniques of cryoenzymology and spectroscopy below 77 K are outside the scope of this chapter but further information can be found in reference (21).

6 Use of the spectrophotometer

6.1 The choices

To some extent, the choice of instrument to be used for obtaining the spectrum of a biochemical sample will be dictated by what is available. For recording an absolute spectrum of a routine sample, such as a dye or chromophore in clear solution, almost any optical configuration could be used, namely single beam, split-beam, dual-wavelength or diode array. If, on the other hand, the investigator wishes to record, say, a difference spectrum of chromophores in a highly turbid membrane or cell suspension, or perhaps measure time-resolved changes, special techniques and instrumentation will be required. Fast kinetic methods are covered in Chapters 8, 9 and 10. As an illustration of the choices that have to be made in 'everyday' routine spectral scanning of clear solutions and in more challenging situations like turbid samples, a guide to selection of operating conditions (*Figure 14*) and two illustrative protocols will be described. However, it cannot be overemphasized that these are not experimental recipes: every possible permutation of sample, spectrophotometer and information sought will require careful consideration of experimental design. Before presenting these protocols, some fundamental practical aspects must be considered.

6.2 Baselines

A baseline is the wavelength-dependent difference in absorbance either (i) between two cuvettes in a split beam apparatus when their contents are thought to be identical or (ii) between two scans of one cuvette in a dual-wavelength spectrophotometer over a period when the contents are thought not to have changed. Any irregularities in a baseline will be included in subsequent difference spectra and may be superimposed on the spectral regions of interest. In extreme cases, baseline irregularities may be mistaken for spectral peaks and troughs, and a steeply sloping baseline can alter the apparent position of an absorption peak. Therefore, baseline flatness should always be checked before recording a difference spectrum and either presented with the latter or, at least, reported. The following additional measures can be taken to improve baseline flatness.

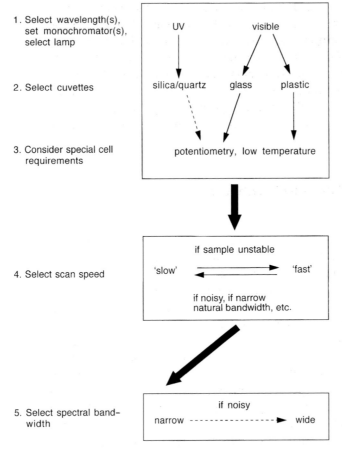

Figure 14 Scheme to show choices necessary for obtaining a routine absorbance spectrum.

6.2.1 Electronic baseline correction facilities

In most modern instruments, any irregularities are 'memorized' by the instrument in an initial scan (normally not plotted) of the cuvette(s) and then subtracted from subsequent scans. Remember, however, that in this case, any dissimilarities in the cuvettes are essentially hidden.

6.2.2 Sample preparation

Give extra care to cleanliness and matching of the two samples, the proper alignment of the cuvette(s) and the identicality of the cuvettes' contents. Where both test *and* reference cuvettes contain samples treated in some way, such as is the case in the recording of the baseline (reduced minus reduced) for a CO difference spectrum (Section 3.5), it is generally preferable to split the sample between the cuvettes after (in this example) reduction, rather than try to prepare two samples identically.

6.2.3 Manual correction

In the absence of the above facilities, the uncorrected baseline can be recorded and subtracted from the difference spectrum manually at, say, 2 nm intervals. The difference spectrum can be assumed to cross the baseline at a known isosbestic point (see below). The method is of course tedious.

6.3 Isosbestic points

An isosbestic point is a wavelength where the absorptions (or ε) of two light-absorbing forms are equal (Greek *iso*, equal; *sbestos*, extinguished), and may be seen, for example, where a difference spectrum crosses an appropriate baseline one or more times. The isosbestic point is useful in both quantitative and qualitative work. Where a clear isosbestic point occurs during the course of a reaction, it is often taken as evidence that only two species are involved, since it is considered unlikely that a third absorbing species would also have an identical ε at these wavelengths. It is more likely that a third species may simply have no absorption at this wavelength. However, the *lack* of an isosbestic point does indicate the presence of a third component. The isosbestic point is more generally used as a reference wavelength to which the absorbance at a nearby wavelength may be referred in (i) setting up a dual wavelength spectro-photometer or (ii) measuring from an absorption spectrum the concentration of a component using $\Delta\varepsilon$. Less frequently, it is a wavelength suitable for quanti-fication of the total amount of the two species present. Finally, in the pres-entation of many stacked spectra, quoting the isosbestic point(s) eliminates the need to draw baselines for each spectrum.

6.4 Wavelength and absorbance calibrations

Factors that can affect the apparent position of an absorption maximum are described later. Even when operating conditions are properly chosen, the poss-ibility remains that the spectrophotometer generates peaks that are a nano-metre or more from the expected position. Wavelength accuracy can be checked using the bright line spectrum (λ_{max} = 656.1 nm) of the deuterium light source; alternatively standards for checking the wavelength calibration are available. Most commonly, these are rare earth oxides or a mixture of oxides such as holmium and/or didymium; such 'filters' should be scanned with air as the reference or blank. Holmium has nine useful maxima between 241.5 and 637.5 nm and didymium five between 573 and 803 nm (see manufacturers' charts). A quick and easy check of both wavelength calibration and accuracy of absorbance measurement is to use a 0.2 mM solution of potassium chromate in 0.05 M KOH. Two broad absorption bands at 275 nm and 375 nm should be seen with molar absorption coefficients, respectively, of 3680 M^{-1} cm^{-1} and 4820 M^{-1} cm^{-1} in a 1 cm cell (18). It is useful to run calibration spectra periodically, dating the scans, to check for degradation of accuracy and calibration.

6.5 Choice and use of cuvettes (cells)

6.5.1 Pattern

The most common type of absorption cell or cuvette is the square cell, open at the top, with a pathlength of 10 mm and working volume of 2–3 ml. Only fluorescence cuvettes are polished on all faces. Manufacturers' catalogues illustrate designs for special purposes, with pathlengths ranging from 1 to 100 mm. The essential feature is parallelism of the two end plates which limit the pathlengths. The magnitude of the other horizontal axis (i.e. perpendicular to the light beam) is dictated by the volume of the sample and the desirability of keeping the light beam clear of the side walls, where intense reflection can occur. 'Semi-micro' cuvettes (i.e. pathlength 10 mm, capacity about 0.65 ml) appear to be convenient in reducing the volume of sample needed, but light may pass around the edge of the sample as well as through it. These should be either blackened (these are commonly available from manufacturers such as Starna) or 'masked', or else a narrow spectral bandwidth must be selected (but see Section 5.2). (In this laboratory highly turbid samples are always recorded in conventional 1 cm square cells.) It is a useful exercise to determine the minimum volume that can be used with each set of cuvettes; this is done by gradually adding a coloured or turbid sample until the absorption reading is constant. The working volume can sometimes be reduced further by including a spacer below the cuvettes.

6.5.2 Mixing

Cuvette lids or, better, ground stoppers prevent evaporation of solvent and/or contamination of the cuvette contents. Tapered Teflon stoppers can be drilled to accommodate the fine needle of a syringe such as those supplied by Hamilton. Mixing of cuvette contents can be problematic, especially when the cuvette is almost full and a lid is fitted; a few glass beads (small enough to lie below the light path) can provide sufficient agitation when a cuvette is inverted. An alternative is to use disposable mixers or stirring paddles (Kartell). Those with small horizontal platforms ('plumpers') can be loaded with up to about 100 μl of an addition, then rotated in the cuvette between thumb and forefinger to provide simultaneous and fairly rapid addition and mixing. For semi-micro cuvettes, the mixers and paddles can be trimmed with a blade.

Where the contents of a cuvette must be continuously stirred during the scanning of a spectrum, a small magnetic bar can be rotated magnetically within the cuvette. Some (e.g. Spinbars from Bel-Art Products) are circular with four small vanes and rotate smoothly inside a 10 × 10 mm cuvette. The small stirrer motor is usually located under the cuvette and the power supply is outside the spectrophotometer. A mu metal shield protects the photomultiplier from 'noise'. High quality dc brush motors driven in a feedback loop to maintain constant speed, not voltage, minimize stirring artefacts, especially important if, for example, oxygen electrode measurements are to be made in parallel with absorbance. Such stirrers are available from The University of Pennsylvania Research Instrumentation Shops.

Although suitable for modest stirring of small volumes, such stirrers are sometimes inadequate for mixing the contents of a larger cell such as that required in potentiometric redox titrations (24). Here, a longer magnetic stirring bar rotates around a horizontal axis just *above* the light beam in the main body of the liquid sample (typically 5–7 ml). The bar can be driven by a button magnet mounted on the shaft of a small 12-V variable-speed electric motor.

6.5.3 Optical material

The choice of glass, silica (quartz) or plastic cuvettes will be primarily dictated by whether they will be used in the UV or visible spectral regions. Only silica is suitable below about 360 nm. Plastic disposable cuvettes are cheap, popular and can give perfectly acceptable results. Their less-than-perfect optical faces may be of little consequence when working with highly light-scattering samples.

6.5.4 Special applications

Examples of cuvettes for special purposes and which must usually be constructed by the user are those for retaining standard e.p.r. tubes in a single beam, for low-temperature work and photodissociation spectra, and for potentiometric titrations.

6.5.5 Cleaning

The importance of clean cuvettes is self-evident. Routinely, all non-disposable cuvettes should be emptied immediately after use, rinsed repeatedly in the solvent (e.g. water), then with clean ethanol or acetone and dried with low pressure air or nitrogen from a cylinder. It is prudent to install a filter (such as those with pore sizes of 0.45 μm used in filter sterilization) in the gas line. Cuvette washers (e.g. Aldrich) wash, rinse, and dry cuvettes. Cotton wool 'buds' can also be useful for dislodging interior, stubborn marks and for drying. The outside optical surfaces should be polished with clean lens tissue. Note that plastic 'squeezy' bottles generally used for solvents contain plasticizers such as butyl phthalate, which can interfere with critical UV spectra.

The Perspex windows of low-temperature cuvettes easily become scratched and will eventually crack. However, provided they are still liquid-tight (at least for as long as it takes to plunge them and their contents into liquid nitrogen) they are quite serviceable in this condition. The opacity of the Perspex is negligible compared with that of the frozen sample.

Neglected cuvettes that cannot be dismantled for cleaning may require soaking overnight in concentrated sulphuric acid containing a few crystals of dichromate or permanganate, or boiling in distilled water containing a laboratory detergent.

6.6 A detailed example: recording of a cytochrome difference spectrum (reduced *minus* oxidized)

The most common method of cytochrome spectral analysis is the recording, with a split-beam spectrophotometer, of a spectrum that represents the difference between an oxidized and a reduced sample. Such spectra may provide data on the identity and the amount of the cytochrome types present, and may be obtained with suspensions of intact cells, subcellular fractions derived therefrom or purified preparations. An example of such a spectrum obtained with pure cytochrome *c* was shown as *Figure 3*. Note, however, that attempts to obtain a spectrum of a cytochrome *in situ*, i.e. in intact cells or mitochondria, will be frustrated by the intense light scattering of the sample. *Figure 15* shows three attempts using different approaches. The bacterium *Zymomonas mobilis* has a low content of cytochromes and is a challenging material for spectrophotometry (25). The absolute spectrum of the suspension used (i.e. recorded using a blank) reveals a high signal, particularly at lower wavelengths, due almost entirely to turbidity. When the spectrum is taken in the dual-wavelength scanning mode with 500 nm as a reference wavelength, the absorbance difference across the spectrum is reduced by more than 10-fold (compare scales in *Figures 15A* and *B*). The true contribution of cytochromes is seen in the difference spectrum, obtained by taking oxidized and reduced spectra in the dual-wavelength scanning mode and subtracting the former from the latter in the instrument software. The cytochrome absorbance peaks represent less than 1% of the turbidity measured in A.

A compromise must be reached in the choice of scan speed, slit width and response time, these factors being inter-related. 'Typical' or starting values for the parameters in such an experiment might be 1–4 nm s^{-1}, 2 nm spectral bandwidth and 0.1 s, respectively.

Full reduction of a sample can generally be achieved by adding a few grains of sodium dithionite ($Na_2S_2O_4$) to a cell or membrane suspension, but may take several minutes for completion. It is important that the dithionite should be fresh. It is useful to split a newly purchased sample into small aliquots in, say, 5 ml screw-capped bottles and to use each for only a month before discarding. Alternatively, the preparation can be allowed to become anoxic by respiration of a suitable oxidizable substrate. Sodium borohydride is also a useful reductant.

Oxidation may be achieved by using an exogenous oxidant or by vigorous aeration of the sample just prior to scanning the spectrum. Difficulty can be experienced if significant levels of endogenous reductants are present or if a respiratory chain is terminated by an oxidase with a particularly high O_2 affinity or is capable of particularly rapid rates of electron flux. In such cases, O_2 may be generated in the sample by adding H_2O_2 and catalase, although preparations of cells or sub-cellular fractions may exhibit sufficient endogenous catalase activity. Alternatively, potassium ferricyanide (ferro/ferricyanide +0.43 V), hexachloro-iridate (iridium (IV) chloride), or ammonium persulphate ($S_2O_8^{2-}/2SO_4^{2-}$, +2.0 V) may be added as oxidants ($E^{\circ\prime}$ values in parentheses). The first two are highly coloured however; aqueous solutions of ferricyanide absorb strongly between

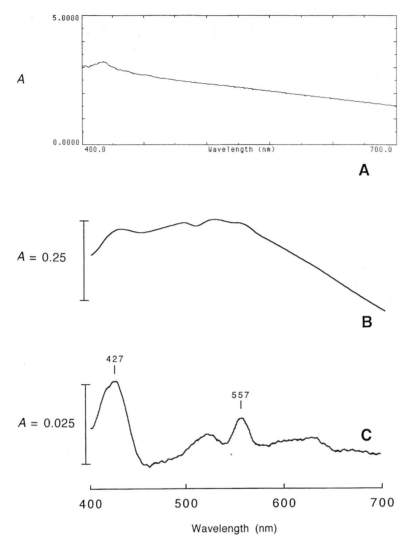

Figure 15 Attempts to record absorbance spectra in a very turbid cell suspension. A shows the 'absolute' spectrum (a scan of a cuvette with the buffer taken as the baseline) of reduced membranes of the bacterium *Zymomonas mobilis,* acquired with a single-beam instrument (Beckman DU 650). B shows ΔA acquired with a dual-wavelength instrument (SDB-4) with 500 nm as the reference wavelength. C is the reduced *minus* 'as prepared' difference spectrum obtained with the same dual-wavelength instrument. The concentration of membrane protein was 10 mg ml^{-1}; sodium dithionite was used as the reductant.

370 and 460 nm (λ_{max} 420 nm), whilst hexachloroiridate solutions have complex spectra in this region with λ_{max} at 488 nm. Ammonium persulphate is colourless. Addition of these oxidants to final concentrations of about 2 mM is usually satisfactory, although it is common practice to add 'a few grains' to 1–4 ml samples. Further details of these procedures and references are given by Jones and Poole (22).

Protocol 1

Recording the absolute UV–vis spectrum of a soluble protein in solution

Equipment and reagents

- Protein in buffer, same buffer for dilution. Impossible to be prescriptive
- Reagents for treatment of protein (e.g. reductant for haem protein)
- Split-beam scanning spectrophotometer (or almost any other kind!)

Method

1   Estimate absorption anticipated at λ_{max}, either from knowledge of protein concentration and anticipated absorbance coefficient, or from preliminary spectral scan.

2   Decide on dilution required or adjustment of absorbance signals to be recorded by choice of pathlength.

3   Establish and set wavelength limits. Select light source (e.g. UV and/or vis).

4   Choose cuvette material (e.g. silica/quartz for UV).

5   Select response time, slit width, scan speed. Use guidelines given in this chapter for 'normal' conditions, follow published protocols, or assess result of present scan and adjust conditions accordingly (see Figure 14).

6   Add buffer to two cuvettes, install in instrument, scan baseline, plot or store.

7   Keep buffer in reference cuvette, add protein to sample cuvette or replace contents of sample cuvette with protein in buffer, install in instrument, scan, plot or store.

8   Compare baseline (step 6) and absolute spectrum of spectrum (step 7) visually or subtract first from second.

Protocol 2

Recording the visible difference spectrum of a membrane-bound chromophore in a turbid sample (cells or membranes)

Equipment and reagents

- Cells or membranes in buffer, same buffer for dilution. Impossible to be prescriptive
- Reagents for treatment of protein (e.g. reductant and oxidant for haem protein)
- Split-beam or dual-wavelength scanning spectrophotometer

Method

1   Estimate absorption anticipated at λ_{max}, either from knowledge of protein concentration and anticipated absorbance coefficient, or from preliminary spectral scan.

2   Decide on dilution required or adjustment of absorbance signals to be recorded by choice of pathlength.[a]

3   Establish and set wavelength limits. Select light source (i.e. vis).

4   Choose cuvette material (generally glass but cheap even scratched cuvettes may not degrade signals).

5   Select response time, slit width, scan speed. Use guidelines given in this chapter for 'normal' conditions, follow published protocols, or assess result of present scan and adjust conditions accordingly (see *Figure 14*).

6   Add identical, well mixed samples of cells or membranes to two cuvettes,[b] install in instrument, scan baseline, plot or store.

7   Keep sample in reference cuvette, add reagent that will generate the required difference (e.g. reductant) to sample cuvette, install in instrument, scan, plot or store.

8   Compare baseline (step 6) and difference spectrum of spectrum (step 7) by subtracting first from second.

[a] It is strongly recommended that the entire wavelength range is scanned at various sample concentrations to avoid signal loss by light scattering at lower wavelengths (see Section 2.3).

[b] In haem protein work, for example, this baseline may be that of the untreated or oxidized sample.

6.7 Post-scan options

6.7.1 Derivative spectra

In its simplest form, a derivative spectrum is a plot of $dA/d\lambda$ versus λ rather than A versus λ. The derivative spectrum of a single, symmetrical peak crosses the baseline at or near the peak in the starting spectrum. More complex forms arise if two peaks are close together. If this first derivative spectrum is subjected to the same derivatization, a second derivative spectrum is obtained in which the original peak is replaced by a trough at the peak position and 'wings' appear above the baseline. Examples were shown in (18) and the technique is discussed in detail by Butler (26). The main application of such derivatives is in resolution of partly overlapping absorption bands, made possible by the progressive band narrowing that occurs as derivatives are taken.

Some spectrophotometers allow derivative spectra to be obtained by scanning two monochromators offset by a small, fixed wavelength interval (as in the old Aminco DW-2a spectrophotometer). This results in a difference signal proportional to $\Delta A/\Delta \lambda$. However, more often, derivative spectra are now obtained by computing data sets obtained in a more conventional scanning mode. Plotting derivatives up to the second is frequently a standard option in modern scanning spectrophotometers.

6.7.2 Data handling

An increasing number of commercially available spectrophotometers are controlled by a computer and allow the user to extract spectral data in two-column format (A and λ) for subsequent manipulation and plotting in publication-ready form. Commercial graphing packages, such as Cricket Graph III, Deltagraph and many more can be used to smooth spectra, perform subtractions and additions of whole spectra, normalizations and calculations of derivatives.

Acknowledgements

RKP is grateful to the large family of research students, postdoctoral researchers and visiting scientists to his laboratory who have put these techniques through their paces in obtaining spectra of microbial haem proteins. Professor Britton Chance has been a source of inspiration for 25 years and introduced us to the Research Instrumentation Shop (RIS) at the University of Pennsylvania School of Medicine. RIS staff Norman Graham and Bill Pennie and Current Designs President Ben Dugan have provided us with machines and limitless help, including the reading of a draft of this chapter. We thank Hewlett-Packard, Molecular Devices and Shimadzu for information on their products. We are very grateful to Mark Johnson for expertly producing many of the figures and Ron Adams for *Figure 7*. Work in this laboratory was supported by BBSRC and The Royal Society/NATO.

References

1. Brown, S. B. (1980). In *An introduction to spectroscopy for biochemists* (ed. S. B. Brown), pp. 1–13. Academic Press, London.
2. Brown, S. B. (1980). In *An introduction to spectroscopy for biochemists* (ed. S. B. Brown), pp. 14–69. Academic Press, London.
3. Bashford, C. L. (1987). In *Spectrophotometry and spectrofluorimetry—A practical approach* (ed. D. A. Harris and C. L. Bashford), pp. 1–22. IRL Press, Oxford.
4. Fewson, C. A., Poole, R. K. and Thurston, C. F. (1984). *Society for General Microbiology Quarterly*, **11**, 87.
5. Koch, A. L. and Ehrenfeld, E. (1968). *Biochimica et Biophysica Acta*, **165**, 262–273.
6. Koch, A. L. (1970). *Analytical Biochemistry*, **38**, 252.
7. Sambrook, J., Fritsch, E. F. and Maniatis, T. (1989). *Molecular cloning, a laboratory manual*, 2nd edn. Cold Spring Harbor Laboratory Press, Cold Spring Harbor Laboratory, NY.
8. Harris, D. A. (1987). In *Spectrophotometry and spectrofluorimetry—A practical approach* (ed. D. A. Harris and C. L. Bashford), pp. 49–90. IRL Press, Oxford.
9. Jackson, J. B. and Crofts, A. R. (1969). *FEBS Letters*, **4**, 185
10. Wood, P. M. (1984). *Biochimica et Biophysica Acta*, **768**, 293.
11. Chance, B., Legallais, V., Sorge, J. and Graham, N. (1975). *Analytical Biochemistry*, **66**, 498.

12. Poole, R. K., Rogers, N. J., D'Mello, R. A. M., Hughes, M. N. and Orii, Y. (1997). *Microbiology*, **143**, 1557.

13. Kavanagh, E. P., Callis, J. B., Edwards, S. E., Poole, R. K. & Hill, S. (1998). *Microbiology*, **144**, 2271.

14. Bashford, C. L. (1987). In *Spectrophotometry and spectrofluorimetry—A practical approach* (ed. D. A. Harris and C. L. Bashford), pp. 115–135. IRL Press, Oxford.

15. Chen, S. S., Yoshihara, H., Harada, N., Seiyama, A., Watanabe, M., Kosaka, H., Kawano, S., Fusamoto, H., Kamada, T. and Shiga, T. (1993) *American Journal of Physiology*, **264**, G375.

16. Thibodeau, E. A. and D'Ambrosio, J. A. (1997) *European Journal of Oral Science*, **105**, 373.

17. Brown, S., Colson, A. M., Meunier, B. and Rich, P. R. (1993). *European Journal of Biochemistry,* **213**, 137.

18. Poole, R. K. and Bashford, C. L. (1987). In *Spectrophotometry and spectrofluorimetry—A practical approach* (ed. D. A. Harris and C. L. Bashford), pp. 23–48. IRL Press, Oxford.

19. Wilson, D. F. (1967). *Archives of Biochemistry and Biophysics*, **121**, 757.

20. Vincent, J.-C., Kumar, C. and Chance, B. (1982). *Analytical Biochemistry*, **126**, 86.

21. Douzou, P. (1977) *Cryobiochemistry.* Academic Press, London.

22. Jones, C. W. and Poole, R. K. (1985). In *Methods in Microbiology* (ed. G. Gottschalk), **18**, 285. Academic Press, London.

23. Chance, B. (1978). In *Methods in Enzymology* (ed. S. Fleischer and L. Packer), **54**, 102. Academic Press, New York.

24. Dutton, P. L. (1978). In *Methods in Enzymology* (ed. S. Fleischer and L. Packer), **54**, 411. Academic Press, New York.

25. Kalnenieks, U., Galinina, N., Bringer-Meyer, S. and Poole, R. K. (1998) *FEMS Microbiology Letters*, **168**, 91.

26. Butler, W. L. (1979). In *Methods in Enzymology* (ed. S. Fleischer and L. Packer), **56**, 501. Academic Press, New York.

Chapter 2

Fluorescence principles and measurement

Arthur G. Szabo

School of Physical Sciences, University of Windsor, 401 Sunset Avenue, Windsor, Canada N9B 3P4

1 Introduction

Fluorescence spectrometry is the most extensively used optical spectroscopic method in analytical measurement and scientific investigation. During the past five years more than 60 000 scientific articles have been published in which fluorescence spectroscopy has been used. The large number of applications ranges from the analytical determination of trace metals in the environment to pH measurements in whole cells under physiological conditions. In the scientific research laboratory, fluorescence spectroscopy is being used or applied to study the fundamental physical processes of molecules; structure–function relationships and interactions of biomolecules such as proteins and nucleic acids; structures and activity within whole cells using such instrumentation as confocal microscopy; and DNA sequencing in genomic characterization. In analytical applications the use of fluorescence is dominant in clinical laboratories where fluorescence immunoassays have largely replaced radioimmunoassay techniques.

There are two main reasons for this extensive use of fluorescence spectroscopy. Foremost is the high level of sensitivity and wide dynamic range that can be achieved. There are a large number of laboratories that have reported single molecule detection. Secondly, the instrumentation required is convenient and for most purposes can be purchased at a modest cost. While improvements and advances continue to be reported fluorescence instrumentation has reached a high level of maturity.

2 Physical principles

2.1 The absorption process

A review of the physical principles of the fluorescence phenomenon permits one to understand the origins of the information content that fluorescence measurements can provide. A molecule absorbs electromagnetic radiation through a quantum mechanical process where the molecule is transformed from a 'ground' state to an 'excited' state. The energy of the absorbed photon of light corresponds to the energy difference between these two states. In the case of light in the ultraviolet and visible spectral range of 200 nm to 800 nm that corresponds to energies of 143 to 35.8 kcal mol^{-1}. The absorption of light results in an electronic transition in the atom or molecule. In atoms this involves the promotion of an electron from an outer shell orbital to an empty orbital of higher energy. In molecules, an electron is promoted from the highest occupied molecular orbital (HOMO) to the lowest unoccupied molecular orbital (LUMO). This process is governed by a number of selection rules, and the transition probability for the absorption of a photon is reflected in the molecule's molar absorptivity, ε (M^{-1} cm^{-1}), that is determined from the Beer–Lambert law.

Figure 1(a) provides an example of the absorption process in the case of a carbon–carbon double bond. On interaction with a photon of appropriate

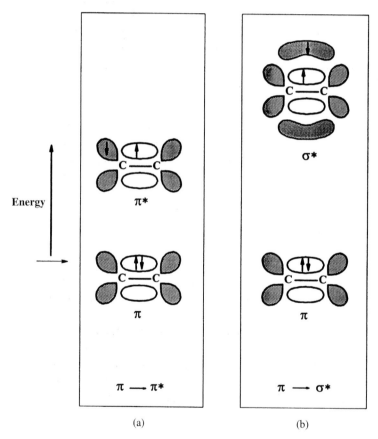

(a)   (b)

Figure 1 (a) A π–π* transition of a C–C double bond, with the lower energy π state showing the π bonding orbital filled with two electrons with antiparallel spin and the unfilled π antibonding orbital (shaded), and the higher energy π* state with the electron spins still antiparallel but with one electron in the π bonding orbital and one in the π antibonding orbital. (b) A π–σ* transition of a C–C double bond with the same lower energy electron configuration as in *Figure 1(a)*, and the higher energy σ* state with one electron in the π bonding orbital and one electron in the σ antibonding orbital (shaded).

energy an electron from the π bonding orbital of the ground electronic state is promoted to a π^* antibonding orbital (LUMO). The energy difference between these two states corresponds to the absorbed photon energy, $\Delta E = hc/\lambda$, where h is Planck's constant, c is the velocity of light in a vacuum and λ is the wavelength of the light. This excitation process is termed a π–π^* transition and the excited state is referred to as a π–π^* state. There are other antibonding orbitals of higher energy, and in this example such an orbital is a σ^* antibonding orbital. The excitation of an electron from the π bonding orbital to the σ^* antibonding orbital would occur when the photon energy corresponds to the energy difference between the ground state and the π–σ^* excited state (*Figure 1(b)*).

One of the selection rules that governs the transition probability is that the spin of the electrons must be retained, i.e. the electron spins in the ground state π bonding orbital are paired and antiparallel and they retain their antiparallel spin in the transition to the excited π–π^* state. The multiplicity, M, of these states describes the spin states of the electrons. Electrons are designated to have a spin quantum number, S, of either $+\frac{1}{2}$ or $-\frac{1}{2}$ and so the multiplicity of the state can be calculated according to *equation 1*:

$$M = 2|\Sigma S| + 1 \qquad\qquad 1$$

In the ground state $\Sigma S = 0$, and so $M = 1$ and this state is designated as a ground singlet state, S_0. In the π–π^* excited state the original electron spins are conserved and $M = 1$. This excited state is then designated as a singlet state. In the example in *Figure 1* the π–π^* state is the first excited singlet state, S_1, and the π–σ^* state is the second excited singlet state, S_2.

2.2 Excited singlet state deactivation processes

A modified Jablonski diagram (*Figure 2*) is a convenient description of the photon absorption processes and the deactivation processes of excited states. Associated with each electronic state are a set of vibrational states that correspond to different vibrational energies of the molecule. In turn associated with each vibrational state are sets of rotational states. The energy difference between the vibrational states corresponds to the infrared frequencies, and the rotational energy level differences correspond to microwave frequencies. At 20°C most molecules exist in the lowest vibrational level of the ground state. The total energy, E, of a particular state of a molecule can be represented as the sum of the electronic energy, E_{el}, the vibrational energy, E_{vib}, and the rotational energy, E_{rot}.

$$E = E_{el} + E_{vib} + E_{rot} \qquad\qquad 2$$

The overall change of energy as a result of the absorption of a photon of light is:

$$\Delta E = \Delta E_{el} + \Delta E_{vib} + \Delta E_{rot} \qquad\qquad 3$$

The absorption of a photon by a molecule is usually to a vibrational level above the 'zeroth' vibrational level of the excited singlet state. The molecular absorption spectrum traces out the energy difference between the ground state

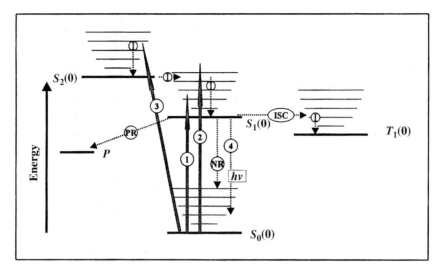

Figure 2 A modified Jablonski diagram showing the different transitions and processes between the excited states. $S_0(0)$, $S_1(0)$ and $S_2(0)$ being the 0 level vibrational energy level of ground state, first singlet state and second singlet state, respectively. $T_1(0)$ is the lowest vibrational level of the first triplet state and P represents the energy level of a stable photoproduct which may be formed. The transitions with the arrows pointing towards the top of the figure represent absorption transitions and they are described in the text. The dashed arrows pointing towards the bottom of the figure represent competing pathways by which the excited molecule deactivates. I represents internal conversion processes; ISC, intersystem crossing; NR, non-radiative deactivation; PR, photoproduct formation; $h\nu$ the radiative or fluorescence process.

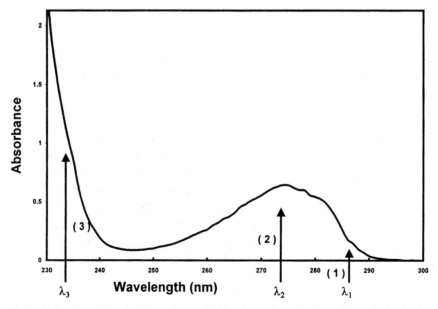

Figure 3 An absorption spectrum representing the three absorption transitions designated 1, 2 and 3, in *Figure 2*.

and the several vibrational levels of the excited singlet states. *Figure 3* is a representation of a typical absorption spectrum, where absorption at λ_1 corresponds to the absorption transition marked 1 (*Figure 2*). Another wavelength, λ_2, would correspond to transition 2 (*Figure 2*) to an upper vibrational level of S_1, and λ_3, corresponds to absorption transition 3 (*Figure 2*) to S_2.

The light absorption process occurs very rapidly (10^{-15} s) to form the excited state, so fast that there is no change in the nuclear coordinates of the molecule during the absorption process (Frank Condon transition). If the absorption is to S_2, the molecule undergoes a very rapid process (10^{-14} s) known as internal conversion from the lowest vibrational level of S_2 to an upper vibrational level of S_1. In solution the molecule rapidly loses the excess vibrational energy through collision with solvent molecules. The net result is that within 10^{-13} to 10^{-14} s of excitation the molecule is in the lowest (0th) vibrational energy level of S_1 ($S_1(0)$). The absorption process has very rapidly created a metastable state that can undergo a number of competing processes that were not available to the ground state.

2.2.1 Intersystem crossing

The process known as intersystem crossing occurs when one of the electrons undergoes a spin inversion from either $+\frac{1}{2}$ to $-\frac{1}{2}$ or vice versa. This results in an electronic state of lower energy (single electrons in different orbitals prefer to have parallel spin) with a multiplicity $M = 3$, which is thus designated as a triplet state, T_1. The rate constant that is associated with the intersystem crossing process is k_{ISC}.

2.2.2 Photoproduct formation

Since S_1 is an electronic state that has a different electronic distribution from S_0, the ground state, the molecule may undergo changes in molecular orbital overlap. A new bonding configuration may result in the formation of a new molecule or photoproduct. This is a photochemical process that has an associated rate constant, k_{PR}, that competes with other S_1 deactivation processes.

2.2.3 Non-radiative deactivation

Another process of internal conversion can occur when the 0th level of S_1 overlaps with a very high energy vibrational level of S_0. The electron moves back into the unfilled bonding orbital, and the molecule then relaxes very rapidly through a series of vibrational steps to the lowest level of S_0. Non-radiative relaxation to an upper vibrational level of S_0 can also occur through interactions with solvent. The excess energy of the molecule is converted into heat energy. These non-radiative deactivation pathways of S_1 to S_0 are grouped into a single rate process described by the non-radiative rate constant, k_{NR}.

2.2.4 Radiative deactivation (fluorescence)

The molecule can also return to one of several vibrational/rotational energy levels of S_0 by the emission of a photon of light that corresponds to the energy difference between $S_1(0)$ and a particular vibrational level of S_1 (transition 4 in *Figure 2*). The emission of light from an upper singlet state to the ground state is known as fluorescence. In this radiative process the electron in the π^* anti-bonding orbital makes a quantum jump back into the half filled π bonding orbital emitting a photon of light whose energy corresponds to the energy difference between these two states. The rate constant that is associated with the fluorescence radiative process is defined as k_R. Just as the absorption of light occurs between the lowest vibrational state of S_0 ($S_0(0)$) and one of several vibrational levels of S_1, the fluorescence radiative process takes place from $S_1(0)$ to any of the vibrational levels of S_0.

2.2.5 Competitive deactivation processes

All of these different processes of deactivation of S_1 compete with one another. Later other rate processes of deactivation of S_1 will be described that will also be in competition with the processes discussed above. These different kinetic processes are summarized in *Table 1*.

Table 1 Summary of deactivation processes of the excited singlet state of a molecule, M

Absorption of a photon: after vibrational relaxation to $S_1(0)$	$M(S_0)$ + $h\nu$	$\rightarrow$ $M(S_1)$				
Intersystem crossing:	$M(S_1)$ $\rightarrow$ $M(T_1)$	k_{ISC}				
Photochemistry:	$M(S_1)$ $\rightarrow$ Product	k_{PR}				
Non-radiative relaxation:	$M(S_1)$ $\rightarrow$ $M(S_0)$	k_{NR}				
Radiative (fluorescence):	$M(S_1)$ $\rightarrow$ $M(S_0)$	+ $h\nu$	k_R			
Quenching:	$M(S_1)$ + [Q]	$\rightarrow$ $M(S_0)$	+ [Q]	$k_Q[Q]$		
Resonance energy transfer	$M(S_1)$ + Acceptor	$\rightarrow$ $M(S_0)$	+ Acceptor	k_{ET}		

Where $M(S_1)$ corresponds to the electronically excited molecule, M.

It is this competition between the different processes that provides an important aspect of the information content of fluorescence spectroscopy and makes it so useful in the determination of the properties of molecules. If the molecule that is excited undergoes changes in its structure or environment, then usually one of the rate constants, k_{ISC}, k_{NR}, or k_{PR} is affected with a concomitant change in the relative probability of the fluorescent process.

2.2.6 Phosphorescence

Phosphorescence is also a radiative process. It describes the radiative process originating from T_1 to S_0. The required spin inversion of one of the electrons when it moves from the LUMO to the half filled HOMO is a forbidden process. When observed it occurs on a timescale much longer than the fluorescence process. In most cases it can only be observed in frozen glass solutions at low temperature (77 K).

3 Fluorescence parameters

3.1 Fluorescence spectrum

The light emitted by a molecule from $S_1(0)$ corresponds to the energy difference between $S_1(0)$ and different vibrational levels of S_0, i.e. $S_0(n)$. There will be a spectrum of energies of the fluorescence photons and this energy spectrum (*Figure 4*) is what is measured on a fluorescence spectrometer. The wavelength, λ_0 corresponds to the radiative transition from $S_1(0)$ to $S_0(0)$. This is the highest energy radiative transition and is the energy difference between these two states. At λ_{MAX}, the wavelength corresponding to the highest fluorescence intensity, the transition from $S_1(0)$ to $S_0(n)$ represents the most probable radiative transition between these two electronic states.

3.2 Fluorescence quantum yield, Φ_F

The quantum yield of fluorescence, Φ_F is defined according to:

Φ_F = number of photons emitted/total number of photons absorbed

Obviously the maximum value of Φ_F is 1, that is all molecules that have been excited to S_1 return to S_0 by the emission of a light photon. It can be shown that Φ_F is related to the several rate constants discussed above:

$$\Phi_F = \frac{k_R}{k_R + k_{NR} + k_{ISC} + k_{PR}} \qquad 4$$

Other rate processes to be described later will be included in this relationship. The denominator of this expression is related to A, the absorbance of the

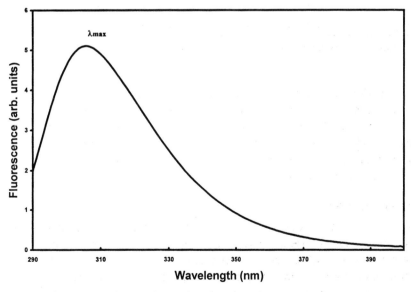

Figure 4 A typical fluorescence spectrum representing the energy distribution of the photons from $S_1(0)$ to different vibrational levels of the ground state, S_0.

molecule. If I_0 is the intensity of exciting light in photons s^{-1} then the integrated intensity of fluorescence, I_F (photons s^{-1}) is:

$$I_F \propto \text{(total number of photons absorbed / s, } I_A). (\Phi_F) \qquad 5$$

From the Beer–Lambert law:

$$A = \varepsilon cl \quad \text{and} \quad \varepsilon cl = \log (I_0/I) \qquad 6$$

where I is the transmitted light intensity in photons s^{-1}.

$$I_A = I_0 - I = I_0(1 - 10^{-\varepsilon cl}) \qquad 7$$

and

$$I_F = I_0(1 - 10^{-\varepsilon cl}) (\Phi_F) \qquad 8$$

Thus the integrated fluorescence signal depends on the incident light intensity, the absorbance of the sample at the wavelength of excitation, λ_{EX}, and Φ_F.

3.3 Singlet and radiative lifetime

If one excites a sample with an infinitely narrow pulse of light creating a population of excited singlet state molecules $M(S_1)$, one can write a differential equation that relates the concentration of $M(S_1)$ at any time, t, to the various deactivation processes:

$$\frac{-d\, M(S_1)}{dt} = (k_R + k_{NR} + k_{ISC} + k_{PR})\, M(S_1) \qquad 9$$

and on integration this leads to:

$$M(S_1)(t) = M(S_1)(0)\, e^{-\Sigma kt} \qquad 10$$

where $M(S_1)(t)$ is the concentration $M(S_1)$ at any time t; $M(S_1)(0)$ is the concentration of $M(S_1)$ at time $t = 0$; and $\Sigma k = k_R + k_{NR} + k_{ISC} + k_{PR}$. Since $I_F \propto M(S_1)$ then one can write the expression:

$$I_F(t) = I_F(0)\, e^{-\Sigma kt} \qquad 11$$

By definition the lifetime of the excited singlet state, $M(S_1)$, is:

$$\tau_S = \frac{1}{k_R + k_{NR} + k_{ISC} + k_{PR}} \qquad 12$$

showing that the singlet excited state has a finite lifetime that is dependent on the efficiency of the several competing deactivating processes. The radiative lifetime of S_1 is defined according to:

$$\tau_R = 1/k_R \qquad 13$$

Thus if the only deactivation process of S_1 was fluorescence then $\tau_S = \tau_R$, and $\Phi_F = 1$. An important relationship is:

$$\Phi_F \tau_R = \tau_S \qquad 14$$

When one measures τ_S and Φ_F then the intrinsic molecular rate constant, $k_R = \tau_R$, can be calculated.

4 Fluorescence spectrometers

The above section has outlined the physical parameters that describe the fluorescence process. One can measure the fluorescence spectrum, $F(\lambda)$, the singlet excited state lifetime, τ_S, and determine the Φ_F. These parameters can be interpreted in terms of the structure, environment and dynamics of the molecule of interest. In this section, the different optical and electronic components comprising an instrument that can measure $F(\lambda)$ and Φ_F will be described. This instrument is generally known as a steady state fluorescence spectrometer, since it integrates the fluorescence intensity over a given time period. Time-resolved fluorescence instrumentation that is used to measure the excited singlet state decay times is described in Chapter 3.

Figure 5 is a block diagram of the essential components of a fluorescence spectrometer. All fluorescence spectrometers comprise a light source, an excitation wavelength selector (the wavelength of excitation is determined from the molecule's absorption spectrum), a sample holder, a second wavelength selector to isolate the fluorescence wavelength, and a photodetector that converts any incident fluorescence photons into an electronic signal. Another important

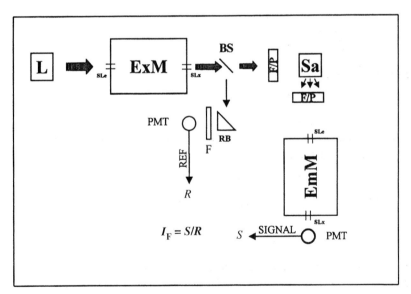

Figure 5 A block diagram of a typical fluorescence spectrophotometer. L represents the light source; ExM and EmM, the excitation and emission monochromators, respectively; BS is a quartz glass beam splitter; F/P is a filter or polarizer; Sa is the sample holder; Sle and Slx are the entrance and exit slits of the monochromators; RB is the rhodamine B sample used in the reference channel; F; is the red filter in that channel; PMT, represents the photomultiplier tubes which detect the fluorescence and reference fluorescence signals, *S* and *R*, respectively; the shaded arrows represent the excitation light beam; the solid arrow pointing towards RB represents *c.* 10% of the light reflected from BS; and the arrows from Sa represent a portion of the fluorescence light intensity that is emitted in a sphere surrounding the sample.

component, the reference channel, is a system to create an electronic signal that is proportional to the incident light intensity on the sample.

4.1 The light source

When a solution sample is excited to S_1 the fluorescence that emanates from the sample propagates in all directions. The fluorescence photons that pass through the emission monochromator (EmM, *Figure 5*) are only a small fraction of the total fluorescence intensity. It is necessary therefore to use a light source that has a relatively high intensity light output. In very simple fluorescence spectrometers used for routine dedicated analyses, a Mercury arc lamp may be satisfactory, but it only produces light at a limited number of wavelengths in the ultraviolet and visible spectral range (e.g. 253.7, 302.2, 313.2, 435.8, 546.1 nm). The continuum lamps used in absorption spectrometry, deuterium and quartz/iodine sources have too low an intensity and hence the fluorescence intensity is also very low. Most fluorescence spectrometers use a high pressure xenon arc lamp as an excitation source since it produces a useful high intensity continuum light spectrum from 200 nm to 1000 nm (*Figure 6*). These lamps are designated according to the electrical power consumed, i.e. 150 W, 450 W, 1000 W. In instruments recently made available, a 75 W xenon arc is used, but it is pulsed, so that the peak intensity of each pulse is high.

4.2 Wavelength selectors

In simple spectrometers that have a dedicated use, optical glass filters are often used. There are a very large number of bandpass filters available commercially

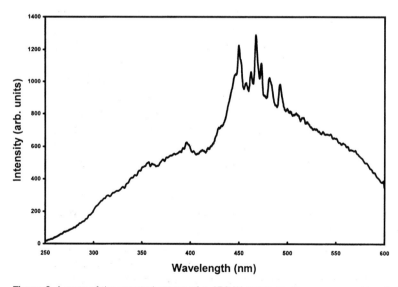

Figure 6 A scan of the spectral output of a 450 W xenon arc lamp measured by placing an optically dense sample of rhodamine B in the sample holder. The fluorescence passed through a 610 nm cut-off filter, and EmM was set at 620 nm. Excitation bandpass, 2 nm; emission bandpass 2 nm.

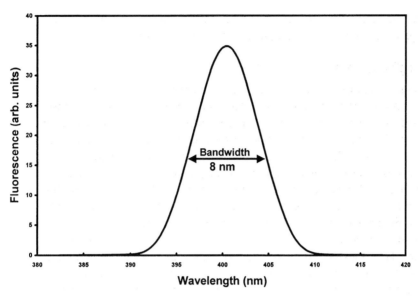

Figure 7 A trace of the monochromator spectral bandpass, with an emission slit width of 2 mm, and a reciprocal dispersion, D^{-1} of 4 nm mm^{-1}.

that can be used to isolate a particular line of a light source, such as a mercury arc, a portion of the continuum spectrum of a xenon arc, or a portion of the sample's fluorescence spectrum. Besides selecting for the maximum transmission wavelength of a filter, one should also consider the half bandwidth that defines the range of wavelengths that the filter will transmit. In addition one must be cognisant that the intensity of transmitted light through a filter will be reduced as the bandwidth is reduced.

The excitation wavelength, λ_{EX}, or fluorescence wavelength, λ_{EM}, of light is often selected through a monochromator. The dispersion element in a monochromator may be either a grating or a quartz prism. A grating monochromator is preferred since the reciprocal dispersion, D^{-1}, is constant over all λ, while for a prism this parameter varies with λ and is cumbersome to account for. Hence a quartz monochromator is rarely used in most standard laboratory instruments. D^{-1} is the parameter that relates the effective bandwidth of λ passing through a monochromator to the slit widths (mm) of the monochromator and has units of nm mm^{-1} (*Figure 7*). The value of D^{-1} depends on the quality of the monochromator, but typically it ranges from 2 nm mm^{-1} to 8 nm mm^{-1}.

One further point to note is that a grating monochromator disperses light in a series of orders. Thus a monochromator setting at 500 nm will also pass light whose wavelength is 250 nm since the second order of dispersion of light at λ_1 will occur at $2\lambda_1$. Placing an appropriate cut-off filter after a monochromator, i.e. a Pyrex filter in this example, can efficiently remove this second-order dispersion.

4.3 Sample excitation components

Light from the xenon arc is collected with appropriate mirrors in the lamp housing and then focused through a lens into a light beam. The appropriate λ_{EX} is determined from the absorption spectrum of the sample and is selected using either a filter or monochromator (ExM, *Figure 5*). In most laboratory instruments a monochromator is used to select λ_{EX} and entrance and exit slit widths set the effective bandwidth of λ. Choice of λ_{EX} and slit widths are discussed below. An excitation double monochromator is an advantage when one is working with samples that are highly light scattering, such as cellular suspensions, micelles, or membrane preparations. This dramatically reduces the stray light that passes through the monochromator and falls on the sample. The use of a double monochromator will result in a lower light throughput and may necessitate using a higher power light source.

The light beam passing through ExM is collected by a lens that collimates the light beam. The next element that is usually encountered is a thin piece of quartz glass, placed at an angle of 45° to the light beam. It can be appreciated that any fluctuation or 'jitter' in the light intensity of the xenon arc will cause a corresponding fluctuation in the sample fluorescence signal, since the fluorescence intensity is proportional to the intensity of the excitation. In addition the light output of a xenon arc decreases as the lamp ages. Further, if one wishes to compare the fluorescence intensity at two or more λ_{EX} then the intensity spectrum of the light source at the two wavelengths needs to be accounted for. The beam splitter reflects a small fraction (*c.* 10%) of light into a 'reference' channel having a cell containing a concentrated ethylene glycol solution of one of the rhodamine dyes, usually rhodamine B (RB). The high concentration of RB (*Protocol 1*) results in the solution having an absorbance >2.5 from 250 to 600 nm. Effectively all light at $\lambda < 600$ nm incident on this cell is absorbed, and most within the first mm of the solution. The dye fluorescence occurs at $\lambda > 600$ nm and a red cut-off filter placed after the dye cell permits only the dye fluorescence to pass. Since the RB absorbance is so high any variation in lamp intensity will be directly reflected in dye fluorescence intensity. The fluorescent photons from the dye fall onto a photodiode or photomultiplier that subsequently generates an electronic signal, *R*, that is proportional to the reference fluorescence intensity.

4.4 Sample compartment

The sample cells used in fluorescence spectroscopy have all four sides clear, since in most instruments the fluorescence photons are detected at an angle of 90° to the propagation direction of the excitation beam. The 90° geometry is used in order to minimize any interference in the fluorescence light detection by the excitation light beam. In most cases the cuvettes are made from fused silica (quartz) but if the excitation and fluorescence wavelengths are above 300 nm plastic disposable cuvettes may be used.

The deactivation processes of $S_1(0)$ are often highly temperature dependent

Protocol 1

Preparation of reference solution

Equipment and supplies

- High purity rhodamine B
- Ethanol
- Ethylene glycol, spectral grade
- Methanol, spectral grade
- 25-ml scintillation vial with a screw cap
- Polyfilm
- Pasteur pipette
- 1-cm square quartz stoppered cuvette or 1-cm triangular quartz stoppered cuvette
- Sonication bath
- Optical tissues
- Latex gloves

A. Solution preparation

1 The rhodamine dyes are considered to be toxic. It is important to wear gloves while handling the dyes and to avoid all spills. Weigh 15 mg of rhodamine B into the scintillation vial.

2 Add 0.5 ml of ethanol to aid the solubility of the rhodamine. Add 4.5 ml of ethylene glycol and cap the vial. Shake the solution and then hold it in a warm sonication bath until there is no evidence of any particulate matter in the sample. The sonicator assists in the solution of the rhodamine.

3 Totally fill the cuvette with the rhodamine solution using a Pasteur pipette. Stopper the cuvette, and wrap the stopper and top of the cuvette with a small piece of polyfilm. Wipe the outside of the cuvette with a tissue wet with methanol and then a dry tissue.

4 Place the cuvette into the reference cell holder of the instrument.

5 The solution should be replaced approximately every six months of steady use.

B. Cleaning the optical elements of the reference channel

1 It is a good idea to clean the optical elements of the reference channel. The beam splitter and red cut-off filter should be carefully wiped with a tissue wet with methanol followed by wiping with a dry tissue.

and so the cuvette should be contained in a cell holder that can be thermo-statted. The fluorescence photons are then collected with a lens and focused onto the fluorescence wavelength selector.

A sample holder that can hold four cuvettes in a rotating turret is an advantage. In this way other samples and blank solutions can be maintained at the same temperature and one can simply rotate between the four cuvettes.

4.5 Fluorescence path optical components

In simple fluorimeters where a limited number of analytes are measured a bandpass or cut-off filter may be used to select the fluorescence photons. If a cut-off filter is used, then light of all λ above the cut-off λ will be detected. This will permit the measurement of very low fluorescence light intensities and correspondingly low concentrations of the analyte. In standard laboratory fluorimeters where a large variety of different fluorescence measurements will be performed, a grating monochromator (EmM, *Figure 5*) is used to select λEM. The fluorescence from the sample is focused onto the entrance slit of EmM and after passing through the exit slit falls onto the photocathode of a photo-multiplier tube (PMT).

4.6 Fluorescence instrumentation electronics

The PMT converts the photon intensity into an amplified electronic signal that can be processed with electronic circuitry in either an analogue or digital mode. The photon energy falling on a photocathode of a PMT causes the ejection of 'photoelectrons' from the bi-alkali metal surface. These electrons in turn cause the ejection of electrons from a dynode, and this process is repeated at several subsequent dynodes until a very large number of electrons reach the anode of the PMT. The electron current is proportional to the number of photons falling on the photocathode surface and hence the fluorescence intensity. The anode current then passes into electronic circuitry where it is translated into a voltage signal. This analogue signal, S, is proportional to the average photon density falling on the PMT. By appropriate electronic circuitry the signal S can be divided by the internal reference signal, R, to provide the ratio, $I_F = S/R$ and thus I_F will be independent of any variation of excitation light intensity. The signal, I_F, can then be relayed to a readout device such as a chart recorder. More commonly, today's instruments are interfaced to a personal computer where the electronic signals can be stored, processed or displayed.

4.6.1 Photon counting

Many of the current spectrofluorophotometers operate in a digital or photon counting mode. In these instruments, the electronic circuitry permits recording of individual fluorescence photons as an anode pulse. This has several positive advantages. Most importantly thermionic emission of electrons from any of the dynodes of the PMT that also results in an anode signal will be rejected by the electronics of the instrument. These pulses will have a lower voltage than an anode pulse originating from a fluorescence photon that falls on the photo-cathode. This latter anode pulse will have a narrow band of a characteristic voltage, and pulses of lower voltage can be electronically discriminated and not recorded. By contrast, in an analogue instrument all thermionic emission from the dynodes is recorded and hence the background or 'dark' signal will be much higher than in a photon counting instrument. Since one is counting photons, it can be readily appreciated that the instrumental sensitivity will be higher in these instruments, as single photon events will be detected. In photon counting, the detection system is much more stable and less affected by small variations in the PMT voltage. A disadvantage is that stray light sources or light leaky monochromators may cause a background signal from such parasitic light. One additional disadvantage is that high light intensity levels result in a significant non-linearity, as two photons arriving simultaneously at the photocathode will only be counted as a single photon, or depending on the electronics will be rejected entirely. However, this situation can be easily avoided in a number of convenient and different ways. The enhanced sensitivity and lower background of a photon counting instrument outweighs any disadvantages.

4.6.2 Computer interfacing

Most instruments sold today that function in either an analogue or photon counting mode are interfaced to a personal computer. The computer software from the supplier should control the entire operation of the spectrofluoro-photometer, save the setting of the slits. The software will control the stepping motors for changing the monochromator λ, the sample PMT and reference PMT voltages, and in some cases the rotation of the sample turret and any light polarizers (*vide infra*) in the instrument. The computer interfacing will be used to set all instrument parameters required for the recording of a fluorescence spectrum, including the spectral wavelength range, the scan rate (nm s^{-1}) and the related integration time at each λ position (photon counting instrument). Because the fluorescence spectrum is recorded and stored in the computer memory, it subsequently can be conveniently manipulated. Blank or reference spectra can be subtracted; spectra can be normalized, smoothed according to some spline function and integrated, and multiplied or divided by different factors for spectral comparison; and differences between spectra can be re-viewed. Finally, the output of such instruments can be conveniently formatted to whatever representation one requires.

5 Fluorescence spectra

5.1 Inner filter effect

After the sample's absorption spectrum has been measured, one can select an ex-citation wavelength, λ_{EX}, that falls within the absorption envelope. There will be other factors that may influence the selection of λ_{EX}. If the sample is homogene-ous and contains a single molecular chromophore, then any wavelength may be appropriate. In the case of samples that contain more than a single molecular chromophore, such as in a protein (*vide infra*), then the choice of λ_{EX} will depend on whether preferential excitation of one of those chromophores is desired.

An important parameter of the sample that must be considered is the absorbance of the sample at λ_{EX}. *Equation 8* shows that the rate of fluorescence emission, I_F, is determined by the rate of light absorbance. In dilute solution *equation 8* can be expanded in terms of a series and where εcl is small then it can be approximated that:

$$I_F = I_0(\varepsilon cl)\, \Phi_F \qquad\qquad 15$$

This approximation can lead to an error known as the inner filter effect. The solution at the front of the cell is exposed to a higher excitation light intensity

Table 2 Percentage error due to inner filter effect

Absorbance	Percentage error in I_F
0.01	1.1
0.05	5.5
0.10	10.6
0.20	20.0

than the sample near the opposite side, because of the absorbance of the intervening solution. This will introduce an error and non-linearity in the relation between I_F and A. The terms in the series expansion of *equation 8* that have been neglected in *equation 15* can be evaluated and the percentage error in the fluorescence signal owing to the inner filter effect can be calculated (*Table 2*). One usually adjusts the concentration of the sample so that A at λ_{EX} is <0.1 in order to minimize inner filter errors. If it is necessary to perform an experiment at a certain concentration of the sample and this results in a very high absorbance at the absorption maximum, λ_{AMAX}, then a λ_{EX} can be selected on the low energy side of the absorption spectrum where $A < 0.1$.

5.2 Light scattering

All samples will scatter the exciting light to some extent. There are two scattering phenomena that the operator must be aware of. One is elastic or Rayleigh scattering and this depends on the particle size of the solute or any suspended material. The intensity of Rayleigh scattering, I_{RS}, is proportional to r^6/λ^4, where r is the particle radius. Scattering from the sample results in a large 'scatter' signal at emission wavelengths, λ_{EM}, close to λ_{EX}. *Figure 8* demonstrates the origin of this effect. In this example λ_{EX} and λ_{EM} are 5 nm apart. If the bandpass of each monochromator is 4 nm (typical of many measurements), there is an overlap in the bandpass curves and light wavelengths in the shaded area will contribute to the signal in the emission detector owing to sample scattering.

Obviously, one way to reduce such a scatter signal would be to reduce the bandpass of both ExM and EmM. However, this significantly reduces the signal that one is measuring. Thus a compromise needs to be reached between scatter contribution and fluorescence signal intensity. As stated above most fluorescence spectral measurements use bandwidths of 4 nm. Therefore, one usually begins the recording of the fluorescence spectrum from a λ_{EM} that is 10 nm higher than λ_{EX}. In a solution of a single chromophore this scattering artefact can be easily avoided by choosing an λ_{EX} close to λ_{AMAX}. In such a case, one should choose a starting wavelength for the fluorescence measurement, λ_I, that is on the low energy (red) edge of the absorption spectrum. Otherwise another inner filter error is encountered where there is appreciable self-absorption of the fluorescence in the absorption envelope. Measuring the signal from a solvent blank cannot correctly account for elastic scattering as there is no scattering solute in the blank. In the case of fluorescence measurements of cell suspen-

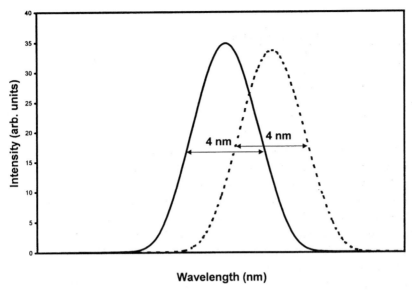

Figure 8 A demonstration of the overlap of excitation and emission monochromator bandpass when λ_{EX} and λ_{EM} are close to one another thus explaining the observation of Rayleigh scattering of λ_{EX}.

sions, dispersed particulate samples, and micelles or vesicle samples, Rayleigh scattering is a significant problem. A double monochromator that has lower stray light and a narrower bandpass can reduce this effect. A method to correct for the scattering of such samples has been proposed and applied with some success (1, 2).

The second type of scattering that cannot be avoided is inelastic Raman scattering of the solvent. In this phenomenon some of the excitation light energy is abstracted by vibrational modes of the solvent molecules. In the case of water or hydroxylic solvents, the most dominant vibration that absorbs this energy is the O–H stretch. The energy of the O–H vibration is observed near 3300 cm^{-1}. The Raman signal from the solvent will be observed at a wavelength that is 3300 cm^{-1} lower in energy than λ_{EX}. The λ of Raman scattering, λ_{RA}, can be calculated in the following way:

$$\frac{1}{\lambda_{RA}} = \frac{1}{\lambda_{EX}} - 0.00033 \qquad\qquad 16$$

For example with λ_{EX} = 290 nm a Raman scatter peak will be observed at λ_{RA} = 321 nm. *Figure 9*, includes a spectrum of an aqueous blank that shows this Raman band and the effect of the Raman signal is seen on the sample fluorescence spectrum. In the latter case it appears as a weak shoulder on the high energy side of the fluorescence wavelength maximum, λ_{FMAX}. When the sample fluorescence is intense the contribution of the Raman band will be negligible. A simple test to determine if a spurious peak is due to a Raman scattering signal is

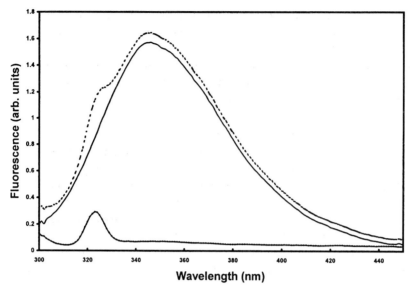

Figure 9 The dashed curve is the fluorescence spectrum of a protein sample with $\lambda_{EX} = 290$ nm. The solid curve with circles is the spectrum of the buffer blank, showing a maximum signal at 321 nm, corresponding to the Raman scatter of the water. The solid curve is the signal resulting from subtracting the buffer blank from the raw spectral data.

to shift λ_{EX} by 10 nm to a lower wavelength. The Raman band will shift in the same direction.

5.3 Instrumental settings

The first step is to set ExM at λ_{EX} considering the factors discussed above. In most cases an excitation bandpass of 4 nm is satisfactory. If the sample fluorescence is weak and Rayleigh scattering can be avoided then the fluorescence signal can be increased by increasing the slit width of ExM. By varying λ_{EM} of EmM in a regular manner the fluorescence signal, I_F, plotted against λ_{EM} gives the sample fluorescence spectrum $F_M(\lambda)$ (*Figure 9*). A 4 nm bandpass is typical of most measurements unless there are very narrow vibrational band features in the fluorescence spectrum that one wishes to resolve. The fluorescence scan is initiated at λ_I that is usually 10 nm $> \lambda_{EX}$ in order to avoid a scattered light component in the spectrum. A wavelength range of between 100–200 nm is scanned depending on the sample fluorescence spectrum. It is good practice to record the fluorescence spectrum to λ_F where the fluorescence intensity is reduced to a value close to zero. Recall, the recorded signal has been ratioed by the reference signal, R. Since most instruments now use stepping motors to scan the monochromators the step size becomes important. In the interest of good spectral resolution a step size of no greater than 1 nm should be selected. The integration time chosen for each step should be sufficient to obtain a good signal to noise, S/N, ratio. An integration time of 0.5 s or 1 s is usually sufficient. If a very wide wavelength range is being scanned then the total time to acquire the total spectrum becomes a consideration. This becomes a highly important factor if the sample undergoes any appreciable photodecomposition (k_{PR}) as this will reduce or modify the fluorescence signal.

It is always necessary to record a solvent/buffer blank under identical instrumental conditions. This can be subsequently subtracted from the sample fluorescence spectrum for quantitative purposes.

5.4 Fluorescence spectral corrections

When fluorescence spectra having different shapes and λ distributions are to be compared quantitatively it is necessary to correct for the wavelength-dependent sensitivity of the EmM/PMT combination. If the spectra of a solute under different sample conditions have identical shape and only vary in intensity, then no instrumental correction is necessary. PMTs do not have constant photon sensitivity across the useful UV–VIS spectral range. A typical photon sensitivity curve is shown in *Figure 10* for a Hamamatsu R955 PMT, a PMT that is used in many instruments. It would be desirable is to have a flat wavelength response. Obviously, this sensitivity versus λ variation will distort the true fluorescence spectrum of the solute and lead to errors in quantitation, especially where spectral integration is performed. Superimposed on the PMT λ sensitivity variation is a smaller effect of the light throughput dependency on λ of the monochromator.

There are three methods of generating spectral correction curves to account for these sensitivity variations. One is to measure the fluorescence spectra of standard compounds whose true fluorescence spectra have been determined and reported (3). Then the sensitivity factors could be calculated by comparing the experimentally measured spectra with the reference spectrum.

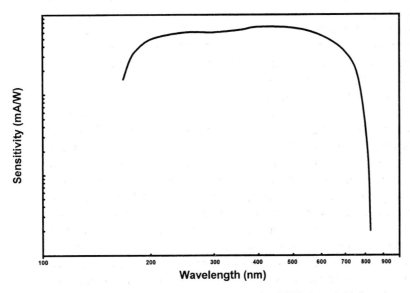

Figure 10 The spectral sensitivity curve of a Hamamatsu R955 photomultiplier tube.

A second method that is often used is to illuminate a solid scatterer, such as MgO, which is held in the sample holder, using a calibrated standard tungsten lamp available from the National Institute of Standard Technology (US). Such lamps act as a black body radiator that has a smooth light output intensity variation with wavelength. Light from the scatterer passes through EmM and is detected by the PMT. The measured spectral intensity curve is compared to the lamp calibration values and a reliable spectral correction curve can be generated and stored in the instrument computer for use in correcting spectra. A limitation of this procedure is that such tungsten lamps have negligible output at $\lambda < 360$ nm. Thus a correction curve can be determined from 360–800 nm with this lamp system. A further consideration is the necessity of purchasing a calibrated lamp, housing and power supply at an appreciable cost, in order to determine a correction factor curve that needs to be measured only rarely.

A method that has shown to be quite satisfactory and relatively straight-forward is described in *Protocol 2*. In this method a diffuser plate or an MgO scatterer plate is placed in the sample holder and then both ExM and EmM are scanned synchronously from 250–600 nm. This will produce a spectrum of the xenon arc source that has been distorted by the sensitivity variation of the emission detection system. The reflector is replaced by a concentrated sample of rhodamine B in a triangular cuvette. The cuvette is oriented so that the hypotenuse is on the face opposite to EmM. This is a similar configuration to the cell in the reference channel of the instrument. A red cut-off filter is placed in front of the entrance slit to EmM and λ_{EM} is set to 620 nm. Then ExM is scanned from 250–600 nm and the spectrum is recorded. This spectrum is a close representation of the true xenon lamp intensity profile. Often neutral density filters will have to be placed in the emission beam in order to prevent signal saturation of the PMT. The two curves thus generated can be compared and the correction factor curve, $C(\lambda)$, calculated:

$$C(\lambda) = \frac{F_{RB}(\lambda_{EX}, 620 \text{ nm})}{F_{MM}(\lambda_{EX}, \lambda_{EM})} \qquad 17$$

where $F_{RB}(\lambda_{EX}, 620 \text{ nm})$ is the recorded spectrum of rhodamine B with EmM constant at 620 nm and λ_{EX} is varied, and $F_{MM}(\lambda_{EX}, \lambda_{EM})$ is the recorded spectrum using a diffuser plate of MgO and both ExM and EmM are varied synchronously.

If $F_M(\lambda)$ represents the measured uncorrected sample fluorescence spectrum then a corrected sample fluorescence spectrum can be obtained by a multiplication of $C(\lambda)$ and $F_M(\lambda)$ according to

$$F_M(\lambda)_{CORR} = C(\lambda)F_M(\lambda) \qquad 18$$

Thus the fluorescence signal at each measured λ of the sample is multiplied by the correction factor at that λ. Again this process can be easily achieved in the data acquisition computer.

Protocol 2

Generation of spectral correction factors (250–600 nm)

Equipment and supplies

- A quartz diffuser plate mounted in a holder so that the plate is oriented at an angle of 45° to the excitation light propagation direction or a metal plate coated with an MgO film. This plate should be mounted in a holder so that it makes an angle of 35° to the excitation light propagation direction.

- A triangular quartz cuvette filled with a concentrated ethylene glycol solution of rhodamine B (see *Protocol 1*)
- Quartz neutral density filters
- Fine wire mesh screens
- 610-nm glass cut-off filter

A. Diffuser set-up and signal levels

1 Place the diffuser plate or MgO plate into the sample holder. In the case of the diffuser plate, the incident light face should be on the side opposite to the EmM entrance slit. In this way light that passes through the plate and is scattered by it will fall on the entrance slits, but reflected light will be avoided. For the MgO plate the incident face should be on the same side as the entrance slit of EmM.

2 Adjust the slits of ExM to a 1-nm bandpass. The bandpass of EmM should be 4 nm or whatever bandpass is commonly used for fluorescence spectral determination. Place a Neutral Density (ND) filter with ND = 1 in front of EmM. Set both monochromators to 400 nm and open the shutters of the instrument. Note the signal level of the emission detector. The level should be well below that where the PMT is being saturated with light. If saturation is observed then place additional ND filters or fine wire mesh screens in front of the EmM entrance slits until a modest signal level is achieved.

B. Instrument settings for $F_L(\lambda)$

1 Set the instrument so that both monochromators will run synchronously.

2 Set both monochromators to 250 nm.

3 Set the wavelength scan range to 250–600 nm.

4 Set the instrument so that the signal detected is not ratioed by the reference signal, R.

5 Set a scan rate of 2 nm s^{-1} if the instrument monochromators are not controlled by stepping motors. In the case where the monochromators are controlled by stepping motors set the step size to 0.5 nm and the integration time to 0.2 s. This step size will be such that most sample fluorescence spectra can be corrected.

C. Record spectrum $F_L(\lambda)$

1 Record the spectral light signal between 250–600 nm with both monochromators scanning synchronously. This spectrum designated $F_L(\lambda) = F_{MM}(\lambda_{EX}, \lambda_{EM})$ should be similar in shape to the spectrum of the xenon lamp. It is distorted by the λ sensitivity of the emission detection system.

Protocol 2 continued

D. Rhodamine sample

1 Replace the diffuser plate by the triangular cuvette of rhodamine. The incident face should be opposite to the entrance slit of EmM. In this way the only light incident on EmM will be rhodamine B fluorescence. Reflected light is directed in the opposite direction.

E. Instrument settings for $F_T(\lambda)$

1 Remove the ND filters and any wire mesh screens and replace them with the 610-nm cut-off filter.

2 Set ExM to 250 nm and EmM to 620 nm.

3 The wavelength scan range and scan rate should be the same as for the previous spectrum.

4 Set the instrument so that only ExM will scan.

5 Record the spectral light signal between λ_{EX} 250–600 nm. This spectrum is considered to be the true lamp spectrum, $F_T(\lambda) = F_{RB}(\lambda_{EX}, 620 \text{ nm})$. Since EmM and the PMT only 'saw' a single wavelength (620 nm) any λ bias was removed from the emission detector.

F. Data handling and $C(\lambda)$

1 Using the computer software divide $S_T(\lambda)$ by $S_L(\lambda)$.

2 Normalize the correction factor curve that is obtained so that the value at 400 nm is 1. Most software packages have a normalization command. The curve thus obtained (*equation 17*) can be used to correct all fluorescence spectra between 250 and 600 nm.

3 A fluorescence spectrum is corrected according to *equation 18*.

5.5 Excitation spectra

A useful fluorescence spectral measurement especially when studying mixtures of substances is an excitation spectrum. λ_{EM}, a wavelength representing a point on the fluorescence spectrum, is held constant and then λ_{EX} is scanned by changing the λ of ExM. This scan is usually carried out across the wavelength range of the absorption spectrum of the sample. The spectrum that is traced out should correspond closely with the absorption spectrum of the molecule that is responsible for the fluorescence (*Figure 11*). This spectrum reveals whether the sample is homogeneous, and whether all fluorescence features result from a single molecule. For example in *Figure 12(a)* the fluorescence spectrum shows a pronounced shoulder at long wavelengths. By measuring the excitation spectrum at two different fluorescence wavelengths, one at the high energy (blue) (λ_B) side and the other at the low energy (red) (λ_R) side of the spectrum, the shape of the absorption spectrum of the molecule or molecules from which this fluorescence originates can be obtained. *Figure 12(b)* shows the result where the

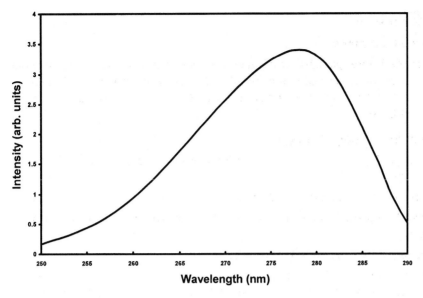

Figure 11 The excitation spectrum of a protein sample with the λ_{EM} = 340 nm.

fluorescence spectrum in *Figure 12a* results from a mixture of two molecules. Each molecule has a different absorption and fluorescence spectrum. If the fluorescence originated from the excitation of a single chromophore, then the excitation spectra will have identical shape. The intensity will be proportional to the fluorescence intensity at each λ_{EM}. Excitation spectra can also provide information as to whether FRET is occurring between two chromophores in a sample, since the excitation spectrum will include signals from the donor absorbance.

5.6 Quantum yield measurement

In many studies using fluorescence for analytical purposes, knowledge of the absolute quantum yield Φ_F is not necessary. Usually one is only comparing fluorescence intensities of different samples and so the relative fluorescence intensity offers a satisfactory measurement. In many other studies one wants to compare the fluorescence efficiencies of the samples in order to relate the fluorescence to the structural properties and interactions of the molecule. This is the situation when biomolecules such as proteins are being studied. It is necessary then to determine Φ_F of the fluorescent chromophore. This is also the case if one is interested in the photophysical parameters of a chromophore. When combined with the measured singlet lifetime, τ_S, according to *equation 14*, then the radiative rate constant, k_R can be calculated. This permits the evaluation of the rate constant of the non-radiative deactivation processes, k_{NR}, of the excited singlet state of the chromophoric molecule being studied. Since k_R is an intrinsic rate constant of a chromophore, direct comparison of Φ_F of the

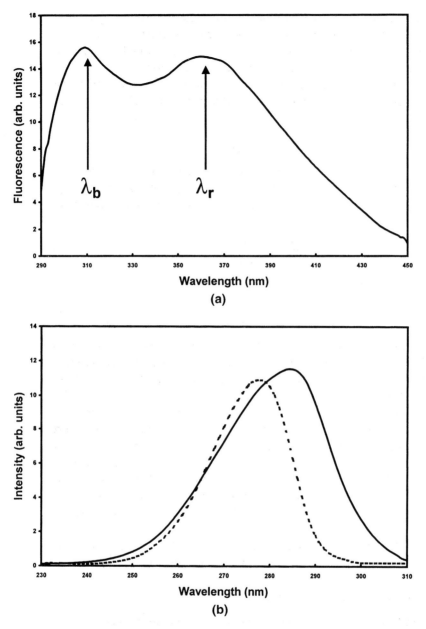

Figure 12 (a) A fluorescence spectrum showing two spectral maxima. λ_r is the fluorescence wavelength setting selected to determine the excitation spectrum responsible for the low energy fluorescence maximum; λ_b is the fluorescence wavelength setting selected to determine the excitation spectrum responsible for the high energy fluorescence maximum. (b) The solid line is the excitation spectrum when EmM was set at λ_r; the dashed line is the excitation spectrum when EmM was set at λ_b for the sample whose fluorescence is shown in (a).

chromophore under different conditions permits the correlation of Φ_F with molecular structural properties and inter or intramolecular interactions.

The measurement of the Φ_F of a sample is generally carried out by comparing the sample fluorescence to that of a standard molecule whose Φ_F has been determined using an absolute method. The basis of this 'secondary' method of comparison of Φ_F of the sample to that of a standard is the following

$$S_M(\lambda) \propto \Phi_F(M)A_M \qquad\qquad 19$$

$$S_S(\lambda) \propto \Phi_F(S)A_S \qquad\qquad 20$$

where $S_M(\lambda)$ and $S_S(\lambda)$ are the integrated areas of the corrected fluorescence spectra of the sample, M, and the standard, S, respectively; and Φ_F (M) and Φ_F (S) are their respective quantum yields; and A_M and A_S are their absorbancies at λ_{EX}. Thus

$$\frac{S_M(\lambda)}{S_S(\lambda)} = \frac{\Phi_F(M)\,A_M}{\Phi_F(S)\,A_S} \qquad\qquad 21$$

which on rearranging gives

$$\Phi_F(M) = \frac{S_M(\lambda)}{S_S(\lambda)} \cdot \frac{A_S}{A_M} \cdot \Phi_F(S) \qquad\qquad 22$$

The determination of Φ_F of a molecule by a primary method is a very difficult task and has been the subject of several reviews (4, 5). The quantum yields of several molecules under rigorous experimental conditions have been determined (6, 7). A selection that is generally useful is listed in Table 3. When such a standard molecule is chosen in order to measure Φ_F of the molecule of interest, the exact sample conditions for the standard must be followed. Temperature control is also critical owing to the temperature dependency of k_{NR}. One of the most often used standards is quinine sulphate in 1 N H_2SO_4 at concentrations $<10^{-4}$ M. At 25°C its Φ_F is 0.546 ($\lambda_{EX} = 365$ nm). For protein fluorescence, where tryptophan is the fluorescent chromophore, N-acetyltryptophanamide (NATA) at pH 7, 20 °C is used as a secondary standard. The accepted value of Φ_F of NATA is 0.14 (8).

When the Φ_F of the sample is determined, $F(\lambda)$ of the sample and the standard should cover a similar λ range. Obviously, both the sample and standard

Table 3 Fluorescence quantum yield standards

Standard/conditions	Spectral range (nm)	Excitation / (nm)	Φ_F
N-Acetyltryptophanamide/ aqueous buffer, pH 6–7, 20°C	300–450	280–300	0.14
Quinine sulphate/ 1 N H_2SO_4, 25°C	380–580	260–390	0.546
2-Aminopyridine/ 0.1 N H_2SO_4, 20°C	315–480	250–300	0.66
9,10-Diphenylanthracene/ ethanol, 20°C, degas	390–530	320–380	0.95

should absorb at the same λ_{EX}. Since it is important to minimize inner filter effects the absorbance of both the sample and the standard should be <0.05 at λ_{EX}. This becomes one of the largest sources of error in measuring Φ_F since there is a large percentage error in measuring such low absorbancies. A means of reducing this error is to measure the absorbance of the sample and standard in a 5 cm cylindrical cuvette whose outside diameter is 7 mm. The absorbances of these solutions (c. 2 ml volume) are measured in this cuvette, and then the solutions are transferred to a 1-cm fluorescence cuvette for the fluorescence measurements. In this way an absorbance that is five times that in a 1 cm cuvette is measured and the error in the absorbance is significantly reduced.

The fluorescence spectra of both the sample and standard should be scanned over an identical scan range, and it is important that $F_S(\lambda)$, the fluorescence spectrum of the standard, should have reached a value close to zero on its red edge. The solvent blank is subtracted from these spectra. Since it is unlikely that the sample and the standard will have identical $F(\lambda)$, the spectra should be corrected for the λ sensitivity of the PMT. The resulting corrected spectra, $F(\lambda)_{CORR}$ are then integrated across the same spectral range, i.e. from λ_I to λ_F. This integration can usually be conducted in the instrument computer. Alternately, the integration can be accomplished using Simpson's rule or a trapezoidal approximation over narrow λ ranges (1–5 nm). Prior to the availability of computer interfaced instruments, the integration was accomplished by cutting out the spectral recording on a chart paper and weighing the paper cut-out.

When one is measuring Φ_F of a fluorophore in an aqueous solution and the standard is in a non-aqueous solvent then a refractive index correction is recommended. This correction is applied according to:

$$\Phi_F = \Phi_F{}' \cdot \frac{n^2(\text{sample})}{n^2(\text{standard})} \qquad 23$$

where $\Phi_F{}'$ is the measured quantum yield without correction for refractive index difference, Φ_F is the corrected quantum yield value, $n^2(\text{sample})$ is the square of the refractive index of the solvent of the sample solution, and $n^2(\text{standard})$ is the square of the refractive index of the solvent of the standard solution. If the sample had been dissolved in benzene and the standard was quinine sulphate the correction would have been 27%.

6 Fluorescence applications

The application of fluorescence measurements to a large variety of chemical, biochemical, biological and physical problems is extensive. The majority of fluorescence applications involve the use of extrinsic fluorescence probes. These are chromophoric molecules that are attached to or adsorbed onto another molecule and their fluorescence is measured. These molecules probe the properties of the substances to which they are attached. The catalogue and manual published by Molecular Probes (9) is an excellent collection of a large number of different examples of the use of fluorescence, especially as they relate to bio-

Protocol 3

Measurement of Φ_F of a sample

Equipment and reagents

- Purified sample
- Purified standard
- Purified solvents

- Four 1-cm quartz cuvettes
- A 5-cm quartz cylindrical cuvette (outside diameter = 0.7 cm)

A. Selection of standard

1 A standard material is selected from a list of quantum yield standards (6, 7 and *Table* 3). The standard should have an absorption spectrum in the same range as that of the sample. The fluorescence spectra of the standard and sample should span a similar λ range.

B. Excitation wavelength selection and absorbance measurement

1 Select λ_{EX} for the sample and if there is a single chromophoric species in the sample then this wavelength should be one where the absorbance does not change rapidly. The standard should also absorb at this λ.

2 Measure and record the absorbance of the sample and standard at this λ_{EX}, subtracting any solvent blank contribution, using the 5-cm cylindrical cuvette. Adjust the concentration of the sample and the standard so that A(5 cm) $\approx$ 0.25. The absorption spectrophotometer bandpass in these measurements should be nearly the same as the bandpass of ExM, that is typically 4 nm.

3 Transfer the solutions to separate 1-cm rectangular quartz fluorescence cuvettes. Two cuvettes should also contain the respective blanks for the sample and standard.

C. Spectral measurement

1 Set the bandpass of ExM and EmM to an appropriate value, typically 4 nm.

2 Measure $F(\lambda)$ of the sample and standard over the identical λ range. It is important that the fluorescence of both are close to zero at the long λ of the scan. The initial or starting wavelength, λ_I, should be selected so that the fluorescence of both solutions is as close to zero as possible. This may be difficult to achieve in highly scattering solutions.

3 The scan rate or step size/integration time should be set to minimize the noise on the spectrum.

4 Measure the background signal of the solvent blanks for the two solutions.

D. Data handling and calculations

1 Subtract the solvent blank spectra from the fluorescence spectra of the sample and standard.

2 Multiply these blank subtracted spectra by the correction factor curve, $C(\lambda)$ (*Protocol 2*).

3 Integrate $F(\lambda)_{CORR}$ over an identical λ range to obtain $S_M(\lambda)$ and $S_S(\lambda)$.

4 Calculate $\Phi_F(M)$ of the sample by substituting the appropriate values into *equation 22*.

5 If the solvents used are different then the value of $\Phi_F(M)$ should be corrected for the refractive index (*equation 23*).

chemical, chemical and biological investigations. Intrinsic fluorescence probes are those chromophoric species that occur as natural constituents of the substances under investigation. An area where intrinsic fluorescent chromophoric components have been extensively utilized is in the field of protein fluorescence (see Chapter 9).

6.1 Fluorescence resonance energy transfer

A highly useful application of fluorescence spectroscopy is the technique known as fluorescence resonance energy transfer (FRET). This is treated in more detail in Chapter 3. The basic principle is to have a pair of chromophores attached to a molecule. One chromophore, the donor, absorbs the excitation energy and transfers this excitation energy to the second chromophore, the acceptor, through a dipole–dipole interaction mechanism. The Forster equation describes the physical terms that determine the efficiency of this process. FRET is used as a 'spectroscopic ruler' to estimate the distance between attachment points of the donor and acceptor on a large macromolecule such as a protein. It can also be useful in providing information on distance distributions owing to the backbone dynamics of the molecule.

6.2 Fluorescence anisotropy

Fluorescence anisotropy measurements and their applications are covered in Chapter 3. Plane polarized exciting light can be obtained by placing a Polaroid sheet or a Glan Taylor prism in front of the sample holder. The polarization or anisotropy of the fluorescence of the sample is then determined by placing another polarizer in front of EmM. The fluorescence that is polarized parallel, $I_\parallel$, and perpendicular, $I_\perp$, to the excitation polarization direction is then measured. The fluorescence anisotropy, r, of the sample is defined according to

$$r = \frac{I_\parallel + I_\perp}{I_\parallel + 2I_\perp} \qquad 24$$

The denominator is proportional to the total fluorescence intensity. r is in the main a measure of the rotational motion or dynamics of the molecule, the latter being very useful in obtaining flexibility parameters of local segments of a macromolecule.

6.2.1 Polarization effects on fluorescence intensity

If one is measuring the Φ_F of a macromolecule such as a protein, or a large molecular complex, it is advised that two additional optical components be included in the spectrophotofluorimeter. The light exiting ExM is not randomly polarized, i.e. it is unpolarized but it will have a polarization bias in either the vertical or horizontal plane. As a result the macromolecular fluorescence will be anisotropic owing to slow rotational tumbling and the fluorescence will be biased in a particular polarization direction. This can be corrected by placing a depolarizer (a wedge plate) optic in the excitation light beam after the beam splitter, which results in the excitation light being unpolarized. A polarizer is placed in front of the slits of EmM and oriented at an angle of 35° to the vertical plane. The fluorescence passing this polarizer will be proportional to the total intensity and will avoid any polarization bias.

Another point to be aware of is that holographic grating monochromators have a polarization anomaly known as the Wood's anomaly. At certain wavelengths the light is polarized entirely in the horizontal plane. This can cause important spectral artefacts such as false maxima or false shoulders. The anomaly can be avoided by using an excitation depolarizer and emission polarizer as described above.

6.3 Protein fluorescence

One research area that has benefited from extensive fluorescence investigations is the field of protein structure–function studies. There are three aromatic amino acids that absorb light in the ultraviolet spectral range, phenylalanine (Phe), tyrosine (Tyr) and tryptophan (Trp). Because of their molar absorptivity the fluorescence of Tyr (ε_{276} = 1405 M^{-1} cm^{-1}) and Trp (ε_{280} = 5579 M^{-1} cm^{-1}) have been extensively used to study proteins.

Trp, an indole amino acid, has received by far the greatest attention. There are several reasons for this. Usually the number of Trp residues in a protein are limited and there are several proteins containing only a single Trp residue. The absorption spectrum of Trp with its higher ε compared to Tyr has an extended absorbance beyond the normal absorbance of Tyr. This permits the preferential excitation of the Trp residues in a protein. Because of the electronic properties of the Trp excited state, its fluorescence spectrum is highly dependent on the nature of its environment. Selective intramolecular interactions can lead to fluorescence enhancement or quenching. The origin of these effects are now reasonably well understood permitting a discussion of the local molecular structure in the vicinity of the Trp residue. Obviously if there are several Trp residues in a protein it is difficult to make specific interpretations of their individual behaviour. However, the fluorescence of such proteins can still be useful if all one wishes to determine is an association constant. The literature on Trp fluorescence and its application to the study of protein structures, interactions and dynamics is vast and the information content of Trp fluorescence is considerable.

Time-resolved fluorescence studies have shown that ground state conformational heterogeneity of the Trp residue can be revealed. This heterogeneity has been attributed to the fluorescence of different rotamers of the indole ring around the C_α-C_β bond (10, 11). Recently it has been shown how Trp analogues when substituted into proteins in place of the normal Trp residue can provide useful information on molecular details of interacting segments of protein–protein and protein–nucleic acid complexes (12). These analogues have extended ultraviolet absorption beyond the normal Trp absorbance and hence can be selectively excited in the presence of a large number of normal Trp residues.

It is beyond the scope of this chapter to review this literature. The reader is referred to several review articles that provide a useful summary (13–15).

It is useful, however, to discuss a few practical aspects of measuring Trp fluorescence and some of the information that can be obtained. One of the first factors that it is necessary to consider is the number of Tyr residues in a protein that contains a single Trp residue. In most cases where the number of Tyr residues is small, $\lambda_{EX} = 295$ nm will predominantly excite the single Trp residue and only Trp fluorescence will be observed. This is because Tyr absorbance at 295 nm is very small ($\varepsilon_{295} = 18$ M^{-1} cm^{-1}) but not zero. When there are a large number of Tyr residues in a protein, the sum of the absorbance of these residues makes a modest contribution to the total absorbance and in these cases the fluorescence generated with $\lambda_{EX} = 295$ nm will contain Tyr spectral components (16). Also FRET from Tyr to Trp may occur and affect the integrated fluorescence intensity. In order to virtually eliminate the Tyr contribution to the fluorescence, λ_{EX} can be changed to 300 nm. However, this generates another problem, especially when one measures Φ_F, since at typical protein sample concentrations of 1–5 $\times$ 10^{-6} M the absorbance at 300 nm is very low.

There is also another cautionary note to consider when the Φ_F of proteins is being measured. In order to selectively excite Trp fluorescence λ_{EX} is usually chosen to be 295 nm or greater in order to minimize the Tyr absorbance. An assumption that is usually made is that the molar absorptivity at 295 nm of the particular protein under the different conditions of the experiments is the same. For example if the thermal unfolding or chemical denaturation of a protein is being studied using fluorescence hardly any attention is given to the fact that ε_{295} is in fact changing. This λ is on the very steep falling slope of the protein absorption spectrum. Changes in ε_{295} are not readily apparent on casual inspection of the absorption spectrum. As a Trp residue that may originally have been buried in the interior of a protein becomes solvent exposed, ε_{295} will change. When performing chemical denaturation experiments this can be accounted for. However, corrections for these changes in thermal unfolding are difficult to estimate.

6.4 Fluorescence quenching

6.4.1 Intramolecular quenching

In proteins the Trp fluorescence is often quenched by intramolecular interactions with different amino acids. These quenching processes contribute to the magnitude of the non-radiative relaxation rate constant, k_{NR}, of the excited state. Cysteine and especially disulfide bridges will quench Trp fluorescence provided they are within 6–10 Å of the Trp residue. This quenching process is thought to be due to an electronic exchange process.

Histidine may quench Trp fluorescence through a proton transfer mechanism. There is evidence (17) that position 4 of the indole ring acquires more electron density in the excited singlet state and can accept a proton from a proton donor such as histidine.

The neutral amide function of glutamine or asparagine, as well as the amide backbone of the protein, can quench Trp fluorescence. This is another short-range electron transfer process. The short lifetime of one of the Trp fluorescence decay components may be due to the close proximity of the indole ring to the amide backbone in that particular rotamer state.

When the Trp residue is located in a hydrophobic environment of the protein its Φ_F is enhanced and importantly the fluorescence spectral maximum is shifted to a much higher energy than that normally seen for a solvent exposed residue.

6.4.2 Intermolecular quenching

A very informative experiment that can be performed with a protein is the quenching caused by the addition of a quenching molecule or ion to the solution. Three species are often used as quenchers, acrylamide, iodide ion, I^-, and cesium ion, Cs^+. The former is a neutral molecule while the latter two are negatively and positively charged, respectively. The concentrations of quencher that are used range from 0.01 to 0.5 M. In the case of proteins, acrylamide quenching has been extensively used to determine the degree of exposure of Trp residues. The experiment is conducted by adding aliquots of a concentrated acrylamide solution to the protein and measuring the fluorescence of the solution after each addition. This quenching process is described according to

$$M(S_1) + [Q] \rightarrow M(S_0) + [Q] \, k_Q[Q] \qquad\qquad 25$$

and the second-order rate constant, k_Q, is the kinetic term that is a measure of the process. In the absence of external quencher, Q,

$$\Phi_F^\circ = \frac{k_R}{k_R + k_0} \qquad\qquad 26$$

where

$$k_0 = k_{NR} + k_{ISC} + k_{PR} \qquad\qquad 27$$

and Φ_F° is the solute quantum yield in the absence of added quencher. In the presence of quencher

$$\Phi_F = \frac{k_R}{k_R + k_0 + k_Q[Q]} \qquad\qquad 28$$

which gives the ratio

$$\frac{\Phi_F^\circ}{\Phi_F} = \frac{k_R + k_0 + k_Q[Q]}{k_R + k_0} \qquad\qquad 29$$

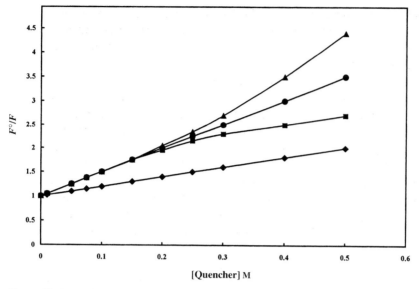

Figure 13 A set of typical Stern Volmer fluorescence quenching plots. ●, a linear Stern Volmer plot; ▲, a plot showing upward curvature owing to a high concentration of quencher surrounding the fluorophore; ■, a plot showing downward curvature owing to the presence of two fluorescent chromophores being quenched with different efficiencies; ◆, a linear Stern Volmer plot representing the case where the chromophore is less accessible to the quencher.

and since the unquenched singlet lifetime, $\tau_S = 1/(k_R + k_0)$, *equation 29* can be simplified to

$$\frac{\Phi_F^\circ}{\Phi_F} = 1 + k_Q \tau_S [Q] \qquad 30$$

This is the well known Stern Volmer relationship that relates the fluorescence yield of the sample to the concentration of the added quencher. A plot of Φ_F°/Φ_F or, since $\Phi_F \propto F$, F°/F versus [Q] should yield a straight line with an intercept of 1, and the Stern Volmer slope, $K_{SV} = k_Q\tau_S$ (*Figure 13*).

Quenching by the addition of an external quencher is a diffusion controlled process. As a first approximation the value of K_{SV} reveals the degree of exposure to the solvent of a Trp residue. A high value suggests a solvent exposed residue, while a low value indicates that the residue is buried inside the protein. However, this is not strictly correct since this assumes that τ_S is the same for both cases. A more accurate estimation of the degree of exposure of Trp is to calculate k_Q from $K_{SV} = k_Q\tau_S$. In this case the value of τ_S should be the intensity weighted decay time when multi-exponential decay kinetics are observed (that is often the case even for a single Trp protein). The intensity weighted decay time is

$$<\tau_S> = \frac{\Sigma\alpha_I \tau_I^2}{\Sigma\alpha_I \tau_I} \qquad 31$$

where α_I and τ_I are the pre-exponential term and decay time of the *i*th fluorescence decay component, respectively. It is important to use the intensity weighted lifetime to calculate values of k_Q since in a steady-state measurement the fluorescent component that makes the largest contribution to the total fluorescence, the largest value of $\alpha\tau$, will be the one that dominates in the fluorescence quenching experiment.

In a protein containing more than one Trp residue the Stern Volmer quenching curve often shows downward curvature (*Figure 13*). This would result because the less accessible Trp residues will be quenched after the more exposed residue, and will have a lower value of k_Q. However, the direction of curvature will depend on the relative values of k_Q and τ_S, especially in the case of a single chromophore when upward curvature is observed. This is due to the fact that at a high quencher concentration diffusion controlled kinetics are not observed, and there will be an appreciable fraction of the chromophore that is surrounded by the quencher at the instant of excitation. That molecule will be immediately quenched and F will be less than that expected for a diffusion controlled process.

In acrylamide quenching experiments it is important that $\lambda_{EX} \geq 290$ nm. Acrylamide has an appreciable absorption at 285 nm. At the high concentrations of acrylamide that are normally used (0.01–0.5 M) a large proportion of excitation light at 285 nm will be absorbed by the acrylamide. This is an inner filter effect that will render the data and analysis invalid.

When quenching by I^- or Cs^+ is the experimental procedure, the ionic strength of the solution must be held constant. In quenching experiments mM concentrations of quencher are used. Constant ionic strength of the solutions are maintained by the appropriate addition of salt solutions.

Protocol 4

Acrylamide quenching of protein fluorescence

Equipment and supplies

- High purity spectral grade acrylamide
- High purity buffer for solution of the protein
- Volumetric pipettes
- 1-cm quartz fluorescence cuvettes

A. Solution preparation

1   A stock solution of 5 M acrylamide is prepared in the sample buffer.

2   2.00 ml of a protein sample is prepared to give an absorbance at $\lambda_{EX} \approx 0.05$.

B. Fluorescence and titration measurement

1   Record the fluorescence spectrum of the protein and of the buffer blank.

2   Add aliquots of acrylamide to the protein sample and measure the fluorescence spectrum of the sample after each addition of acrylamide. The first addition should be 5 µl of stock acrylamide, to give [acrylamide] = 0.0125 M. A second 5-µl aliquot is

Protocol 4 continued

added followed by a third aliquot of 10 μl. Subsequently five separate aliquots of 20 μl are added, followed by two separate aliquots of 40 μl each. This will provide a final [acrylamide] ≈ 0.5 M.

3 The same aliquots of acrylamide should be added to 2.00 ml of the buffer blank and the background signal measured.

C. Data handling and analysis

1 The buffer blank signal is subtracted from the corresponding sample fluorescence spectrum.

2 The integrated areas of each blank subtracted spectrum, $S_M(\lambda)$ are determined. The integrated area of the initial sample, [acrylamide] = 0 M, gives $F°$. Each of the other integrations are the corresponding values of F at the different concentrations of acrylamide.

3 Calculate the ratio $F°/F$ for each acrylamide concentration.

4 Correct the acrylamide concentration for the dilution factor of each sample according to

$$[\text{acrylamide}] = [\text{acrylamide stock}]V_A/(2000 + V_A) \text{ M} \qquad 32$$

where V_A is the cumulative volume of acrylamide stock added for each sample measured.

5 Plot the ratio $F°/F$ versus [acrylamide] for each sample.

6 After inspection of the plot, determine the slope and intercept of those points that appear to obey a linear relationship. For this a linear regression analysis is usually performed.

7 The slope gives K_{SV}, the parameter related to the accessibility of the Trp residue. The intercept should be close to 1.

8 From time resolved fluorescence experiments determine the intensity-weighted singlet decay time, $<\tau_S>$, and calculate k_Q from $K_{SV} = <\tau_S>k_Q$.

7 Conclusion

The description above is a summary of the important practical aspects of steady state fluorescence spectroscopic measurements. Further details and applications can be obtained from a number of reviews and publications, a selection of which are referenced below (18–21).

References

1. Teale, J. (1969) *Photochem. Photobiol.* **10**: 363–374.
2. Lenz, B. R., Moore, B. M., and Barrow, D. A. (1979) *Biophys. J.* **25**: 489–494.
3. Melhuish, W. H. (1972) *J. Res. Nat. Bur. Stand.* **76A**: 547.
4. Demas, J. N. and Crosby, G. A. (1971) *J. Phys. Chem.* **75**: 991

5. Birks, J. B. (1977) *Standardization in spectrophotometry and luminescence measurements* (ed. K. D. Mielenz, R. A. Velapoldi and R. Mavrodineanu). US Department of Commerce.

6. Miller, J. N. (1981) *Standards in fluorescence spectroscopy.* Chapman and Hall, London.

7. Parker, C. A. (1968) *Photoluminescence of solutions.* Elsevier, London.

8. Chen, R. F. (1967) *Anal. Lett.* **1**: 35

9. Haugland, R. P. (1996) *Handbook of fluorescent probes and research chemicals* (6th Edn). Molecular Probes, Eugene.

10. Dahms, T. E. S., Willis, K. J., and Szabo, A. G. (1995) *J. Am. Chem. Soc.* **117**: 2321-2326.

11. Willis, K. J., Neugebauer, W., Sikorska, M., and Szabo, A. G. (1994) *Biophys. J.* **66**: 1623-1630.

12. Ross, J. B. A., Szabo, A. G., and Hogue, C. W. V. (1997) *Methods of Enzymology* **278**: 151-189 (ed. L. Brand and M. L. Johnson). Academic Press, New York.

13. Eftink, M. R. (1991) *Protein structure determination: Methods of biochemical analysis* **35**: 127-207 (ed. Suelter, C. H.). John Wiley, New York.

14. Szabo A. G. (1989) *The enzyme catalysis process* (ed. A. Cooper, J. L. Houben, and L. C. Chien), pp. 1213-1239. Plenum, New York.

15. Millar, D. P. (1996) *Curr. Opin. Struct. Biol.* **6**: 637-642.

16. Willis, K. J. and Szabo, A. G. (1989) *Biochemistry* **28**: 4902-4908.

17. Saito, I., Sugiyama, H., Yamamoto, A., Muramatsu, S. and Matsuura, T. (1984) *J. Am. Chem. Soc.* **106**: 4286-4287.

18. Millar, D. P. (1996) *Curr. Opin. Struct. Biol.* **6**: 322-326.

19. Eftink, M. R. (1998) *Biochemistry (Mosc.)* **63**: 276-284

20. Lee, Y. C. (1997) *J. Biochem.* **121**:818-825.

21. Soper, S. A., Warner, I. M., and McGown, L. B. (1998) *Anal. Chem.* **70**: 477R-494R.

Chapter 3

Time-resolved fluorescence and phosphorescence spectroscopy

Thomas D. Bradrick* and Jorge E. Churchich[†]
* Optical Spectroscopy Section/LBC/NHLBI, Bldg. 10, Rm 5D14,
MSC 1412, National Institutes of Health, Bethesda, MD 20892–1412, USA
[†] Dept. of Applied Biology and Chemical Technology, Hong Kong
Polytechnic University, Hong Kong, PRC

1 Introduction

In this chapter, we outline the basic theory and methodology for making time-resolved fluorescence and phosphorescence measurements. We begin with a brief discussion of the intrinsic time dependence of fluorescence and phosphorescence decays, and also introduce several important photophysical concepts. Energy transfer measurements, which are important for determining molecular distances, are then addressed, followed by the convolution integral (which describes the luminescence decay that is actually observed in pulsed excitation experiments) and the relationships that are used to determine fluorescence lifetimes from phase/modulation data. Polarized fluorescence measurements, which are an important tool for following molecular motions, are then discussed, and the fluorescence portion of the chapter concludes with an overview of data analysis and brief descriptions of the instrumentation that is used in making time-resolved fluorescence measurements. The remainder of the chapter is then devoted to a discussion of phosphorescence spectroscopy.

2 Background

2.1 Basic photophysics and time dependence of fluorescence and phosphorescence decays

The intrinsic time dependence of fluorescence decays is derived in Section 3.3 of Chapter 2. There it is shown that if a population of excited singlet-state molecules is generated instantaneously, its size decreases exponentially with time, as does the intensity of the emitted photons. The fluorescence lifetime τ_S is typically used to describe the rate of decay, where $\tau_S^{-1} = k_R + k_{NR} + k_{ISC} + k_{PR}$ (*equation 12*, Chapter 2) and k_R, k_{NR}, k_{ISC} and k_{PR} are the rate constants for the various parallel unimolecular de-excitation processes (*Section 2.2* and *Figure 2*, Chapter 2).[1] If there are several non-interacting fluorophore species in solution, their excited states will decay independently and a sum of exponentials will be needed to describe how the sample's fluorescence intensity decreases with time.[2] The general expression for the intrinsic fluorescence decay $F(t)$ is therefore

$$F(t) = \sum_i \alpha_i \exp(-t/\tau_i) \qquad\qquad 1$$

where the sum is over all the different fluorophore species 'i' in solution, and we now express as α_i their individual pre-exponential factors. The α_i depend on the species' respective k_R values, concentrations and extinction coefficients, as well as on the exciting light intensity and geometric factors (among other things). In carrying out fluorescence lifetime measurements, some of these variables are typically not quantified; as their product is thus not known before hand, it is simply expressed as a constant that is determined when the data are analysed (see Section 2.6).

The steady-state fluorescence intensity $I(\lambda)$ at emission wavelength λ is related to the time-resolved decay parameters through

$$I(\lambda) \propto \sum_i \alpha_i(\lambda)\tau_i \qquad\qquad 2$$

where we now write α_i as a function of the emission wavelength. The fractional intensity f_i of the ith decay component is given by

$$f_i = \frac{\alpha_i \tau_i}{\sum_i \alpha_i \tau_i} \qquad\qquad 3$$

Analogous to the fluorescence quantum yield (*equation 4*, Chapter 2), we can define a quantum yield for intersystem crossing (Section 2.2.1, Chapter 2), which is the fraction of initially excited molecules that makes it from S_1 to T_1 and is given by

$$\Phi_{ISC} = \frac{k_{ISC}}{k_R + k_{NR} + k_{ISC} + k_{PR}} = k_{ISC}\,\tau_s \qquad\qquad 4$$

Phosphorescence is a longer-lived emission than fluorescence, with the former having lifetimes varying from microseconds ($\mu s = 10^{-6}$ s) to seconds, as compared with picoseconds (ps $= 10^{-12}$ s) to nanoseconds (ns $= 10^{-9}$ s) for the latter. Even though phosphorescence emission is a two-step process (inter-

---

[1] Excited singlet-state molecules may also participate in a number of bimolecular reactions such as energy transfer, excited-state complex (excimer) formation, and excited-state proton transfer, all of which result in non-exponential fluorescence decays. A discussion of the kinetics of these process-es is beyond the scope of this book, however. The interested reader may wish to consult (1).

[2] 'Other' fluorescent species need not be different molecules. Depending on its environment, the fluorescence decay of a single tryptophan residue at some position in a protein can be com-plex, as several alternate processes, such as electron and proton transfer, and the existence of dif-ferent microenvironments for different rotational orientations of the indole side chain, may result in the residue having several different decay times.

system crossing followed by a radiative return to S_o from T_1), generally k_{ISC} is very much greater than the rate of triplet state de-excitation, and creation of the triplet state population is essentially complete before emission from it proceeds. For the unimolecular decay of the triplet state this means that its population and the phosphorescence intensity also decay exponentially with time. The rate of decay may be described using the phosphorescence lifetime or decay time, which is given by $\tau_P = (k_P + k_{QP})^{-1}$, where k_P and k_{QP} are the respective rate constants for the radiative and non-radiative relaxation of T_1.

The phosphorescence quantum yield is the fraction of absorbed photons that is emitted as phosphorescence and so is the product of the fraction that makes it from S_1 to T_1 (i.e. Φ_{ISC}), and the fraction of the triplet state molecules that returns to S_o by emitting a photon:

$$\Phi_P = \frac{k_P}{k_P + k_{QP}} \Phi_{ISC} = k_P \tau_P \Phi_{ISC} \qquad 5$$

The ratio of the quantum yields for phosphorescence and fluorescence is then

$$\frac{\Phi_P}{\Phi_F} = \frac{k_{ISC} k_P / k_R}{k_P + k_{QP}} \qquad 6$$

In the event that $k_{QP} \ll k_P$, *equation 6* becomes

$$\frac{\Phi_P}{\Phi_F} \approx \frac{k_{ISC}}{k_R} \qquad 7$$

(i.e. $\Phi_P \approx \tau_S k_{ISC} = \Phi_{ISC}$). The intersystem crossing rate constant (k_{ISC}) is sensitive to spin orbit coupling by heavy atoms (2).

2.2 Fluorescence and phosphorescence energy transfer and sensitized luminescence

The theory of energy transfer via dipole–dipole interactions was originally developed by Förster (3, 4) to explain singlet–singlet transfer: $D(S_1) + A(S_o) \rightarrow D(S_o) + A(S_1)$. It is also applicable to triplet–singlet transfer (5, 6): $D(T_1) + A(S_o) \rightarrow D(S_o) + A(S_1)$. (Here, D is the donor, A is the acceptor and their electronic states are indicated in parentheses.) The strength of the interaction is usually expressed in terms of a 'critical radius,' R_o. At this distance of separation between donor and acceptor, the probability that energy transfer will take place is 50%.

According to Förster's theory

$$R_o^6 = 8.8 \times 10^{-25} \, \kappa^2 \Phi_o J / n^4 \quad (\text{cm}^6) \qquad 8$$

where J is the overlap integral

$$J = \int f_D(\bar{v}) \varepsilon_A(\bar{v}) d\bar{v} / \bar{v}^4 \qquad 9$$

κ^2 is a dimensionless orientational factor, Φ_o is the quantum yield of the donor in the absence of energy transfer, and n is the refractive index of the medium. In *equation 9* the emission spectrum of the donor (f_D) is normalized to unity (i.e. $\int f_D(\bar{v}) d\bar{v} = 1$) and ε_A is the molar extinction coefficient of the acceptor. Both f_D

and ε_A are plotted versus wavenumber ($\bar{v}$), which is the reciprocal of the wavelength (λ, in cm). R_o values for a number of donor–acceptor pairs free in solution (for which κ^2 is 2/3) may be found in (7); otherwise, they must be calculated using *equations 8* and *9*.

The efficiency of energy transfer is given by

$$E = \frac{\Phi_D - \Phi_{DA}}{\Phi_D} = \frac{\tau_D - \tau_{DA}}{\tau_D} \qquad 10$$

where Φ_D and Φ_{DA} are the donor's quantum yields and τ_D and τ_{DA} are its decay times in the absence and presence of the acceptor, respectively. (The interchangeability of quantum yield and lifetime follows from *equation 14* of Chapter 2). Because of the transfer efficiency's sensitivity to the donor–acceptor distance R

$$E = \frac{R_o^6}{R_o^6 + R^6} \qquad 11$$

energy transfer measurements can be used to determine molecular distances. If the equipment is available, it is generally easier to do this accurately by measuring emission lifetimes than it is by measuring quantum yields. The reader may wish to consult (8) for additional details.

Molecules in their triplet states may also participate in collisional transfer processes known as triplet–triplet energy transfer: $D(T_1) + A(S_o) \rightarrow D(S_o) + A(T_1)$. This, however, takes place through an electron exchange mechanism that involves overlap of the electronic clouds of the donor and acceptor molecules. Consequently, the interaction energy cannot be determined from optical measurements of the donor. (Triplet–triplet energy transfer by dipole–dipole interaction would not be expected to occur over long distances since the electronic transitions in both the acceptor and donor are spin-forbidden.) Radiationless energy transfer between triplet levels becomes appreciable when the acceptor concentration reaches the level of 10 mM or higher. The sensitized phosphorescence yield of the acceptor resulting from $D(T_1) \rightarrow A(T_1)$ transfer is given by the equation

$$\Phi_S = \frac{\Phi_{PA}}{\Phi_{PD}^\circ - \Phi_{PD}} \qquad 12$$

where Φ_{PD}° and Φ_{PD} are the phosphorescence quantum yields of the donor in the absence and presence of the acceptor, respectively, and Φ_{PA} is the phosphorescence quantum yield of the acceptor. When the forbidden nature of a transition in the donor results in a corresponding increase in the lifetime of its excited state (as is the case for the triplet state of many molecules), energy exchange over long distances can be high (provided there is a good overlap of the spectra concerned) as the distance over which the triplet-state molecule can diffuse is relatively great.

2.3 Observed time dependence of fluorescence

2.3.1 Introduction to convolution integral

In Section 2.1 we stated that the fluorescence intensity due to a population of excited-state molecules decreases exponentially with time following their instantaneous excitation. In practice, however, excitation is never instantaneous but is achieved either by using pulses that have a significant duration compared to the fluorescence lifetimes one is measuring (pulsed excitation, Section 2.3.2), or else by continuously illuminating the sample with light whose intensity varies sinusoidally with time (traditional phase/modulation, Section 2.3.3). Both types of excitation distort the observed fluorescence decay away from being exponential, as the decaying excited-state populations created by earlier parts of the excitation pulse are subsequently replenished by its later parts. This distortion complicates obtaining the fluorescence decay times (τ) and relative amplitudes (α) from the collected data. In the case of pulsed excitation, the observed fluorescence is described by an expression called the convolution integral.

2.3.2 Pulsed excitation and the convolution integral

An excitation pulse L that has finite duration is treated as consisting of a succession of instantaneous (so-called 'δ-function') pulses of varying 'heights' or intensities, each of which generates a subpopulation of excited-state molecules that immediately begins to decay. The relative sizes of these different subpopulations and thus their relative fluorescence intensities are proportional to the intensities of the δ-function pulse components that generated them. The observed fluorescence intensity at a particular time t is then the sum of the contributions each of these sub-populations makes:

$$F_{obs}(t) = \lim_{\Delta t' \to 0} \sum_i L(t'_i)\Delta t' F(t - t'_i) = \int_0^t L(t')F(t - t')dt' \qquad 13$$

Here, $L(t'_i)$ is the height and $\Delta t'$ the width of the excitation pulse's δ-function component at time t'_i. $F(t)$ (the intrinsic fluorescence decay) is given by *equation 1* and its argument $t - t'_i$ accounts for the fact that the elapsed time varies between when the different subpopulations are created (at times t'_i) and when the fluorescence intensity is measured. The (relative) 'height' of L at t'_i appropriately 'scales' the magnitude of the resulting fluorescence response. The effect of convolving an exponential decay with an excitation pulse of finite width will be seen in Section 2.6.4, where we describe the analysis of pulsed-excitation data. A fuller discussion of the convolution integral may be found in (9).[3,4]

----

[3] Photomultiplier tubes, which are commonly used to convert light into an electronic signal for further processing (see Section 3.2), typically distort both the measured excitation pulse profile and observed fluorescence decay by broadening them. For that reason, the measured $L(t')$ is often called the instrument response function as it is actually a convolution of the excitation pulse and the photomultiplier response. It is, in fact, the 'spread' in the signal caused by the photomultiplier, rather than that due to a laser pulse, that most distorts the observed fluorescence decay away from being exponential and requires we use the convolution integral description.

[4] On the much longer timescale over which phosphorescence is emitted, nanosecond pulses that populate S_1 'look' like δ-function pulses. In that case, the triplet state population is still formed virtually instantaneously and the phosphorescence decay remains undistorted and decays exponentially. Usually the phosphorescence decay profile does not have to be described by a convolution integral unless one is using a xenon flashlamp for excitation, for which the pulses are a few tens of microseconds wide, with a weak 100 μs tail.

2.3.3 Phase/modulation relationships

Traditional phase/modulation fluorometry involves measuring fluorescence life-times by continuously illuminating the sample with light whose intensity varies sinusoidally. The resulting fluorescence intensity also varies sinusoidally at the same frequency as that of the exciting light, but is delayed or 'phase shifted' by an angle θ with respect to the latter (*Figure 1*). Also, the 'modulation' of the fluorescence ($m_F = b/a$, the ratio of the amplitude of oscillation to the average intensity; see *Figure 1*) is reduced compared to that of the exciting light ($m_L = B/A$; see *Figure 1*).

The phase shift and demodulation both depend on the fluorescence lifetimes and amplitudes and on the modulation frequency ω according to (10)

$$\tan\theta = N_\omega/D_\omega$$
$$m = \sqrt{N_\omega^2 - D_\omega^2}$$

14

where

$$N_\omega = \sum_i \frac{f_i \omega \tau_i}{1 + \omega^2 \tau_i^2}$$
$$D_\omega = \sum_i \frac{f_i}{1 + \omega^2 \tau_i^2}$$

15

and the sums are over the number of exponential terms 'i' that are being used to model the fluorescence decay. f_i is the fractional intensity of the ith decay component (*equation 3*).

Thus, by measuring the phase shift θ and the demodulation m, one can determine the fluorescence decay times (τ) and amplitudes (α). Such measurements are typically made sequentially over as broad a frequency range as possible in order to minimize the error in the extracted parameters, and to enable one to distinguish between different multi-exponential fluorescence decays (see

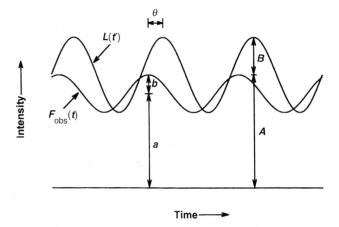

Figure 1 Illustration of exciting light intensity $L(t')$ and fluorescence response $F_{obs}(t)$ for a phase/modulation experiment. B and b are the respective amplitudes of oscillation of the two waves, while A and a are their respective average intensities. θ is the angle by which $F_{obs}(t)$ is delayed or phase shifted with respect to $L(t')$.

Section 2.6.5). (Note: MHz modulation frequencies are required for measuring nanosecond fluorescence lifetimes.) The reader who desires additional details about phase/modulation fluorometry may wish to consult (10).

2.4 Decay associated spectra (DAS) and discrete lifetimes versus lifetime distributions

Due to space limitations, we can only make brief mention of DAS and lifetime distributions. DAS (decay associated spectra) are the emission spectra associated with the different decay species and are described in (11). Examples of applications may be found in (12–14). The treatment of fluorescence lifetimes as distributions (rather than as the discrete values implied in Section 2.1) is discussed in (15–18). Commercially available software packages (see Section 2.6.1) are typically able to carry out the necessary analysis to obtain DAS from decay measurements made at multiple emission wavelengths, or to obtain lifetime distributions.

2.5 Polarized excitation and emission anisotropy decay

2.5.1 Introduction to emission anisotropy

Implicit in the previous discussion of fluorescence decay was that the fluorescing sample was being excited using unpolarized light. In fluorescence anisotropy decay measurements, the sample is excited using vertically polarized light and the horizontal and vertical polarization components of the fluorescence decay are separately recorded. The observed decrease in the intensity of either component is due both to the decrease in the size of the population of excited-state molecules, and the tumbling motion of the molecule, which reorients the transition dipole moment between absorption and emission and so changes the intensity of the light observed using a particular emission polarizer orientation. Thus, the polarized fluorescence decay components contain information about the motional dynamics of the molecule.

The time-dependent emission anisotropy $r(t)$ is defined as

$$r(t) = \frac{I_{\parallel}(t) - gI_{\perp}(t)}{I_{\parallel}(t) + 2gI_{\perp}(t)} \qquad 16$$

where $I_{\parallel}(t)$ is the emission polarization component with orientation parallel to that of the exciting light (i.e. vertical), $I_{\perp}(t)$ is the emission polarization component with orientation perpendicular to that of the exciting light (i.e. horizontal) and g is a factor that corrects for any bias the emission monochromator and detector may have towards one orientation over another. (Typically, monochromators preferentially transmit vertically polarized light. Systems for making time-resolved measurements may come equipped with a 'depolarizer' placed in front of the entrance slits of the emission monochromator that negates the effects of the bias and causes g to equal unity. Otherwise, g is determined by exciting the fluorophore with horizontally-polarized light and taking the ratio of the resulting vertical and horizontal emission intensities.) The denominator of *equation 16* ($I_\parallel(t) + 2gI_\perp(t)$) is the total fluorescence intensity of the sample.

The expected time dependence of the emission anisotropy has been worked out for several important models, including those detailed in the following sections.

2.5.2 Freely-rotating spherical molecule

In this case, the time-dependent fluorescence anisotropy is given by

$$r(t) = r_0 \exp(-t/\phi) \tag{17}$$

(A globular protein with a fluorophore (either intrinsic or extrinsic) rigidly attached so that the latter's motion is dominated by the tumbling of the protein, would be an example of this as the protein is completely free to rotate.) Here, r_0 is the initial anisotropy (i.e. r at $t = 0$) and ϕ is the rotational correlation time. The latter quantity is a measure of how fast the molecule tumbles, which, for a spherical molecule, is given by

$$\phi = \frac{\eta V}{kT} \tag{18}$$

where η is the micro-viscosity of the sphere's environment, V is the sphere's hydrodynamic volume, k is Boltzmann's constant and T is the absolute temperature. **A good 'rule of thumb' is that a protein's rotational correlation time in nanoseconds should be about half its molecular weight in kDa.** (e.g. we expect a 16-kDa globular protein to have a ϕ of about 8 ns.) Protein unfolding and the concentration-dependent association of protein subunits are examples of processes that would be expected to cause ϕ to change and so can be examined using time-resolved emission anisotropy.

2.5.3 Rod-shaped molecule with restricted motion

For a rod-shaped molecule with its transition dipole moment oriented along the rod's long axis and whose motion is confined to a cone, with one end of the rod fixed at the cone's apex, the emission anisotropy is given by

$$r(t) = (r_0 - r_\infty) \exp(-t/\phi) + r_\infty \tag{19}$$

Because the molecule's motion is restricted, the anisotropy does not decay to zero but reaches some final constant value, r_∞. *Equation 19* is typically used to model the anisotropy decay of a rod-shaped fluorophore such as 1-(4-trimethylammoniumphenyl)-6-phenyl-1,3,5-hexatriene (TMA-DPH) embedded in a phospholipid membrane, where it reports on the order and dynamics of the latter. The so-called 'order parameter' S of the membrane is defined as

$$S = \left(\frac{r_\infty}{r_0}\right)^{1/2} = \frac{\cos\theta(1 + \cos\theta)}{2} \tag{20}$$

where θ is the half angle of the cone. A more general model for restricted motion is

$$r(t) = \sum_i r_i \exp(-t/\phi_i) + r_\infty \tag{21}$$

A more complete description of the theory of emission anisotropy may be found in (19) (which also contains specific examples of the technique's use as well as the appropriate equations for phase/modulation measurements).

2.5.4 Examples

A study of the effects of melittin on phospholipid membranes that included time-resolved fluorescence anisotropy measurements of 1,6-diphenyl-1,3,5-hexatriene (DPH, the parent compound of TMA-DPH) incorporated into the membranes is described in (20). (Note: TMA-DPH is now the preferred of the two probes as its cationic tertiary amine helps to anchor one end of the probe molecule to the bilayer–water interface.)

References (21–23) describe time-resolved fluorescence anisotropy studies (using TMA-DPH and DPH as fluorescent probes) of the effect of immune complexes on macrophage membrane fluidity, the effect of succinate and phenyl succinate on the fluidity of the inner mitochondrial membrane, and the role of cholesterol in modulating the fluidity of renal cortical brush border and basolateral membranes, respectively.

A time-resolved fluorescence anisotropy study of the base motions in DNA, made using the intrinsic fluorescence of the bases themselves is described in (24). Large amplitude motions were found on the picosecond timescale that are sensitive to solvent viscosity. This provides support for the idea that base motions contribute substantially to the observed fluorescence depolarization in DNA and is consistent with a mechanism that involves concerted motions in the interior of the polymer.

Reference (25) details a study in which time-resolved fluorescence intensity and anisotropy decay measurements were combined with stopped-flow mixing to monitor the local and global dynamics of *E. coli* dihydrofolate reductase during refolding. A general review of such (so-called) 'double kinetic' experiments is given in (26).

Recent contributions that both time-resolved fluorescence intensity and anisotropy decay studies have made to our understanding of the structure and dynamics of proteins and nucleic acids are reviewed in (27).

2.5.5 Caveats

It is very important that the lifetime of the fluorophore (either intrinsic or extrinsic) being used to measure the anisotropy decay be appropriate to the timescale of the motions under examination. Too short a lifetime will mean that the fluorescence intensity will 'die away' before much rotational motion takes place; the anisotropy will remain 'high' over the course of the fluorescence intensity decay and the latter's components will contain limited information about the motional dynamics. On the other hand, too long a lifetime means that the fluorescence will become too quickly depolarized on the timescale of the intensity decay, so that again the signal will contain limited information about the dynamics. Generally, tryptophan fluorescence (with a lifetime of a couple of nanoseconds) can be reliably used to measure ϕ values of up to about 25 ns.

(The required accuracy for φ depends, of course, on the purpose of the measurement.) Protein motions with φ values greater than this will probably require that a longer-lived extrinsic fluorophore (e.g. pyrene) be attached to the molecule. Motions on the microsecond to second timescales can be monitored using phosphorescent probes (see Section 4.2).

In the event that one wishes to measure only the fluorescence decay times of a sample, but is forced to make the measurement with polarizers present, it is imperative that the emission polarizer be oriented at an angle of $54.7° \sim 55°$ with respect to the vertical excitation orientation (the so-called 'magic angle') in order to correct for the presence of the polarizers. Otherwise, as the 'apparent' decay rates of the horizontal and vertical fluorescence components are due to both the depletion of the excited state population and to fluorophore motion (as we mentioned in Section 2.5.1), the measured 'lifetimes' will be incorrect.

Proteins are generally excited at 295 nm in order to avoid (i) energy transfer from tyrosine to tryptophan residues, which can occur at shorter wavelengths (e.g. at 280 nm where tryptophan has its maximum absorbance) and which depolarizes the fluorescence, and (ii) depolarization of tryptophan fluorescence due to relaxation from its ^1L_b to ^1L_a electronic states. (At shorter wavelengths electrons are excited to both levels, after which electrons in the higher level relax radiationlessly to the lower, which also depolarizes the fluorescence.) One would like to maximize the initial anisotropy in order to obtain the most sensitivity for polarized fluorescence measurements.

Due to a process called photoselection, r_o can have a maximum value of 0.4. A measured value for r_o that is significantly less than this strongly suggests that some fast depolarization process is taking place that is beyond the equipment's temporal resolution. Energy transfer, excimer formation, fast local libration and radiationless relaxation between two different electronic states (such as that from ^1L_b to ^1L_a in tryptophan) are some examples of processes that result in a significant decrease in r_o.

2.6 Data analysis

2.6.1 Introduction to least-squares analysis of data

After the data from a time-resolved fluorescence measurement have been collected, they must be analyzed to obtain the lifetimes (τ) and amplitudes (α) and, in the case of polarized fluorescence, the anisotropies (r) and rotational correlation times (φ). For results from pulsed-excitation experiments, this involves performing a non-linear least-squares fit of the convolution integral (*equation 13*) to the experimental data. (The analysis of phase/modulation data is described in Section 2.6.5.) The goal is the same as that in linear least-squares fitting (or linear regression analysis) in that the parameters of the model function (the slope and y-axis intercept of a straight line in the case of linear regression, the α and τ values in the case of fluorescence intensity decay) are adjusted so that the fitted line evenly 'cuts' through the distribution of noise in the data and balances the (squared) vertical distances from the fitted line to the data points above and below it. These vertical distances or differences are called residuals; mathematically, the best fit is defined as being that which minimizes the sum of their squared values. (The residuals are squared as the maximum likelihood of a normal distribution is then at their centres.) The individual residuals are often weighted according to their uncertainties (if known), so that the 'fit' is not unfairly influenced by values that possess a relatively high degree of uncertainty. If the sum of the squares of the weighted residuals is 'rescaled' according to the 'number of degrees of freedom,' we have the definition of χ^2, which has an optimal value of unity.

A general discussion of the use of least-squares fitting in fluorescence measurements may be found in (28). The global analysis of fluorescence data is discussed in (29). Commercially available time-resolved fluorimeters are typically sold with data analysis software included. Available stand-alone packages include the Globals Unlimited suite, which is capable of analysing both time- and frequency-domain data, stopped-flow kinetics, etc. The Center for Fluorescence Spectroscopy at the University of Maryland (USA) also offers software for frequency- and time-domain fluorescence lifetime analysis.

2.6.2 Goodness-of-fit criteria

Strictly speaking, one does not know before hand how many exponentials will be required to adequately describe a particular fluorescence decay, so that one has to perform several separate fits, each using a different number of such terms. We therefore need criteria to use in choosing which model best describes the data. The standard ones are the forms of:

(1) the residuals;

(2) the autocorrelation function values; and

(3) the χ^2 value.

Note: the autocorrelation function 'looks' for correlations between residuals separated by various amounts of time. Specifically, its ith value expresses the amount of correlation between pairs of residuals separated by $i - 1$ time points (or recorder channels). If there is no correlation amongst the residuals, then the autocorrelation function values should be completely random. There are typically only half as many (plus one) autocorrelation function values as there are residuals, owing to the way the former are calculated. Also, by definition, the first value is equal to unity, as each residual is completely correlated with itself. Typically, autocorrelation values have not been calculated in the case of non-linear least-squares fitting of phase/modulation data.

2.6.3 Initial guesses.

In the case of linear regression analysis, the problem has an 'analytic' solution: finding the slope and y-intercept of the best-fitting straight line is reduced to 'plugging' the x- and y-coordinates of the data points into two formulae. There is no such analytic solution for the problem of non-linear regression; the best-fitting values for the parameters have to be found by an iterative process that proceeds from some starting point. One begins the fitting procedure using initial 'guesses' for the parameter values, and the fitting program employs one of several techniques to try to reach the best-fitting values by 'stepping' towards them from the initial guesses. The values for the initial guesses need to be chosen with care as the fitting program can have difficulty reaching the best-fitting values if the former are too far away from the latter. For a single exponential fit, one can estimate the lifetime from the rate at which the fluorescence decays at long times (i.e. far away from the excitation pulse and the distortion it causes). For tryptophan fluorescence in proteins, a clustering of values around 1, 3 and 7 ns occurs, which can serve as initial guesses. For other fluorophores, published values can be used. The fitting procedure is less sensitive to the initial guesses for the amplitudes, so arbitrary values for them often work.

2.6.4 Analysis of pulsed excitation data

1. Non-polarized (or total fluorescence intensity) decay data. We illustrate this with an example.

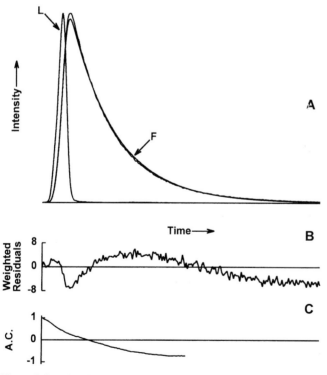

Figure 2 Results of a single-exponential fit to synthetic data F generated by convolving a double exponential decay with the laser pulse profile L. In panel A, the single-exponential fit can also be seen as it falls slightly below F at the fluorescence peak. Panel B shows a plot of the resulting weighted residuals versus time while panel C shows the plot of the autocorrelation (A.C.) function values for the residuals.

Figures 2 and 3 show how the appearance of the residuals and autocorrelation function for a pulsed excitation experiment typically depend on the appropriateness of the fitting function. In panel A of both figures, L shows an actual excitation pulse profile (more properly, the instrument response function) that was generated by an argon ion laser that pumped a dye laser circulating rhodamine 6G, the tuned output of which was frequency-doubled to 295 nm (nanometer = 10^{-9} m = 10 Å) by passage through a β-barium borate crystal (cf. Section 3.1.1). The pulse was recorded using a photomultiplier after being scattered by a dilute silica suspension in distilled water. Synthetic data (F in the figures) were generated by convolving a double exponential decay ($\alpha_1 = 2.5$, $\tau_1 = 2.5$ ns, $\alpha_2 = 0.5$, $\tau_2 = 5.0$ ns) with the laser pulse profile. Random Gaussian noise, appropriately scaled for single photon counting (cf. Section 3.3.1), was then added to this 'signal'. The resulting peak intensity was about 45 500 counts.

Figure 2 shows the results of fitting the convolution of a single exponential to this decay, which yielded best-fitting values of $\alpha = 2.8$ and $\tau = 3.0$ ns. This lifetime is close to the intensity-weighted average lifetime of 3.2 ns (given by $\Sigma f_i \tau_i$, where the f_i are the fractional intensities (*equation 3*)), which is usually to

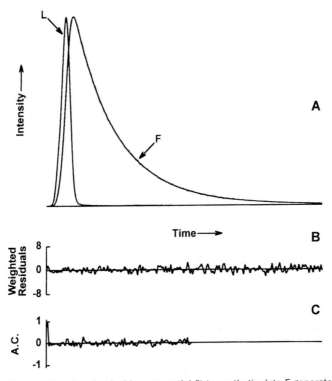

Figure 3 Results of a double-exponential fit to synthetic data F generated by convolving a double exponential decay with the laser pulse profile L. Panel B shows a plot of the resulting weighted residuals versus time while panel C shows the plot of the autocorrelation (A.C.) function values for the residuals.

be expected. The fit can also be seen in panel A as it falls somewhat below the 'data' at the peak of the fluorescence curve. Otherwise, the two seem to agree fairly well, especially at long times. The weighted residuals (*Figure 2B*) indicate, however, that the single exponential fit deviates from the data essentially everywhere along the decay curve, and the autocorrelation of the residuals (*Figure 2C*) is also far from being random, which confirms that a definite trend exists in the residuals. The χ^2 value of 14.88 also indicates a poor fit. (Recall that the aim of the fitting procedure is to 'adjust' the model function so that it evenly 'cuts' through the noise in the data. If the model is fundamentally inadequate, no amount of adjustment will enable one to obtain a good fit with a random distribution in the residuals.)

Figure 3 shows the results of the double exponential fit that was attempted next. In this case the fit appears to completely overlay the data in panel A and this impression is borne out by the completely random distribution of the residuals around the zero line (panel B). The random distribution of the autocorrelation function values (*Figure 3C*) confirms that there is indeed no trend in the residuals, while a χ^2 value of 0.84 (close to unity) further attests to the goodness of the fit. A fit of a triple exponential decay was also attempted (results not shown). A χ^2 value of 0.85 indicated that the addition of another term was not warranted, and this judgement was reinforced by the fact that two of the three exponentials had nearly identical lifetimes. (The general rule is that, for ~300 degrees of freedom (i.e. ~305 data points), the addition of another exponential term is not justified if it results in a less than 10% improvement in χ^2.) Thus, the decay is taken to be double exponential, the results for which were $\alpha_1 = 2.55$, $\tau_1 = 2.52$ ns, $\alpha_2 = 0.45$ and $\tau_2 = 4.61$ ns. These values are in excellent agreement with the generating function concerning the short lifetime component. The amplitude and lifetime values of the long lifetime component are each about 10% in error, presumably because that 'species' only contributed about 28% of the total fluorescence intensity at the chosen emission wavelength, and the two lifetimes only differ by a factor of two. Much greater accuracy can be obtained by carrying out a global analysis that includes fluorescence decays measured at other emission wavelengths (i.e. DAS, Section 2.4) at which the longer lifetime component would make a different contribution.

2. Polarized fluorescence data. As the emission anisotropy (*equation 16*) is not a directly measured quantity, polarized fluorescence data cannot be analysed by fitting calculated experimental values for $r(t)$ with a simple convolution of the lamp profile and a model function (e.g. *equation 17* for a rotating sphere). Instead, the polarized data are often analysed in two steps. First, the total fluorescence decay ($I_{\parallel}(t) + 2gI_{\perp}(t)$ in the denominator of *equation 16*, and typically referred to as the 'sum' data, $S(t)$) is fitted as described above. Once the appropriate expression $F(t)$ for the intrinsic fluorescence decay has been obtained, the product $F(t)r(t)\mathbf{L}_{ij}$ is then convolved with the lamp profile and fitted to the 'difference' data $D(t) = I_{\parallel}(t) - gI_{\perp}(t)$ (the numerator of *equation 16*). Here, $r(t)$ is the model function for the anisotropy decay and L_{ij} is an 'association matrix' (made up of ones and zeros) that combines the appropriate intensity and anisotropy decay components, depending on whether the decay is associative or non-associative (30). The expression for $r(t)$ can be modified as required until a good fit is obtained, where χ^2 and the form of the residuals and autocorrelation again serve as the fitting criteria as described previously. Alternatively, $I_{\parallel}(t) = F(t)[1 + 2r(t)L_{ij}]/3$ and $I_{\perp}(t) = F(t)[1 - r(t)L_{ij}]/3g$ convolved with the lamp pulse profile may be fitted simultaneously to their respective polarized fluorescence decay components in a global analysis.

2.6.5 Analysis of phase/modulation fluorescence decay data

In *Figure 4* we show as an example the results from an analysis of synthetic phase/modulation data. In this case, *equations 14* and *15* were used to generate phase angle and modulation values for a double exponential fluorescence decay

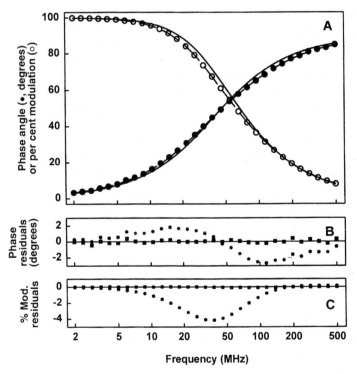

Figure 4 Results of single- and double-exponential fits to synthetic phase/modulation data for a double-exponential decay. In panel A, the solid circles (●) are the phase angle values (in degrees) and the open circles (○) are the percentage modulation values (modulation × 100%), both plotted as a function of the modulation frequency of the exciting light. The solid and dashed lines show the respective results of single- and double-exponential analyses of these data. Panels B and C show the residuals from those fits to the phase (B) and % modulation (C) data. For both types of residuals, the solid circles (●) indicate the residual values from the single-exponential analysis, while the solid squares (■) indicate the values from the double-exponential analysis.

for which $\alpha_1 = 0.33$, $\tau_1 = 2.0$ ns, $\alpha_2 = 0.67$ and $\tau_2 = 5.0$ ns. Gaussian random noise with standard deviations of 0.2° and 0.005 (10, 18) for the phase angle and modulation, respectively, were added to the calculated values. These standard deviations were also used as the inverse weighting factors in the non-linear least-squares analysis of the 'data.' (Note: for phase/modulation data the weighting factors, which are the reciprocals of the standard deviations, are estimated from repeated measurements so that, depending on their accuracy, the optimal χ^2 value may not be unity. The appropriate number of decay components is simply that which minimizes χ^2.)

The 'data' are shown in *Figure 4A*, where the closed circles are the phase angle values and the open circles are the per cent modulation (modulation $\times$ 100%) values, both plotted as functions of the modulation frequency (in MHz). The solid lines (which clearly do not everywhere overlay the data) show the results of simultaneous fits of the single-exponential expressions ($\theta = \arctan(2\pi f \tau)$ and $m = (1 + 4\pi^2 f^2 \tau^2)^{-\frac{1}{2}}$) to the data ($\chi^2 = 30.1$), which yielded a value of 4.1 ns for the single lifetime. This value is close to the intensity-weighted average lifetime $\langle \tau \rangle$ of 4.5 ns, which is usually to be expected. The residuals for this fit (closed circles in panels B and C for the phase and per cent modulation residuals, respectively), clearly show trends, which reinforces our impression from panel A that this decay is not well modelled by a single exponential. The dashed lines in *Figure 4A* (which do seem to fit the data well) show the results of a double-exponential fit to the data ($\chi^2 = 0.57$), which gave values of $\alpha_1 = 0.345$, $\tau_1 = 2.04$ ns, $\alpha_2 = 0.655$ and $\tau_2 = 5.02$ ns. These are in excellent agreement with the values used to generate the data, and the goodness of the fit is indicated by the random distribution in the residuals (solid squares for the phase and per cent modulation residuals in panels B and C, respectively). Usually, the autocorrelation function values are not calculated for fits to phase/modulation data. The reader may consult (10) for the equations used in analysing polarized phase/modulation data.

3 Equipment for time-resolved fluorescence measurements

Experimental set-ups for time-resolved fluorescence measurements can be broken down into the following component blocks:

(1) an excitation source;

(2) a sample chamber with optics for focusing the exciting light and fluorescence;

(3) a detector for converting fluorescence into an electronic signal; and

(4) the recording electronics for processing the detector's output.

Below, we briefly describe the options available for (1), (3) and (4). We also mention several commercially available systems as examples.

3.1 Excitation sources

3.1.1 Lasers

Lasers are a source of coherent, monochromatic, highly-collimated radiation. UV tunability requirements for protein and DNA fluorescence measurements mean that the typical set-up usually involves either a mode-locked argon-ion or Nd:YAG laser synchronously pumping a dye laser.[5] Different dyes give laser output at different wavelengths. Rhodamine 6G, for example, can provide pulses at 590 nm, which can then be frequency-doubled to 295 nm for exciting tryptophan fluorescence. Frequency doubling is a non-linear optical effect (called second harmonic generation) that results from the interaction of the laser beam's intense electric field with the doubling crystal. β-Barium borate (BBO) crystals can provide a doubling efficiency of greater than 15%. (Note: third harmonic generation using the tunable, modulated output of titanium–sapphire lasers is beginning to be used to provide short UV pulses.) The repetition rate of the pump–dye laser combination is typically high enough that the time between successive pulses is much shorter than the time needed to accommodate a complete fluorescence decay. A technique called cavity dumping is used to select out, for instance, one in twenty pulses for subsequent doubling. This is accomplished by placing an acousto-optic deflector inside the dye cavity that is periodically triggered with a radio frequency burst. The reader who desires additional information about lasers may consult (31–34).

3.1.2 Flash lamps

A flash lamp consists of two electrodes (separated by a narrow gap) contained in a housing and surrounded by a gas under pressure. A device called a thyratron rapidly switches voltage pulses to the electrodes so that an electrical discharge takes place across the gap ~50 000–100 000 times per second (Hz). Each discharge excites the atoms/molecules of gas in the region so that they give off a pulse of light, the profile of which (in terms of either time or wavelength) depends on the pressure and composition of the gas. These pulses typically have a breadth of several nanoseconds, with a tail.

3.1.3 Synchrotron radiation

Synchrotron radiation can also be used to provide excitation pulses for performing time-resolved fluorescence measurements. We will limit our comments, however, as the necessary support facilities required for a synchrotron radiation source restrict them to being located at major universities and national laboratories. Such facilities include: the Aladdin Biofluorescence Center (ABC) at

---

[5] In mode-locking, the combination of different standing wave 'modes' that the laser cavity supports is fixed so that superposition of the different modes gives short pulses. Although argon-ion and Nd:YAG lasers do not themselves produce the shortest mode-locked pulses, their periodically-modulated output can synchronously pump or drive a dye laser (whose mode-locking has been matched to that of the pump laser), which then produces much shorter pulses than the pump laser.

the Synchrotron Radiation Center of the University of Wisconsin, Madison and the National Synchrotron Light Source (NSLS) at Brookhaven National Laboratory (both in North America); the Synchrotron Radiation Source (SRS) at Daresbury (UK); and the Laboratoire pour l'Utilisation du Rayonnement Electromagnétique (LURE) at the Université de Paris-Sud at Orsay (France). (Links to the websites of other facilities may be found at the LURE website.) The interested reader may find (35) provides additional useful information.

3.2 Detectors

Commercially available time-resolved fluorescence spectrometers typically use photomultiplier tubes (PMTs) to convert light into an electronic signal for further processing. As was pointed out in *Footnote 3*, however, PMTs distort the fluorescence decay profile by broadening it; this places a limit on how short a fluorescence decay time a photomultiplier-based time-resolved system can measure. PMTs are described in Section 4.6 of Chapter 2 ; the reader can also find additional information in (36).

Other, 'faster' detectors that can be used in place of a conventional PMT include microchannel plate photomultiplier tubes (MCP-PMTs) (31) and streak cameras (37). Because of their expense, the use of these devices is usually confined to 'home built' fluorimeters found in dedicated fluorescence laboratories, and is therefore not discussed here.

3.3 Recording electronics

The type of recording electronics used depends on whether pulsed excitation (either pulse sampling or time-correlated single photon counting) or phase/modulation measurement are being made. Therefore, we discuss the electronics in the context of the method.

3.3.1 Pulse methods

For the case of pulsed excitation (as described in the discussion of the convolution integral, Section 2.3.2), a full fluorescence decay profile cannot be captured following excitation by just a single pulse; the emission intensity is too dim, and recording devices are generally too slow to allow that. Instead, a composite decay profile is assembled from recorded 'portions' of many (typically hundreds of thousands of) consecutively generated fluorescence decays. Because the final result is a composite 'picture,' it is very important that the excitation source provides stable pulses. If, for example, the excitation pulse profile changes between when the fluorescence decay and excitation pulse profiles are recorded, then the convolution of the latter with the proper expression for the intrinsic fluorescence decay $F(t)$ (*equation 1*) will not fit the recorded decay and reliable lifetimes etc. will not be extracted.

In both of the pulse excitation techniques we describe below, the excitation pulse is also used to provide a triggering signal that the recording electronics uses to synchronize events. In the case of flash lamp excitation, this is commonly accomplished using an 'antenna pick-up'. For laser pulse excitation, a portion of the beam may be reflected into a photodiode whose output provides the triggering pulse. Often there has to be a delay between when the triggering signal arrives at the recording device and when the latter can first begin to process the incoming fluorescence signal from the detector. The arrival of the fluorescence signal typically has to be delayed to allow for this, which can be accomplished by simply increasing the length of cable connecting the detector and the recorder. (It takes an electronic signal about 1 ns to travel one foot (~0.3 m).) What happens next depends on the recording technique being used.

1. Pulse sampling. In this technique, a device called a boxcar averager may be used to record the signal coming from the detector. One can think of a boxcar averager as having an electronic 'shutter' that is opened for a brief period t_g at some time t_d following the reception of the triggering signal at t_t (see *Figure 5*). (This process is called gating, as the effect is like that of opening and closing a gate to control something's entry. Note: the 'width' of t_g in *Figure 5* has been greatly exaggerated in order to make it clearly visible. Typically, the time course of a fluorescence decay will be divided up into ~256 or 512 such segments or 'channels,' rather than the two dozen or so suggested by the figure.)

For a pre-set number of consecutive pulses, the shutter will always open at the same time t_d after triggering and for the same duration t_g to allow a small 'slice' (denoted by the hatched area under the decay curve in *Figure 5*) of the detector's output signal to enter the recorder. There it is used to charge a capacitor, whose charge increases with each repetition of the pulse. If enough pulses are sampled in this way, random noise in the measured decay segment

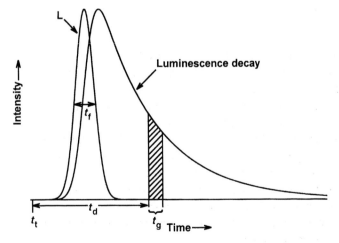

Figure 5 Events occurring during excitation of a sample in the case of pulse sampling (fluorescence) and phosphorescence measurements. t_f is the width at half height of the excitation pulse, t_d is the delay from the arrival of the triggering signal at t_t to the beginning of observation and t_g is the gate width of the detector, during which time the system measures the hatched portion of the decay curve.

will tend to average out. Once the pre-set number of samples has been reached, the capacitor is discharged through a resistor, whose voltage is digitized and stored. The whole process then begins again, with the shutter now opening after a new delay time of $t_d' = t_d + t_g$. In this way, the shutter's 'window' is scanned across the decay profile, which allows the latter to be recorded in discrete, consecutive segments.

Alternatively, the voltage being applied to the detector can be sequentially gated and the signal from the latter stored in a multi-channel analyser (MCA) as described in (38). (Note: the main disadvantage that pulse-sampling methods have is that the PMT is operated in the linear or low-gain analogue mode, so that its output changes if its high-voltage supply 'drifts.')

Boxcar averagers and their use in signal recovery are described in (39). PTI (Photon Technology International) manufactures a series of fluorescence life-time spectrometers (models C-70 to C-73) that use a stroboscopic detection technique (i.e. a gated PMT, as described in the previous paragraph) and either nanosecond flashlamp or nitrogen/dye laser excitation.

2. Time-correlated single photon counting (TCSPC). In this technique, the photon flux is kept low enough and the PMT voltage high enough that the arrival of individual photons can be detected (see Section 4.6.1 of Chapter 2). As in pulse sampling, the lamp pulse provides a signal that initiates a timing or 'clock' sequence. The electronics then wait for the pulse from the arrival of the first photon, which stops the 'clock.' (The 'clock' or chronometer is actually a capacitor that is charged at a constant rate during the period between the start- and stop-pulses, so that the total charge is proportional to the elapsed time. After the arrival of the stop-pulse, the charge is measured and digitized, and one 'count' is added to the appropriate 'bin' or channel of a multi-channel analyser (MCA) as determined by the just-measured arrival time of the first photon. The module that performs all this is called a TAC, or time-to-amplitude converter.) After the arrival of the first photon, the electronics are oblivious to pulses caused by any other photons until after the next excitation pulse, at which point the process is repeated.

Fluorescence decay in a population of excited-state molecules is a random process, so that the arrival time of the first photon is also random, although it is more likely to occur at the peak of the fluorescence decay than it is along the latter's 'tail.' Because of this inherent statistical nature, over many excitation cycles one gradually builds us a statistical profile of the likelihood of emission versus time that corresponds to the decay profile. For a large enough number of counts, the uncertainty in the number of counts in a particular channel is equal to the count number's square root, which gives us the weighting factors to be used in the least-squares analysis of the data.

Another important component of a TCSPC system is a device called a constant fraction discriminator (CFD), whose primary function is to compensate for amplitude variations in the triggering pulses and yield reliable timing.

A comprehensive treatment of TCSPC can be found in (40), which contains introductory chapters on fluorescence and the basic principles of the technique as well as chapters dealing with light sources, photomultipliers, electronics, data analysis, time-resolved emission spectra (including DAS) and time-resolved fluorescence anisotropy. The reader may also wish to consult (41). Commercially available TCSPC systems include Quantum Northwest's QNW 1000 time-resolved fluorimeter and Edinburgh Instruments' FL900 fluorescence lifetime spectrometer.

3.3.2 Phase/modulation equipment

In traditional phase/modulation fluorometry, the output of a xenon arc lamp or laser is modulated at the desired frequency F. The gain of the fluorescence detector (e.g. a PMT or MCP-PMT) is then modulated at a slightly off-set frequency $F + \delta F$ (where δF is typically 25, 40 or 800 Hz), so that when the excitation and emission signals are recombined they mix and one observes the phenomenon called 'beats', in which the product of the signals is modulated at the frequency off-set, δF. (Beating is the same phenomenon one perceives by listening to two vibrating tuning forks of slightly different frequency.) The phase and modulation are related to δF which can easily be measured as the value of the latter is low.

ISA's (Spex's) Fluorolog Tau-3 lifetime system and Spectronic Instruments' SLM-AMINCO 48000 DSCF spectrofluorimeter both use xenon flash lamp excitation (typically 150–450 W) and have modulation frequencies of up to 310 MHz. (Most systems can also be operated as steady-state fluorimeters). SLM also manufactures a multi-harmonic system based on a pulse- modulated continuous wave light source.

4 Phosphorescence

4.1 Phosphorescence of proteins

The triplet lifetimes of aromatic amino acid residues in proteins show a strong dependence on the viscosity of the matrix. In rigid glasses consisting of mixtures containing glycerol/water 50:50 v/v at 77 K, the phosphorescence lifetime of free tryptophan is typically around 6 s (42). Smaller values of the phosphorescence lifetime of proteins, excited at 295 nm, imply specific interactions: e.g. electron transfer with a group in the immediate vicinity (43) or inter-tryptophanyl energy transfer. The phosphorescence spectra of proteins are dominated by tryptophanyl groups that show vibronic structures with a 0–0 vibrational band centred at around 412 nm. The well-resolved vibrational structure of the phosphorescence emission allows one in some cases to distinguish between the emission spectra of tryptophanyl groups positioned in environments of different polarity (44).

For phosphorescence measurements, the samples are typically excited using a pulsed xenon flash lamp. A commercially available instrument (the Perkin-Elmer LS-50B luminescence spectrophotometer) is suitable for phosphorescence measurements designed to record emission spectra and determine phosphorescence decay times. The events occurring during the excitation of the sample with a pulsed xenon source in the phosphorescence mode are the same as those illustrated in *Figure 5* for pulse sampling of fluorescence. In this case, the xenon lamp gives an excitation pulse with a width t_f at half-peak intensity of around 10 μs. During excitation, the phosphorescence increases to a maximum value of I_0 and then decays exponentially to zero. The gating of the sample photomultiplier is delayed by a time (t_d), so that it no longer coincides with the lamp flash. The gate width of the photomultiplier detector is controlled by the gate time (t_g) as indicated in the figure. If a delay greater than 0.1 ms is selected for the phosphorescence measurements, no fluorescence signal is detected.

In typical phosphorescence experiments, the delay time is varied between 0 and 900 ms, whilst the gate time changes from 0.01 to 500 ms. Both t_d and t_g can be varied in multiples of 0.01 ms. The phosphorescence of proteins can be recorded in rigid matrices at 77 K (liquid nitrogen temperature) or in aqueous solutions at room temperature (45). The central requirement for the observation of protein phosphorescence in solution is that dissolved oxygen be reduced to a sufficiently low level, since oxygen can efficiently quench the tryptophan triplet state. Removal of oxygen from the protein solution can be achieved by repeated application of moderate vacuum followed by the inlet of argon. Anoxia can also be attained and maintained by the use of an enzyme system consisting of glucose oxidase, glucose and catalase (46) as outlined in *Protocol 1*.

A check on the thoroughness of deoxygenation is provided by the dependence of the phosphorescence lifetime on the amount of oxygen present in the system. Complete removal of O_2 from a solution of alkaline phosphatase (10 μM)

Protocol 1

Phosphorescence sample deoxygenation

Reagents

- Glucose oxidase and catalase (Sigma) and glucose (Fisher)
- Protein buffer, pH 7

Method

1   Prepare stock solutions of glucose oxidase (20 mg/ml) and catalase (2 mg/ml) in protein buffer.[a]

2   To the desired amount of lyophilized protein, add 3 mg of glucose for every ml of final sample volume. Add the desired amount of protein buffer and place the solution in a cuvette.

3   For every ml of protein/glucose solution, add 1 μl each of the glucose oxidase and catalase stock solutions, stopper the cuvette and mix.

[a] These solutions should keep for a couple of weeks when stored at 4 °C.

at 25 °C yields a phosphorescence lifetime of 1.5 s (47) when excited in the spectral region coinciding with the absorption of the protein. The phosphorescence lifetime is decreased to 1.2 s in the presence of nanomolar concentrations of oxygen.

It is well established that oxygen quenches both excited singlet and triplet states of aromatic compounds. Thus, indole derivatives in solution at room temperature display a ratio for the rate constant k_{NR}/k_{QP} of approximately two (48, 49). However, the quenching of alkaline phosphatase phosphorescence by oxygen proceeds at a rate $k_{QP} = 1.2 \times 10^6$ M^{-1} s^{-1} (47). The magnitude of this rate constant can only be explained in terms of hindered diffusion of the quencher through the protein matrix. Apparently, one of the tryptophanyl residues of alkaline phosphatase is buried within a shell formed by α-helical rods and β-pleated sheets. This particular agglomerate (the so-called 'knot' structure (50)) prevents the penetration of small molecules like oxygen.

The technique of phosphorescence spectroscopy in either rigid glasses or aqueous solutions can be used to study the spectroscopic properties of tryptophanyl residues in wild-type and mutant forms of proteins. As an example, samples of bovine brain inositol monophosphatase containing three, two and one tryptophanyl groups/monomers were examined by phosphorescence spectroscopy at 77 K in rigid glasses. The locations of the three tryptophanyl residues in the wild-type protein have been provided by X-ray crystallography (51). Trp5 is located at the N-terminus, Trp87 is separated by 12 Å from the metal chelating centre of Mg^{2+}, and Trp219 is near the catalytic centre. Mutated forms of the enzyme containing one (double) and two (single) mutant tryptophanyl residues/monomer were constructed (52). The wild-type enzyme and variant W219F display similar phosphorescence spectra (*Figure 6A*) and the emission decays monoexponentially with a phosphorescence lifetime of 4.0 s. By contrast, the phosphorescence emission of the double mutant is blue shifted and the 0–0 vibronic transition is centred at around 409 nm. On the other hand, the double mutant (W219F, W5F) exhibits a shorter decay component (2.75 s) which also fits monoexponential kinetics. The substantial decrease in the phosphorescence lifetime is interpreted to mean that the degree of rigidity of the environment surrounding Trp87 has been perturbed as a result of replacement of Trp5 and Trp219 by the amino acid phenylalanine. The conformational dynamics of the wild-type and mutant forms of inositol monophosphatase were also examined by measuring resonance energy transfer. Because of the luminescence properties of the ion Tb^{3+} (53, 54), the lanthanide bound to the catalytic site of inositol monophosphatase shows a strong sensitized luminescence upon excitation of the tryptophanyl groups of the protein. Two major bands centred at 490 and 545 nm, and two minor bands at 580 and 620 nm are detected upon excitation at 295 nm (*Figure 6B*).

The efficiency of energy transfer for the wild-type enzyme is greater than that observed for either the single mutant (W219F) or the double mutant (W219F, W5F) (*Table 1*). This behaviour was expected since Trp219 is near the metal binding centre (7 Å); it contributes significantly to the sensitized lumin-

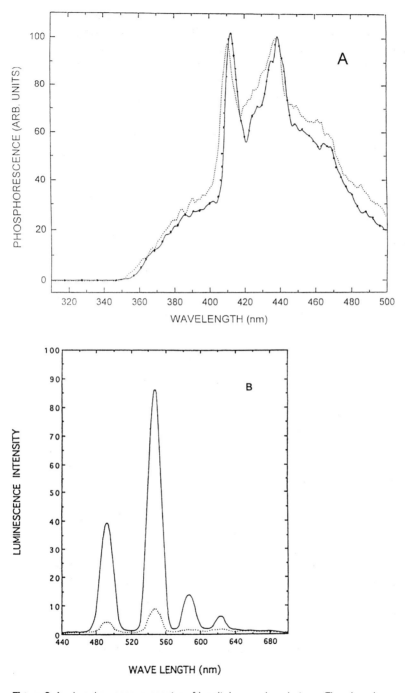

Figure 6 A: phosphorescence spectra of inositol monophosphatase. The phosphorescence spectra were recorded at 77 K with samples dissolved in the solvent system, glycerol–Tris–HCl (pH 7.5) (1:1 v/v). The protein concentrations are 16 μM. Results were obtained for wild-type protein (——), single mutant (W219F) (●) and double mutant containing Trp87 (...). The delay time was 0.1 ms and the gate time was 20 ms. B: Luminescence

escence because the critical transfer distance is 3 Å. On the other hand, Trp87 and Trp5, separated by 12 and 22 Å, respectively, from the metal binding site, show weaker sensitized luminescence. However, the significant decrease in the sensitized luminescence of the double mutant as compared to the single mutant can be attributed to two independent factors: i.e. change in the distance of the donor–acceptor pair (Trp87–Tb^{3+}) and change in coordination efficiency as demonstrated by the binding studies (*Table 1*).

The reader may find (55) and (56) provide additional useful information about phosphorescence measurements.

Table 1. Binding of Tb^{3+} to myoinositol monophosphatase (1.2 μM) at pH 7.5 in 10 mM Tris–HCl

Sample	K_d (μM)	Number of binding sites	Efficiency[a] of energy transfer
Wild-type	1.6	1	0.016
W219F	1.7	1	0.004
W291F, W5F	50	1.3	0.0005

[a]Calculated using *equation 12*. For the wild-type enzyme, $J_{(v)} = 0.73 \times 10^{-19}$ M^{-1} cm^3, $R_0 = 3.1$ Å and R = 6 Å.

4.2 Time-dependent phosphorescence anisotropy

In Section 2.5 we described the use of time-resolved fluorescence anisotropy for monitoring protein motion on the nanosecond timescale. For motion on much longer timescales, time-resolved phosphorescence anisotropy can be used instead. The latter technique has been employed, for example, to examine the rotational motion of membrane-bound proteins labelled with the triplet probe eosin (57, 58). Prior to making the measurements, the protein is labelled with eosine-maleimide as described in *Protocol 2.*

An oxygen-consuming reaction, involving glucose oxidase, glucose and catalase (*Protocol 1*) is then employed to create anaerobic conditions and $I_\parallel(t)$ and $I_\perp(t)$ are measured. The anisotropy data are generally analysed using as a model either a cylinder or ellipsoid undergoing restricted rotation and wobbling (*equation 21*). In the case of eosin-labelled band 3 in erythrocyte membranes, three rotational correlation times (ϕ_i) were determined, ranging from 0.03 to values greater than 1 ms (59). The phosphorescence lifetime of eosin-tagged protein is 2.5 ms.

The presence of several rotational correlation times was interpreted to mean that several classes of band 3 proteins differing in rotational motion exist in ghosts and intact cells. A typical anisotropy decay from (59) is shown in *Figure 7*.

spectra of inositol monophosphatase (2 μM) in the presence of 10 μM $TbCl_3$ before (—) and after addition of 10 mM $MgCl_2$ (...). The samples were dissolved in 10 mM Tris–HCl buffer (pH 7.5) at room temperature. The excitation wavelength was 295 nm. The delay time was 0.1 ms, gate time was 1 ms and the slit widths for excitation and emission were 5 nm.

Protocol 2

Eosine labelling of proteins for time-resolved phosphorescence measurements

Equipment and reagents

- 100 μM solution of protein in protein buffer
- Dimethylformamide (DMF) for preparing eosine stock solution
- Eosine-5-maleimide (Molecular Probes (Cat. no. E-118))
- Protein buffer (10–100 mM phosphate, Tris or HEPES, pH 7.4), deoxygenated
- Sephadex G-25 column equilibrated with protein buffer, or dialysis tubing

Method

1 Prepare a 20 mM (~15 mg/ml) stock solution of eosine in DMF just prior to its use. Shield the container from light by wrapping it in aluminium foil.

2 Add 100 μl of label solution per 1 ml of protein solution to the latter, dropwise, while stirring, to obtain a label-to-protein molar ratio of 20.

3 Shield the mixture from light and incubate it at room temperature for 2 h.

4 Dialyse the probe/protein mixture extensively against buffer at 4°C or pass it over a Sephadex G-25 column to remove excess label.

5 Determine the degree of labelling by measuring the solution absorbance at 584 nm, where the label extinction coefficient is 8.3×10^4 M^{-1} cm^{-1}.

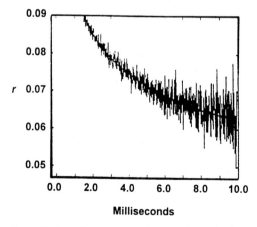

Milliseconds

Figure 7 Phosphorescence anisotropy decay in ghosts. The anisotropy decay of eosin-labelled proteins was monitored as a function of time. The data were fitted according to *equation 21*. Reprinted in part with permission from Matayoshi, E. D. and Jovin, T. M. (1991) *Biochemistry*, **30**, 3527–3538. Copyright 1991 American Chemical Society.

Acknowledgements

T. D. B. would like to express his gratitude to the National Institute of Neurological Disorders and Stroke, the National Institutes of Health (NIH), for its continuing support in the form of a postdoctoral fellowship, and to Dr Jay R. Knutson, Chief, Optical Spectroscopy Section, National Heart, Lung and Blood Institute, NIH, for welcoming him into the latter's laboratory. In addition, he would like to thank Dr Knutson and Dr Solon Georghiou for their critical reading of and helpful comments concerning the fluorescence portion of this chapter. J. E. C. wishes to express his appreciation for the support given by the Hong Kong Technical University during his stay there as a visiting professor.

References

1. Lakowicz, J. R. (1983). *Principles of fluorescence spectroscopy*, p. 383. Plenum Press, NY.
2. McGlynn, S. P., Azumi, T. and Kinoshita, M. (1969). *Molecular spectroscopy of the triplet state*, p. 261. Prentice Hall, NJ.
3. Förster, T. B. (1959). *Discuss. Faraday Soc.*, **27**, 7.
4. Förster, T. B. (1965). In *Modern quantum chemistry,* Part III (ed. O. Sinanoglu), p. 93. Academic Press, NY.
5. Kellog, R. E. (1967). *J. Chem. Phys.*, **47**, 3403.
6. Ermolaev, V. L. and Sveshnikova, E. B. (1964). *Opt. Spectroscopy (English Translation)*, **14**, 320.
7. Berlman, I. B. (1973). *Energy transfer parameters of aromatic molecules*. Academic Press, NY.
8. Cheung, H. C. (1991). In *Topics in fluorescence spectroscopy, vol. 2: principles* (ed. J. R. Lakowicz), p. 127. Plenum Press, NY.
9. Fernandez, S. M. and Sha'afi, R. I. (1983). In *Fast methods in physical biochemistry and cell biology* (ed. S. M. Fernandez and R. I. Sha'afi), p. 1. Elsevier, NY.
10. Lakowicz, J. R. and Gryczynski, I. (1991). In *Topics in fluorescence spectroscopy, vol. 1: techniques* (ed. J. R. Lakowicz), p. 293. Plenum Press, NY.
11. Knutson, J. R., Beechem, J. M. and Brand, L. (1983). *Chem. Phys. Lett.*, **102**, 501.
12. Royer, C. A., Mann, C. J. and Matthews, C. R. (1993). *Protein Sci.*, **2**, 1844.
13. Shen, F., Triezenberg, S. J., Hensley, P., Porter, D. and Knutson, J. R. (1996). *J. Biol. Chem.*, **271**, 4819.
14. Green, S. M., Knutson, J. R. and Hensley, P. (1990). *Biochemistry*, **29**, 9159.
15. Alcala, J. R., Gratton, E. and Prendergast, F. G. (1987). *Biophys. J.*, **51**, 587.
16. Alcala, J. R., Gratton, E. and Prendergast, F. G. (1987) *Biophys. J.*, **51**, 597.
17. Alcala, J. R., Gratton, E. and Prendergast, F. G. (1987). *Biophys. J.*, **51**, 925.
18. Knutson, J. R. (1992). In *Methods in enzymology* (ed. L. Brand and M. L. Johnson). Vol. 210, p. 357. Academic Press, NY.
19. Steiner, R. F. (1991). In *Topics in fluorescence spectroscopy, volume 2: principles* (ed. J. R. Lakowicz), p. 1. Plenum Press, NY.
20. Bradrick, T. D., Dasseux, J.-L., Abdalla, M., Aminzadeh, A. and Georghiou, S. (1987). *Biochim. Biophys. Acta*, **900**, 17.
21. Petty, H. R., Niebylski, C. D. and Francis, J. W. (1987). *Biochemistry*, **26**, 6340.
22. Mutet, C., Duportail, G., Crémel, G. and Waksman, A. (1984). *Biochem. Biophys. Res. Commun.*, **119**, 854.
23. Molitoris, B. A. and Hoilien, C. (1987). *J. Membrane Biol.*, **99**, 165.
24. Georghiou, S., Bradrick, T. D., Philippetis, A. and Beechem, J. M. (1996). *Biophys. J.*, **70**, 1909.

25. Jones, B. E., Beechem, J. M. and Matthews, C. R. (1995). *Biochemistry*, **34**, 1867.

26. Beechem, J. M. (1997). In *Methods in enzymology* (ed. L. Brand and M. L. Johnson). Vol. 278, p. 24. Academic Press, NY.

27. Millar, D. P. (1996). *Curr. opin. struct. biol.*, **6**, 637.

28. Straume, M., Frasier-Cadoret, S. G. and Johnson, M. L. (1991). In *Topics in fluorescence spectroscopy, vol. 2: principles* (ed. J. R. Lakowicz), p. 177. Plenum Press, NY.

29. Beechem, J. M., Gratton, E., Ameloot, M., Knutson, J. R. and Brand, L. (1991). In *Topics in fluorescence spectroscopy, vol. 2: principles* (ed. J. R. Lakowicz), p. 241. Plenum Press, NY.

30. Brand, L., Knutson, J. R., Davenport, L., Beechem, J. M., Dale, R. E., Walbridge, D. G. and Kowalczyk, A. A. (1985). In *Spectroscopy and the dynamics of molecular biological systems* (ed. P. M. Bayley and R. E. Dale), p. 259. Academic Press, London.

31. Small, E. W. (1991). In *Topics in fluorescence spectroscopy, vol. 1: techniques* (ed. J. R. Lakowicz), p. 97. Plenum Press, NY.

32. Rabek, J. F. (1982). *Experimental methods in photochemistry and photophysics*, p. 593. Wiley, NY.

33. Silfvast, W. T. (1996). *Laser fundamentals*. Cambridge University Press, NY.

34. Davis, C. C. (1996). *Lasers and electro-optics: fundamentals and engineering*. Cambridge University Press, Cambridge, UK.

35. Munro, I. H. and Martin, M. M. (1991). In *Topics in fluorescence spectroscopy, vol. 1: techniques* (ed. J. R. Lakowicz), p. 261. Plenum Press, NY.

36. Rabek, J. F. (1982). *Experimental methods in photochemistry and photophysics*, p. 459. Wiley, NY.

37. Nordlund, T. M. (1991). In *Topics in fluorescence spectroscopy, vol. 1: techniques* (ed. J. R. Lakowicz), p. 183. Plenum Press, NY.

38. Lakowicz, J. R. (1983). *Principles of fluorescence spectroscopy*, p. 62. Plenum Press, NY.

39. Rabek, J. F. (1982). *Experimental methods in photochemistry and photophysics*, p. 554. Wiley, NY.

40. O'Connor, D. V. and Phillips, D. (1984). *Time-correlated single photon counting*. Academic Press, London.

41. Birch, D. J. S. and Imhof, R. E. (1991). In *Topics in fluorescence spectroscopy, vol. 1: techniques* (ed. J. R. Lakowicz), p. 1. Plenum Press, NY.

42. Longworth, J. W. (1971). In *Excited states of proteins and nucleic acids* (ed. R. F. Steiner and I. Weinryb), p. 319. Plenum Press, NY.

43. Churchich, J. E. (1966). *Biochim. Biophys. Acta*, **120**, 406.

44. Strambini, G. B. and Gonnelli, M. (1990). *Biochemistry*, **29**, 196.

45. Saviotti, M. L. and Galley, W. C. (1974). *Proc. Natl. Acad. Sci. USA*, **71**, 4154.

46. Englander, S. W., Calhoun, D. B. and Englander, J. J. (1987). *Anal. Biochem.*, **161**, 300.

47. Strambini, G. B. (1987). *Biophys. J.*, **52**, 23.

48. Lakowicz, J. R. and Weber, G. (1973). *Biochemistry*, **12**, 4161.

49. Guzeman, O. L., Kaufman, J. F. and Porter, G. (1973). *J. Chem. Soc., Faraday Trans. II*, **69**, 708.

50. Gregory, R. B. and Lumry, R. (1985). *Biopolymers*, **24**, 301.

51. Bone, R., Spinger, J. P. and Atack, J. R. (1992). *Proc. Natl. Acad. Sci. USA*, **89**, 10031.

52. Gore, M. G., Greasley, P., McAllister, G. and Ragan, C. I. (1993). *Biochem. J.*, **296**, 811.

53. Evans, C. H. (1990). *Biochemistry of the lanthanides*. Plenum Press, NY.

54. Moreno, F., Corrales, S., Garcia Blanco, F., Gore, M. G., Rees-Milton, K. and Churchich, J. E. (1996). *Eur. J. Biochem.*, **240**, 435.

55. Vanderkooi, J. M. (1992). In *Topics in fluorescence spectroscopy, vol. 3: biochemical applications* (ed. J. R. Lakowicz), p. 113. Plenum Press, NY.

56. Schauerte, J. A., Steel, D. G. and Gafni, A. (1997). In *Methods in enzymology* (ed. L. Brand and M. L. Johnson), Vol. 278, p. 49. Academic Press, NY.
57. Brochon, J.-C., Wahl, P. and Auchet, J.-C. (1972). *Eur. J. Biochem.*, **25**, 20.
58. Cherry, R. J. and Schneider, G. (1976). *Biochemistry*, **15**, 3657.
59. Matayoshi, E. D. and Jovin, T. M. (1991). *Biochemistry*, **30**, 3527.

Chapter 4

Introduction to circular dichroism

Alison Rodger and Matthew A. Ismail

Department of Chemistry, University of Warwick, Coventry CV4 7AL

1 Introduction

1.1 Circular dichroism

Circular dichroism (CD) is the ideal technique for studying chiral molecules in solution. It is uniquely sensitive to the asymmetry of the system. These features make it particularly attractive for biological systems. CD is by definition the difference in absorption, A, of left and right *circularly* polarized light (CPL):

$$CD = A_\ell - A_r \qquad\qquad 1$$

CPL has the electric field vector of the electromagnetic radiation retaining constant magnitude in time but tracing out a helix about the propagation direction (Figure 1). Following the optics convention we take the tip of the electric field vector of right CPL to trace out a right-handed helix in space at any instant of time (Figure 2) (1, 2).

CD spectra can in principle be measured with any frequency of electromagnetic radiation. In practice, most CD spectroscopy involves the ultraviolet–visible (UV–visible) regions of the spectrum and electronic transitions, though increasing progress is being made with measuring the CD spectra of vibrational transitions using infrared radiation. We shall limit our consideration to electronic CD spectroscopy since the practical considerations for vibrational CD differ from those for electronic CD.

For randomly oriented samples, such as solutions, a net CD signal will only

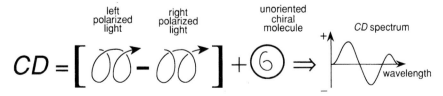

Figure 1 Schematic illustration of CD. (Source: *Circular dichroism and linear dichroism*, A. Rodger and B. Nordén, 1997, by permission of Oxford University Press.)

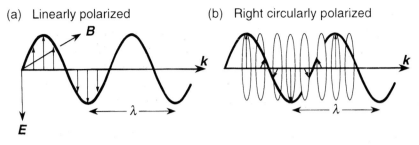

(a) Linearly polarized   (b) Right circularly polarized

Figure 2 (a) Linearly and (b) circularly polarized electromagnetic radiation. The arrows denote direction of the electric field **E**. At each point in space or time, the magnetic field, **B**, is perpendicular to the electric field such that **k** (the propagation direction), **E** and **B** form a right-handed axis system. (Source *Circular dichroism and linear dichroism*, A. Rodger and B. Nordén, 1997, by permission of Oxford University Press.)

be observed for chiral molecules (ones that cannot be superposed on their mirror images (3)). Oriented samples of achiral molecules, such as crystals, will also give a CD spectrum unless the optical axis of the sample aligns with the propagation direction of the radiation. However, such spectra are seldom useful.

CD is now a routine tool in many laboratories. The most common applications include proving that a chiral molecule has indeed been synthesized or resolved into pure enantiomers and probing the structure of biological macromolecules, in particular determining the α-helical content of proteins. *Figure 3* gives an example of a CD spectrum. The key points to remember are that a CD signal is observed only at wavelengths where the sample absorbs radiation, i.e. under absorption bands, and the signal may be positive or negative depending on the handedness of the molecules in the sample and the transition being studied. In this chapter we shall limit our consideration to wavelength scans. Kinetics will be covered in Chapter 10.

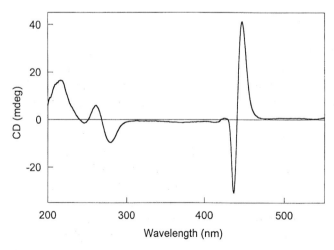

Figure 3 CD spectrum of *trans*-bis-(4-*N*-methylpyridiniumyl)diphenylporphyrin (8 μM) and poly[d(A-T)]₂ (40 μM) in phosphate buffer (1 mM; pH 7) and NaCl (20 mM).

1.2 Optical rotatary dispersion

Solutions of chiral molecules, in addition to absorbing left and right CPL differently, rotate the plane of the polarization of an incident linearly polarized light beam (1, 2). The amount of rotation varies with wavelength, resulting in an optical rotatory dispersion (ORD) spectrum. As Fresnel realized, the phenomenon of ORD is understood when one realizes that a linearly polarized light beam may be expressed as the sum of two CPL beams. The left and right CPL beams travel through a chiral solution at different speeds due to their different refractive indices. Thus when the beams recombine upon emerging from the sample, their phase difference has changed and the resulting linearly polarized beam has a different orientation from the incident beam. Due to the relationship between refraction and absorption (2) a complete CD spectrum may be converted to an ORD spectrum by using the Kramers–Kronig relations. ORD is now little used except as a means of quantifying the chirality of a solution using a single wavelength of light, usually the sodium D line at 589 nm. As discussed below it also has an effect on the units used in CD spectroscopy.

1.3 Chapter outline

CD spectroscopy has become relatively straightforward with the modern instrumentation available. In the next section, how to measure a CD spectrum is thus preceded by a description of the instrumentation. A full understanding of CD requires some knowledge of CD theory; this is the subject matter of Section 3, Section 4 summarizes the units used for CD. The chapter concludes with a discussion of the CD of biomolecules, with particular reference to spectra/structure correlations. More detail on most aspects may be found in the alphabetical *General references* list.

2 Measuring a CD spectrum

2.1 The instrumentation

Measuring a CD spectrum is a routine procedure assuming one has access to a CD instrument. CD instruments are usually referred to as circular dichrometers or spectropolarimeters; however, dichrograph and spectropolarograph better describe the output of modern instruments. If your sample gives a good UV–visible absorbance spectrum then it is highly likely that (if it is chiral) you will get a good CD spectrum. The essential features of a CD spectrometer are a source of (more-or-less) monochromatic left and right circularly polarized light and a means of detecting the difference in absorbance of the two polarizations of light. As the absorbance signal of a typical CD spectrum is a fraction of a per cent of that observed in the normal absorbance spectrum, the normal method of achieving these requirements is to implement a polarization phase-modulation technique.

In most commercial CD spectrometers, a photoelastic modulator (in older instruments a Pockels cell) produces alternatively right and left circularly polarized light with a switching frequency of typically 50 kHz. The light intensity is constant, but upon passage through a sample exhibiting CD an intensity fluctuation (corresponding to the different absorptions of left and right circularly polarized light) that is in phase with the modulator frequency appears. The unabsorbed photons hit a PMT which produces a current whose magnitude depends on the number of incident photons. This current is detected by a lock-in amplifier. Thus, the DC (direct current) component of the PMT current depends on the total absorption of light by the sample (and on lamp intensity and monochromator characteristics), and the AC (alternating current) component relates to the CD (a lock-in amplifier is needed to determine the phase of the AC component which contains information about the sign of the CD). A larger voltage is applied to the PMT when the number of incident (i.e. non-absorbed) photons is smaller, thus *the high tension (HT) voltage on a CD machine is a rough guide to the absorbance of a sample*. The HT voltage can usually be monitored if one wishes to do so. It is advisable to take this option even with routine samples, as it is an easy way to detect sample abnormalities.

The CD spectrometers described above are single beam instruments. A new double beam instrument from On-Line Instrument Systems (OLIS) has recently come on the market, which uses sample and reference beams to enable direct calculation of the CD signal. The instrument utilizes two modulated CPL beams out of phase by 180° which are passed through the sample and detected simultaneously. The absorption of left and right CPL is therefore measured directly. The technique has the advantage that factors such as lamp fluctuations are negated since they affect both beams equally, whereas the CD behaviour is characteristic to each beam. The use of a sample/reference system results in zero drift over time and nearly flat baselines. However, the two beams pass through different parts of the sample and cuvette. This may or may not be important.

The radiation source in UV–visible CD spectrometers is a high energy (150–450 W) xenon arc lamp. In some instruments the lamp is water cooled. This can prove the most problematic part of the whole instrument since if the local water supply is used for a flow through system one is at the mercy of its purity, hardness etc. Alternatively, if a self-contained system is used, then the fairly narrow bore of the tubing that the water flows through requires a reasonably efficient pump, and the water must be kept free from algae etc. and must be cooled. A narrow bore central heating system pump may be the cheapest solution to the pumping problem and using car radiator cooling fluid inhibits the growth of most things if regularly changed. Avoid using clear tubing if possible. Despite the water consumption required, a water flow heat exchanger may be the only cooling option that is sufficiently efficient to cool the water (even for a self contained system) for a 450 W lamp. If at all possible, an instrument with an air cooled lamp is to be preferred. This no longer seems to involve accepting a loss in sensitivity.

Another requirement of the high energy lamps used in CD machines is that the optics of the instrument need to be purged with nitrogen gas to avoid ozone being created and reacting with surfaces of mirrors etc. The nitrogen purging is also required for the sample compartment when running below 200 nm to avoid having O_2 in the sample compartment absorbing the incident radiation. In practice this means a moderate nitrogen flow rate (3–5 cm³/min) at all times, with an increase to 20 or more cm³/min when collecting data below 200 nm.

2.2 The sample

Most CD spectrometers are single beamed instruments and can tolerate a maximum absorbance of about 2 absorbance units before the number of photons passing through the sample become too small to be measured accurately. It is preferable to have the maximum absorbance less than 1.5 (the optimum value is usually approximately 1 unless the dissymmetry factor (see below) is large). It is good practice to check that the CD signal is proportional to the sample concentration by running a spectrum of a diluted sample and checking that the signal scales with concentration. If it does not, then you have evidence of solute–solute interactions in your sample.

As the wavelength and absorbance ranges are the same for both normal absorption and CD it is advisable to run a normal absorption spectrum of the sample first (see *Protocol 1*). *Always* leave the reference beam of the absorbance spectrometer empty for this experiment. In most cases the normal absorption spectrum will indicate absorbance maxima. If these do not align (assuming the instruments are correctly calibrated) with the CD maxima this provides valuable information when it comes to interpreting the spectrum (see below).

2.3 The cuvette

2.3.1 Cuvette width and sample volume

All of the light beam incident upon the cell must pass through the sample and not be clipped or reflected by the walls or base of the cell or the meniscus of the solution, otherwise the measured spectrum is affected by scattered light. Thus the narrow cells often used to minimize sample volume in a normal absorption spectrophotometer *cannot* be used for CD unless the light beam is chopped or focused prior to hitting the cell. While focusing of the light beam is possible, one must ensure that the lenses used for the focusing are not themselves significantly birefringent (CD active) and also that the whole light beam incident on the sample is collected by the PMT. The light beam must not be focused too tightly on the PMT itself otherwise the PMT may be damaged. Business cards are ideal for inserting in the light beam at ~550 nm (the green light most easily detected by our eyes) to see the beam width and height, though note that the beam width is dependent on the instrument slit width, which in turn depends on the lamp energy so may be larger in the UV region than at 550 nm. A typical unmasked, unfocused beam size is about 8 mm by 10 mm.

2.3.2 Cuvette pathlength and shape

For UV–visible CD, high quality quartz cuvettes are required that transmit the full wavelength range of UV–visible light. In the visible region, glass may be used but it is generally advisable to use quartz even here. Plastic cuvettes typically have high intrinsic birefringence so should be avoided. In any case, the need to run a baseline of each cuvette used (see below) removes the usual attraction of disposable plastic cuvettes.

As with normal absorbance spectroscopy, the default cuvette pathlength is usually 1 cm. Most CD spectrometers have cuvette holders of better design than absorption spectrometers and will properly hold a range of cuvette sizes (and shapes). So, if the sample has a large absorbance signal, say greater than 2 absorbance units, try using a shorter pathlength cell. If much of the absorbance is due to solvent or buffer, this is a particularly attractive option if the analyte concentration can be increased.

Either cylindrical or rectangular cuvettes may be used for CD. Cylindrical cells are usually deemed to have lower birefringence (baseline CD) than rectangular cuvettes; however, if UV and CD 'matching' is requested when the cuvettes are purchased, rectangular cuvettes seem to be equally good. Water jacketed cylindrical cells enable the sample to be thermostatted most simply and also take the least sample volume for a given pathlength. With these cuvettes, you must check that the configuration of your light beam and cuvette holder ensure that the light beam passes through the sample and not the quartz walls and cooling water parts of the cuvette. The light beam will almost certainly need masking for this to be achieved.

Rectangular cells have a number of advantages over cylindrical cuvettes for the 1 mm and longer pathlength experiments. They are cheaper, may be used in standard absorption spectrophotometers (so CD and normal absorption data may be collected on exactly the same sample), and may be used for serial titration experiments as ~60% of a rectangular cell can be empty for the first spectrum and then gradually filled. Titration experiments where spectra are collected as a function of concentration, ionic strength, pH etc. often involve adding solution to the cuvette. A simple way to avoid dilution effects is as follows. Consider a starting sample that has concentration x M of species X. Each time y μl of Y is added, also add y μl of a $2x$ M solution of X. The concentration of X remains constant at x M. An infinite number of variations on this theme are possible.

If pathlengths of 0.1 mm or less are required, it is probably best to use demountable cuvettes where the sample is dropped onto a quartz disc that is etched to a predefined depth and then another quartz disk is carefully placed on top. In this case sample recovery is very difficult, so the smaller volume of cylindrical cells makes them more attractive than rectangular ones. A simple cuvette holder for demountable cylindrical UV cells may be created from an infrared cell holder and pieces of rubber (a mouse mat proves ideal) with holes drilled at the appropriate places to allow the holder assembly screws to pass through (*Figure 4*). Any such cell holder must be located perpendicular to the light beam

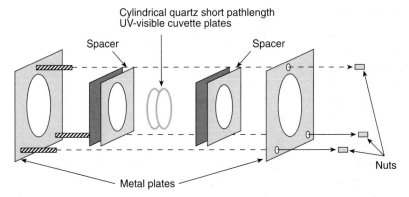

Figure 4 Cell holder for cylindrical quartz UV–visible short pathlength (less than 0.1 mm) cuvettes.

with the whole incident light beam passing through the cell windows when data is being collected.

2.4 The baseline and zeroing

A machine baseline (i.e. CD spectrum of air) measured on a standard CD spectrometer will not be flat. However, instruments seldom have a facility for inputting a baseline for a series of experiments. This is in part because the cuvette used for an experiment will also have its own CD spectrum; even matched cuvettes have slightly different intrinsic spectra. So always collect a baseline spectrum of the solvent/buffer under the same conditions as the sample spectrum using the same cuvette in the same orientation with respect to the light beam. Subtract the baseline spectrum from the sample spectrum to produce the final CD plot. It is usually the case that with small CD signals, even when the baseline is subtracted, the CD is not exactly zero outside the absorption envelope. However, if it is flat outside the absorption envelope, the spectrum may be zeroed by adding or subtracting a constant (either within the spectrometer software or using your chosen data-plotting software).

2.5 The parameters

2.5.1 Wavelength range

CD spectrometers usually scan from longer wavelengths to shorter ones. Select a starting wavelength so that there is at least 20 nm of zero absorbance beyond the normal absorption envelope(s) of interest. When the baseline spectrum is subtracted from the sample spectrum (see Section 2.4), the region outside the absorption envelope should be flat. If it is not, then this probably means either there is a very weak absorbance band that has a large dissymmetry factor (see Section 3.6) and hence a large CD signal compared with its normal absorbance intensity, or, more probably, there is some light scattering of the sample. Sources of light scattering include slightly dirty cuvettes (inside or outside),

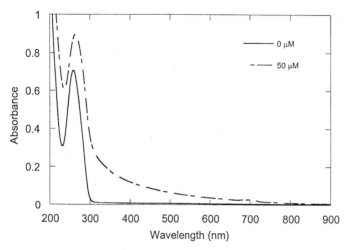

Figure 5 Normal absorption spectra of 100 μM calf thymus DNA in (i) 10 mM NaCl and 1 mM sodium cacodylate buffer (solid line) and (ii) 10 mM NaCl, 1 mM sodium cacodylate and 50 μM spermine (broken line), which is sufficient to condense the DNA and provide light scattering that may appear as an absorbance signal outside the absorption envelope.

undissolved sample, condensation of samples (a particular problem with DNA when highly charged cations are being added; see Sections 5.10 and 5.11), and particulate samples (of particular relevance for membrane proteins; see Section 5.7). Given that the wavelengths of light being scattered are less than 1 μm, one does not necessarily expect to be able to see by eye the presence of such particles. Scattering effects can usually be distinguished from absorption effects, as the scattering more-or-less linearly increases with decreasing wavelength, whereas an absorbance band should increase and then decrease. In addition, if light scattering is the source of a sloping baseline, the normal absorption will show a similar effect, as illustrated in *Figure 5*.

2.5.2 Signal to noise ratio

The signal to noise ratio in a CD spectrum increases with: $\sqrt{n}$, where n is the number of times the spectrum is accumulated (and the data averaged); $\sqrt{\tau}$, where τ is the time over which the machine averages each data point; and $\sqrt{I}$, where I is the intensity of the light beam (which is in turn influenced by the bandwidth, see below). Most CD spectrometers have both short timescale (millisecond to minutes) and long timescale (minutes to hours) baseline variations. If the CD signal is large, both can be ignored. To avoid any problem from short timescale variations, collect data averaged over a number of faster scans rather than one slow one. Longer timescale fluctuations can be more problematic and are usually dealt with by alternating collection of sample and baseline spectra or by assuming it is simply a linear drift that can be removed by subtracting or adding a constant value to each spectrum (Section 2.4). While perhaps appearing to be wishful thinking, the assumption of only linear baseline drift seems to be valid especially with more recent instrumentation.

2.5.3 Parameter sets

CD spectrometers give the operator considerable control over response time (τ), scan speed (s), bandwidth (b) (the wavelength range (error) of the incident light), and data interval (d). To optimize signal to noise effects τ should be selected to be as large as possible subject to $\tau \times s \leq b/2$. If τ is too long for the chosen s and b, then maxima of peaks (both positive and negative) will be cut off and their wavelengths shifted. A control scan using $\tau' = \tau/2$ (or $s' = 2s$) should be used to check that spectra are not being distorted by the chosen parameters. The data interval determines how often a data point is collected. There is no point in having, for example, $b = 5$ nm and $d = 0.1$ nm as all that will be achieved is very long run times and many measurements being performed where essentially the same light intensity and wavelength distribution pass through the sample. It would be much more sensible to measure far fewer points more accurately.

Scan speeds of 100 nm/min, $\tau = 1$ s, $b = 1$ nm and a data step of 0.5 nm (though see Section 5.2 for protein structure determination applications) seem to be a good starting point as a parameter set for most experiments where the samples have the broad band shapes usually found for solution samples. If the bandwidth is fixed, the instrument will be programmed to control the slit width directly. There are occasions where one may wish to adjust the slit width manually. However, such applications are specialized and require care to avoid damaging the instrument.

In general, it is usually advisable to perform a fast preliminary scan to determine whether there is any point in collecting an accurate spectrum and whether the sensitivity scale has been chosen correctly to appropriately display the CD at all wavelengths of the spectrum. Having the display scale exactly right is not important on modern instruments as data can be re-scaled to a great extent (though not usually from the highest to the lowest ranges); however, it is nice to see the spectrum as it is being collected.

2.6 Noise reduction

If the parameters for a run have been chosen correctly then there is no point in smoothing data sets by averaging the signal from neighbouring points. All that is achieved, with an intensity versus wavelength scan, is what would have been more efficiently achieved by having a different parameter set. When a noisy CD spectrum is plotted, however, it is often possible to sketch the spectrum that one 'knows' is really there as illustrated in *Figure 6*. Sometimes what one should do is repeat the experiment by averaging over more scans or increasing τ and reducing the scan speed. However, this is not always practicable if the runs are already long and/or a series of spectra needs to be collected within the lifetime of a sample. In this case, one wishes to have a computational option that will do numerically what your eye and pencil can do. The 'Noise Reduction' option on the Jasco software, for example, is designed for this. It is a Fourier transform smoothing routine that proceeds by transforming the spectrum from the wavelength domain into a frequency domain where the high and low frequency

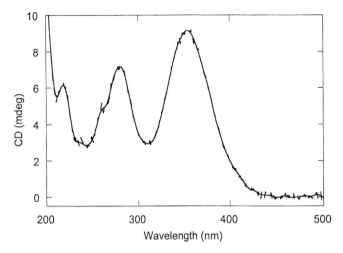

Figure 6 CD spectrum of Hoechst 33258 (20 mM), calf thymus DNA (40 mM), NaCl (20 mM) and phosphate buffer (1 mM; pH 7) (broken line) where it is clear what the noise reduced spectrum (solid line) should look like.

signals (hopefully corresponding to the noise) can be cut out. Transforming back to the wavelength domain ideally gives you the spectrum you 'knew' was hiding under the noise envelope. To ensure this is indeed the case, always over-lay the noise-reduced result on the original data set and use your eye to decide the validity of the transformed spectrum. If it does not 'look right', reject the result and alter the cut-off parameters. The dangers of this type of noise reduction are either the introduction of 'vibronic' structure or the truncation of peaks.

Protocol 1

Measuring a routine CD spectrum

Equipment

- UV–visible absorbance spectrophotometer
- Computer-controlled CD spectrometer
- CD spectrum measurement and data analysis software
- Quartz cuvette of an appropriate type (see Section 2.3)
- cm³ volumes of sample and reference solutions of μM concentrations

A. Sample preparation

1   Determine an appropriate cuvette type for the sample. Use a 1-mm pathlength quartz cuvette for strongly absorbing samples or when only a small volume of the sample is available. Otherwise, use a 10-mm pathlength rectangular quartz cuvette.[a]

2   Run a UV–visible spectrum of the buffer/solvent over the wavelength range of interest. If using a double beam UV–visible spectrophotometer, ensure that the spectrum is collected with the reference beam empty.[b]

3   Prepare a solution of the sample such that the sample absorption at the wavelength of interest is approximately 1 but not more than 1.5 absorbance units. Run a UV/visible absorbance spectrum, again with an empty reference beam.

4   Compare the sample and buffer solvent UV–visible spectra and check that the wavelength maximum of interest is due to sample absorption and not some other component of the system.

B. CD parameter selection

1   Set the wavelength range for the CD spectrum to begin at least 20 nm beyond the absorption envelope of interest and ensure that the scan begins in a region of the spectrum where there is no absorption.

2   Place the quartz cuvette containing the sample solution into the sample compartment and ensure that the light beam is not clipping the meniscus of the solution (increase the volume of liquid in the cuvette if necessary or perhaps raise the height of the cuvette) or the walls or base of the cuvette (in which case lower the cuvette or use a different cuvette type). You may be able to remove the cuvette holder assembly to view down the light beam path. Alternatively use a mirror.

3   Run a fast, approximate scan of the sample to determine an appropriate CD sensitivity level. Use $\tau = 0.25$ s, $b = 0.5$ nm and $s = 500$ nm min^{-1} accumulated over one scan only.

4   For the accurate spectrum, set τ, s and b such that τ is as large as possible subject to $\tau \times s \leq b/2$. Use $s = 100$ nm min^{-1}, $\tau = 1$ s, $b = 1$ nm and $d = 0.5$ nm as a starting point for routine work. Specify up to eight accumulated scans depending on the quality of the spectrum required.[c] In some instances more scans may be required. If you are uncertain of what to choose, select a large number and stop the run when the spectrum 'looks OK'. To ensure proper averaging over the number of scans, it is advisable to stop the instrument while it is rewinding between scans.

5   If collecting CD data for input into the structure fitting program CDsstr (see Protocol 2), set $d = 1$ nm.

C. Measuring the CD spectrum

1   Collect a sample CD spectrum using the above parameters.

2   Run the reference spectrum in the same cuvette as was used for the sample spectrum measurement with the same parameters.[d] As with the sample spectrum, ensure that no clipping of the light beam is occurring.

D. Processing the CD data

1   Obtain the baseline corrected CD spectrum by using your CD data analysis software to subtract the reference spectrum from the sample spectrum.

2   If the spectrum is excessively noisy and adjusting the instrumental parameters to acquire a better spectrum is not practical, apply a noise reduction algorithm (a Fourier transform routine, for example) to improve the spectrum.

3 Overlay the final spectrum on the original spectrum to ensure no peak shape distortion or data mishandling has occurred.

[a] Other cuvette types may be appropriate in certain circumstances. See main text for details.

[b] If you put the solvent or buffer in the reference beam of a double beam normal absorption spectrometer you may get a good spectrum even if the solvent/buffer has a significant absorbance as the instrument will subtract the absorbance of the reference cuvette; however, there is no such compensation mechanisms in a single beam CD spectrometer.

[c] The signal to noise ratio is dependent on a number of factors as detailed in the text. However, increasing the number of accumulated scans is a convenient practical method of improving the quality of the CD spectrum.

[d] By using the same cuvette for both sample and reference measurements, absorption of CPL by the cuvette is eliminated from the final spectrum.

3 Equations of CD spectroscopy

An electronic transition occurs because either the electric field or the magnetic field (or both) of the radiation 'pushes' the electrons to a new stationary state. The effect of the electric field is to cause a linear rearrangement of the electrons; the net linear displacement of charge during any transition is therefore called the *electric dipole transition moment* (edtm) of the transition and is denoted by the vector μ. The magnetic field induces a circular rearrangement of electron density. The net circulation of charge is the *magnetic dipole transition moment* (mdtm), m. In an achiral molecule the net electron redistribution is always planar. It is usually linear (having $\mu \neq 0$, $m = 0$ so being *electric dipole allowed* (eda) and *magnetic dipole forbidden* (mdf)) or circular (having $\mu = 0$, $m \neq 0$ so being *electric dipole forbidden* (edf) and *magnetic dipole allowed* (mda)), but it may also be a planar spiral in molecules of low symmetry.

In a chiral molecule the electron rearrangement during a transition is always helical which requires that μ and m have a parallel component. The Rosenfeld equation for the CD intensity (or CD strength or rotational strength, R) of a transition in a collection of randomly oriented chiral molecules is (4):

$$R = \text{Im}\{\mu \cdot m\} \qquad\qquad 2$$

where 'Im' denotes 'imaginary part of' (the magnetic dipole moment operator contains a factor of $\sqrt{-1}$), μ is the edtm for the transition from the final to the initial state, and m is the mdtm for the reverse transition.

CD may sometimes be used to deduce the structure of a system. This is only really viable when the system can be considered as a collection of chromophores (spectroscopically well-defined subunits of a molecule) each of which is only slightly perturbed by the rest of the system. In the rest of this section we shall consider the coupling of two intrinsically achiral chromophores that are chirally oriented with respect to one another. A range of applications of this theory is given in (4). We have to treat eda/mdf transitions separately from mda/edf transitions since eda transitions require the induction of a magnetic component, whereas mda transitions require an induced electric component. The dependence of the different kinds of induced moments on the geometry of the system is very different, thus when we wish to extract geometric information from CD we must be aware of which situation we have.

The CD spectrum induced into eda transitions is also different if it arises from the coupling of identical chromophores (Section 3.1) rather than from the coupling of non-identical chromophores (Section 3.2). We shall quote the results for each case with little justification. Full derivations using degenerate and non-degenerate perturbation theory may be found in (4). The two cases could be treated as one using a full secular-determinant approach. This would have the advantage of naturally including the near-degenerate case, but has the disadvantage of not leading to transparent geometry/CD correlations.

3.1 Degenerate coupled-oscillator CD

Consider a molecule composed of two identical chromophores A and C. The largest coupling takes place between eda transitions in the two chromophores occurring at the same energy (ε) so they are degenerate (5). The transition moments of two degenerate transitions have the same length but different origins and orientations. The two degenerate edtms couple together to make two equal magnitude opposite handedness transition helices, one from in phase and one from out of phase coupling. Thus we expect to see a CD spectrum with two equal magnitude opposite-signed peaks at energies very close to ε (*Figure 7*).

We choose the axis system and vectors illustrated in *Figure 8* with A in front and C behind. The x-axis lies along $\mathbf{R}_{AC}$, the vector from the A origin to that of C, and z lies along the y–z projection of $\boldsymbol{\mu}^a$, the A chromophore edtm. We also define three angles: α ($0 \leq \alpha < 180°$) the angle between $\mathbf{R}_{AC}$ and $\boldsymbol{\mu}^a$, γ ($0 \leq \gamma <$

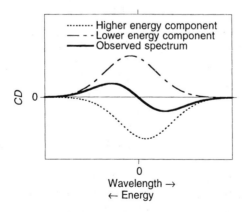

Figure 7 CD resulting from the coupling of one transition in each of two chromophores where both transitions are eda and degenerate (occurring at energy ε). The characteristic form of an excitonic spectrum results from cancellation of overlapping positive and negative bands. (Source: *Circular dichroism and linear dichroism*, A. Rodger and B. Nordén, 1997, by permission of Oxford University Press.)

180°) the angle between $-\boldsymbol{R}_{AC}$ and $\boldsymbol{\mu}^c$, and τ the angle passed through in going from the *y-z projection* of $\boldsymbol{\mu}^a$ to that of $\boldsymbol{\mu}^c$ in an *anticlockwise* direction as illustrated in *Figure 8*. Thus

$$\boldsymbol{R}_{AC} = R_{AC} (1,0,0) \tag{3}$$

$$\boldsymbol{\mu}^a = \mu^a (\cos \alpha, 0, \sin \alpha) \tag{4}$$

$$\boldsymbol{\mu}^c = \mu^c (-\cos \gamma, \sin \gamma \sin \tau, \sin \gamma \cos \tau) \tag{5}$$

We shall refer to the system as being right handed if the three vectors $\boldsymbol{\mu}^c$, $\boldsymbol{\mu}^a$ and $\boldsymbol{R}_{AC}$ (CAR) form a right-handed coordinate system. The angles relate to the vectors as follows (4):

$$\cos \tau = \frac{\hat{\boldsymbol{\mu}}^a \cdot \hat{\boldsymbol{\mu}}^c - \hat{\boldsymbol{\mu}}^a \cdot \hat{\boldsymbol{R}}_{AC} \hat{\boldsymbol{R}}_{AC} \cdot \hat{\boldsymbol{\mu}}^c}{\sqrt{[1 - (\hat{\boldsymbol{\mu}}^a \cdot \hat{\boldsymbol{R}}_{AC})^2][1 - (\hat{\boldsymbol{\mu}}^c \cdot \hat{\boldsymbol{R}}_{AC})^2]}}$$
$$= \frac{\hat{\boldsymbol{\mu}}^a \cdot \hat{\boldsymbol{\mu}}^c + \cos\alpha \cos\gamma}{\sin\alpha \sin\gamma} \tag{6}$$

where ^ denotes a unit vector.

The energies at which one finds the CD signals resulting from the coupling of two degenerate chromophores are:

$$\varepsilon^{\pm} = \varepsilon \pm V \tag{7}$$

where $\varepsilon^{\pm}$ means that ε^{+} takes the upper of the signs of any $\pm$ or $\mp$ in an equation and (4)

$$V = \frac{\hat{\boldsymbol{\mu}}^a \cdot \hat{\boldsymbol{\mu}}^c - 3 R_{AC} \cdot \hat{\boldsymbol{\mu}}^a \hat{\boldsymbol{\mu}}^c \cdot R_{AC}}{R_{AC}^3} \tag{8}$$

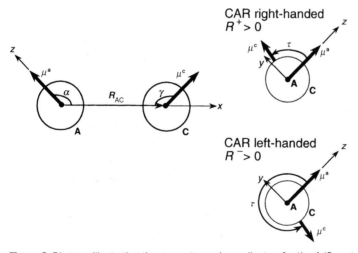

Figure 8 Diagram illustrating the geometry and coordinates for the A/C system described in the text. Note that τ is the angle taken in the anticlockwise direction between the *projections* of the edtms onto the *y–z* plane when the observer is looking down the *x*-axis. $0° \leq \tau < 180°$ if $\boldsymbol{\mu}^c \times \boldsymbol{\mu}^a \cdot \boldsymbol{R}_{AC} > 0$ (i.e. if the three vectors form a right-handed parallelepiped) and $180° \leq \tau < 360°$ if $\boldsymbol{\mu}^c \times \boldsymbol{\mu}^a \cdot \boldsymbol{R}_{AC} < 0$. (Source: *Circular dichroism and linear dichroism*, A. Rodger and B. Nordén, 1997, by permission of Oxford University Press.)

so

$$\varepsilon^{\pm} = \varepsilon \pm \frac{\mu^2 (\sin \alpha \sin \gamma \cos \gamma + 2 \cos \alpha \cos \gamma)}{R_{AC}^3} \qquad 9$$

where μ is the length of $\boldsymbol{\mu}^a$ and $\boldsymbol{\mu}^c$.

The two CD bands have sign and magnitude given by the following vectorial equation (4)

$$R^{\pm} = \pm \frac{\varepsilon}{4\hbar} \{\boldsymbol{\mu}^c \times \boldsymbol{\mu}^a \cdot \mathbf{R}_{AC}\}$$

$$= \pm \frac{\varepsilon \mu^2 R_{AC}}{4\hbar} \sin \alpha \sin \gamma \sin \tau \qquad 10$$

The CD spectrum we expect to see for the coupling of two degenerate transitions on two chromophores is as follows.

1. There are two bands of equal magnitude and opposite sign, R^+ centred at ε^+ and R^- centred at ε^-.

2. The CD strength R depends only on the strength of the transitions in the isolated chromophores and on their orientations relative to one another and to the vector connecting the two chromophores.

3. The sign of the CD of the '+' state is positive if $\boldsymbol{\mu}^c$, $\boldsymbol{\mu}^a$ and $\mathbf{R}_{AC}$ form a right-handed axis system.

4. V is small, so the two bands are centred close to one another and a significant part of the total intensity of each band is cancelled. Hence the spectral form illustrated in *Figure 7*.

5. Although the CD strength is a maximum when τ, α, and γ are all 90° (and the system is achiral), under these circumstances $V = 0$, so the positive and negative CD signals are centred at ε and exactly cancel, resulting in no CD signal as expected for an achiral molecule.

6. According to *equation 10* the CD strength should get larger as the distance between the chromophores increases. However, V concomitantly decreases so the two component bands come closer together and more and more cancellation occurs, thus avoiding the nonsense situation of two chromophores too far apart to interact having an infinitely large CD signal. The extent of cancellation depends upon the shapes of the bands but goes approximately linearly in V thus the CD signal effectively decreases as R_{AC}^{-2}.

7. The CD of two enantiomers may be seen to be equal and opposite by reflecting the system about the plane containing $\boldsymbol{\mu}^a$ and $\mathbf{R}_{AC}$.

Equation 10 describes the magnitude of the area under each CD band resulting from the coupling of $\boldsymbol{\mu}^a$ and $\boldsymbol{\mu}^c$; the actual appearance of the spectrum is also dependent on the band shape, which is usually Gaussian or Lorentzian, so we get spectra as illustrated throughout this chapter rather than single sharp lines at a precise energies.

3.2 Non-degenerate coupled-oscillator CD

When eda transitions occur in two non-identical achiral chromophores, as with the degenerate case, a linear motion of charge at a distance (in another chromophore) has a circular (magnetic) component locally. Thus if the two transitions couple, an induced CD signal (4) is found. The main differences between the non-degenerate and degenerate cases are that for the non-degenerate case:

(1) the two CD bands resulting from the coupling occur at (strictly very close to) the energies of the two non-degenerate transitions, rather than as an exciton couplet where most of the CD intensity is cancelled, and

(2) the coupling between non-degenerate transitions is not as strong as that between degenerate transitions.

The CD induced at ε_a into the transition on A by its coupling with an edtm of different energy in C is (4)

$$
\begin{aligned}
R(\varepsilon_a) &= \frac{-\varepsilon_a \varepsilon_c V}{\hbar(\varepsilon_c^2 - \varepsilon_a^2)} \{\boldsymbol{\mu}^c \times \boldsymbol{\mu}^a \cdot \mathbf{R}_{AC}\} \\
&= \frac{-\varepsilon_a \varepsilon_c}{\hbar(\varepsilon_c^2 - \varepsilon_a^2)} \left\{ \frac{\boldsymbol{\mu}^c \cdot \boldsymbol{\mu}^a - 3\hat{\mathbf{R}}_{AC} \cdot \boldsymbol{\mu}^a \boldsymbol{\mu}^c \cdot \hat{\mathbf{R}}_{AC}}{R_{AC}^2} \right\} \{\boldsymbol{\mu}^c \times \boldsymbol{\mu}^a \cdot \mathbf{R}_{AC}\} \qquad 11 \\
&= \frac{-\varepsilon_a \varepsilon_c (\mu^a \mu^c)^2}{\hbar(\varepsilon_c^2 - \varepsilon_a^2) R_{AC}^2} \{\sin \alpha \sin \gamma \cos \tau + 2\cos \alpha \cos \gamma\} \sin \alpha \sin \gamma \sin \tau
\end{aligned}
$$

A number of general points about non-degenerate coupled-oscillator CD spectra (*Figure 9*) follow from *equations 10* and *11*.

1. The CD strength *decreases* with *increasing* A to C distance according to R_{AC}^{-2}. This is the smallest distance-dependence for any CD mechanism; coupled-oscillator CDs are therefore generally larger than ones arising from other chromophore coupling mechanisms (4). Hence if a transition is both eda and mda, then the coupled-oscillator mechanism generally dominates the observed CD.

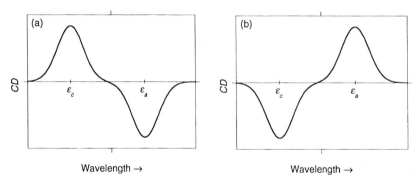

Figure 9 Schematic illustration of non-degenerate coupled-oscillator CD spectra for $\alpha = \gamma = 90°$: (a) $0 \le 2\tau < 180°$ and (b) $180° \le 2\tau < 360°$ (see *Figure 8* for definition of τ). (Source: *Circular dichroism and linear dichroism*, A. Rodger and B. Nordén, 1997, by permission of Oxford University Press.)

2. It is often the case that $\alpha = \gamma = 90°$. When this situation arises, the geometry factor in *equation 11* is $\cos\tau\sin\tau = \frac{1}{2}\sin(2\tau)$ so the CD is a maximum when $\boldsymbol{\mu}^a$ and $\boldsymbol{\mu}^c$ are oriented at $45°$ to one another.

3. If we wish to know the combined effect of many transitions in C on one transition in A, we simply introduce a sum over c into *equation 11*.

4. The CD induced at ε_c is determined by simply permuting all the 'a' and 'c' and 'A' and 'C' labels in *equation 11*, and we see that $R(\varepsilon_c) = -R(\varepsilon_a)$.

3.3 Carbonyl $n{\rightarrow}\pi^*$ CD

Long before any theoretical understanding of the carbonyl $n{\rightarrow}\pi^*$ transition CD had been achieved, it was found empirically that each atom (or group of atoms) of the molecule induces a contribution to the observed CD whose sign is determined by the octant in which the atom lies (6). Any atom that has an equivalent 'partner' in a neighbouring octant has its contribution to the CD cancelled so it is not a net perturber and may be ignored. Thus the CD of an $n{\rightarrow}\pi^*$ transition has its sign determined by the factor (*Figure 10*)

$$-xyz \qquad\qquad 12$$

where (x,y,z) are the coordinates of the unit vector along the line from the carbonyl to the *net perturber* in the coordinate system of *Figure 10*. It was later realized (7) that the relative contributions of two or more net perturbers scale with their polarizability and their distance from the carbonyl group as indicated below.

The equation describing the octant rule dependence of the *dynamic coupling* induced CD of an $n{\rightarrow}\pi^*$ transition is derived as follows (4, 7). One first divides the molecule into an achiral chromophore A composed of the C=O bond and the two carbon atoms to which it is joined (*Figure 11*). The rest of the molecule is divided into chromophores, C_i, which are subunits of the molecule that do not exchange electrons with the rest of the molecule. For a molecule where (apart

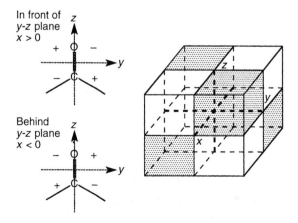

Figure 10 The octant rule. (Source: *Circular dichroism and linear dichroism*, A. Rodger and B. Nordén, 1997, by permission of Oxford University Press.)

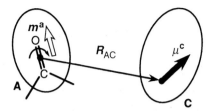

Figure 11 Geometry of a carbonyl A/C system (cf. *Figure 8*). (Source: *Circular dichroism and linear dichroism*, A. Rodger and B. Nordén, 1997, by permission of Oxford University Press.)

from the carbonyl group) all the valence electrons are in sigma bonds, taking the C_i to be bonds is a good approximation (7–10). Alternatively, the C_i may be defined to be atoms such as carbons and nitrogens. Although the bond approximation is the better one, in most cases the difference in *xyz* for the middle of a bond and one of the terminal atoms of the bond is not great. If there are π-bonds or non-bonding electrons in the molecule, then the unit where the electrons are localized is often more than a single bond, e.g. if the molecule has a $-COOH$, $-NH_2$ or $-C_6H_5$ group, these will need to be treated as units.

We may write the CD of the $n \rightarrow \pi^*$ transition in terms of an A optical factor, $f(A)$, and a C optical/geometry factor, $g(C)$, (which includes the energy terms):

$$R = f(A)g(C) \qquad 13$$

The first non-zero term of the CD strength for the carbonyl $n \rightarrow \pi^*$ transition has an A optical factor

$$f(A) = \mathrm{Im}[Q^a_{xy}m^a_z] \qquad 14$$

where m^a_z is the mdtm of the $n \rightarrow \pi^*$ transition, Q^a_{xy} is the x-y component of quadrupole transition moments of other transitions on A and a sum over them is implied. The optical/geometry factor is (*Figure 11*) (4, 8):

$$g(C) = \frac{-6\varepsilon_c}{(\varepsilon_c^2 - \varepsilon_a^2)R^4_{AC}} \mu^c_z \left[\mu^c_x \hat{R}_{ACy} + \mu^c_y \hat{R}_{ACx} - \sum_k^{x,y,z} 5\mu^c_k \hat{R}_{ACx} \hat{R}_{ACy} \hat{R}_{ACk} \right] \qquad 15$$

so

$$R \approx \frac{30\varepsilon_c}{(\varepsilon_c^2 - \varepsilon_a^2)R^4_{AC}} f(A)\mu^c_z\mu^c_z[\hat{R}_{ACx}\hat{R}_{ACy}\hat{R}_{ACk}]$$

$$\approx \frac{30\varepsilon_c}{(\varepsilon_c^2 - \varepsilon_a^2)R^4_{AC}} f(A)\mu^c_z\mu^c_z[xyz] \qquad 16$$

$$\approx \frac{30\alpha(\varepsilon_a)}{R^4_{AC}} f(A)\mu^c_z\mu^c_z[xyz]$$

where $\alpha(\varepsilon_a)$ is the isotropic polarizability of A at ε_a.

Thus, to apply the octant rule in the revised form indicated by *equation 16* first identify the chromophores, C_i, into which the non-carbonyl part of the molecule is to be divided (usually sigma bonds and functional groups as discussed above). Second, determine for each C_i

- its isotropic polarizability (8)
- its distance from the carbonyl chromophore origin (which we take to be the centre of the C=O bond)
- the unit vector from the carbonyl origin to the C_i origin

Equation 16 may then be evaluated for each C_i in terms of the parameter $f(A)$ which we assume is negative and constant for all ketone carbonyls. In some instances other terms in the full expression for the CD must be included (10).

3.4 *d–d* transitions of tris-chelate transition metal complexes

Another class of mda/edf transitions that has been extensively studied are the *d–d* transitions of transition metal complexes (*Figure 12*). A particularly successful empirical sector rule for this system is as follows: the CD of the E band of tris-chelate complexes such as $[Co(propylenediamine)_3]^{3+}$ is larger than that of the A_2 band, and the sign of the E band of the $\triangle$ enantiomer is negative (*Figure 13*). As the A_2 and E bands of *d–d* transitions lie very close in energy (these $\textbf{\textit{D}}_3$

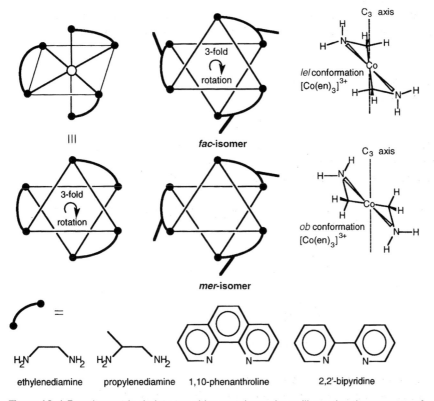

ethylenediamine propylenediamine 1,10-phenanthroline 2,2'-bipyridine

Figure 12 Δ-Enantiomer tris-chelate transition metal complexes illustrating the geometry of the *fac* and *mer* isomers adopted by $[Co(propylenediamine)_3]^{3+}$ and the *lel* and *ob* conformers of $[Co(ethylenediamine)_3]^{3+}$. (Source: *Circular dichroism and linear dichroism*, A. Rodger and B. Nordén, 1997, by permission of Oxford University Press.)

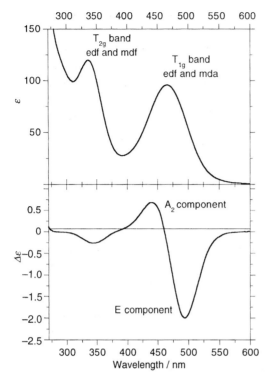

Figure 13 d–d CD spectrum of Δ-[Co((R)-propylenediamine)$_3$]$^{3+}$. (Source: *Circular dichroism and linear dichroism*, A. Rodger and B. Nordén, 1997, by permission of Oxford University Press.)

symmetry tris-chelate molecules are close in geometry to octahedral, in which case the transitions would be degenerate) so they overlap. The isotropic polarizability equations analogous to *equation 16* for this case are (11)

$$R(\text{E}) = \frac{15\alpha(\varepsilon_a)}{4R_{AC}^4} \text{Im}[m_x^a Q_{x^2}^a - m_x^a Q_{y^2}^a - 2m_y^a Q_{xy}^a]x(x^2 - 3y^2)] \qquad 17$$

$$R(\text{A}_2) = \frac{35\alpha(\varepsilon_a)}{4R_{AC}^6} \text{Im } m_z^a[H_{x^3z}^a - 3H_{xy^2z}^a][x(x^2 - 3y^2)(9z^2 - 1)] \qquad 18$$

where $Q_{x^2}^a$ etc. and $H_{x^3z}^a$ etc. are, respectively, quadrupole and hexadecapole transition moments of A (11). The distance dependence of these equations would lead us to expect the E band CD to be larger than the A$_2$ band—i.e. the 'dominant E band rule'.

There has been some controversy about the dominant E band rule: when the CD spectrum of crystalline [Co(ethylenediamine)$_3$]$^{3+}$ was measured it was found that the crystal CD for the A$_2$ component of the T$_{1g}$ d–d band had almost the same magnitude as that of the E band and both were an order of magnitude larger than those observed in the solution spectrum—so the solution spectrum appears to result from the cancellation of two large (A$_2$ and E) bands (4). This raises the question of why the E band in solution spectra is always the larger if the observed solution CD is only a small percentage of the total crystal signal. The obvious solution, namely crystal packing effects, has been eliminated as a possible explanation. The dilemma is probably resolved by realizing that the above equations give the CD intensity for a D_{3d} environment about the metal. In practice, the symmetry is lower and other terms contribute to the CD of each transition; however, these other contributions are equal and opposite for the A_2 and E bands so cancel due to the small energy difference of the two bands (4).

3.5 Optical activity (optical rotation, OR)

Optical rotation (Section 1.2) is the difference in refractive indices of left and right circularly polarized light upon passing through the medium:

$$OR = (n_l - n_r)\frac{\pi d}{\lambda} \text{ radians} \qquad 19$$

where $(n_l - n_r)$ is called the *circular birefringence* and λ and d have the same units (usually dm). OR is usually measured by determining the rotation of linearly polarized light upon passing through the solution. If the linearly polarized light is rotated clockwise (viewed looking into the light source), the OR is called a positive or right (dextro) OR. Optical rotation as a function of λ is called ORD (see above). ORD is an S-shaped curve centred at the CD maximum (so-called anomalous ORD). For a positive CD band the long-wavelength side of the ORD curve is a large positive ORD contribution. The short-wavelength side is less positive or negative. ORD is also non-zero away from absorption bands. Hence α_D values (the ORD at the sodium D line) may be used to characterize the enantiomeric excess of a solution.

3.6 Dissymmetry factor

A measure of the size of a CD signal is given by the dissymmetry factor g for the transition:

$$g = \frac{4R}{cD} \qquad 20$$

where R is the rotational strength of the transition, D is the dipole strength of the transition and c is the speed of light.

4 Units of CD spectroscopy

The relationship between CD signal (in absorbance units) and sample concentration and pathlength is analogous to the Beer–Lambert law for absorbance, namely:

$$\Delta A = (\Delta\varepsilon)Cl \qquad 21$$

where C is the sample concentration in mol dm^{-3}, l is pathlength in cm and $\Delta\varepsilon$ is in units of mol^{-1} dm^3 cm^{-1}. The units used for extinction coefficients are almost always absorbance units mol^{-1} dm^3 cm^{-1}, which gives values that are ten times greater than those in SI units. The moles to which one refers may be either in terms of molecules or, for macromolecules, in terms of residues such as DNA bases or protein amino acids.

A general summary of possible CD units and their interrelationship follows. In absorbance units we write

$$CD = A_l - A_r \qquad\qquad 22$$

Alternatively in molar units it is

$$CD = \varepsilon_l - \varepsilon_r \qquad\qquad 23$$

where ε_r is the molar extinction coefficient for the absorption of right circularly polarized light.

CD is often given in terms of the *ellipticity*, θ. θ is obtained from the ratio of the minor and major axes of the ellipse traced out by the electric field vector of the elliptically polarized light that emerges from a circularly dichroic sample onto which linearly polarized light was incident.

$$\tan \theta = \tanh\!\left(\frac{\pi l}{\lambda}(n'_l - n'_r)\right) \qquad\qquad 24$$

where n'_l is the absorption index for left circularly polarized light. The wavelength, λ, must be in the same units as l. For small ellipticities

$$\theta\,/\,\text{radians} \approx \frac{\pi l}{\lambda}(n'_l - n'_r)$$
$$= \frac{2.303lC}{4}(\varepsilon_l - \varepsilon_r) \qquad\qquad 25$$

Upon converting to millidegrees, as commonly used for CD:

$$\theta\,/\,\text{millidegrees} = 32\,980\,Cl(\varepsilon_l - \varepsilon_r)$$
$$= 32\,980\,(A_l - A_r) \qquad\qquad 26$$

Most CD spectrometers, although they actually measure differential absorbance, produce a CD spectrum in ellipticity, with units θ, in millidegrees, versus λ, rather than ΔA versus λ. Most CD machines will perform the conversion from ellipticity to absorbance units on request.

Another old measure of CD is the *specific ellipticity*

$$[\theta] = \frac{\theta}{C_g d} \qquad\qquad 27$$

where C_g is sample concentration in g cm^{-3} and d is pathlength in dm.

The related molar ellipticity is

$$M_\theta = \frac{[\theta]M}{100} = \frac{100\theta}{Cl} \qquad\qquad 28$$

where M is molar mass in g, C is in mol dm^{-3}, and l is in cm. If dm^{-3} is substituted by 1000 cm^{-3} we see that *molar ellipticity* can have the commonly used units of *deg cm^2 dmol^{-1}*. Thus

$$M_\theta = 3298\,(\varepsilon_l - \varepsilon_r)$$
$$= 3298\Delta\varepsilon \qquad\qquad 29$$

where ε_r is the extinction coefficient for the absorption of right circularly polarized light. A related unit that is often used is the mean residue ellipticity. In this case, as mentioned above, the sample concentration is given in terms of total numbers of residues (using the average molecular mass of an amino acid residue) without reference to their identity, hence it is an average over the types of residue present in the macromolecule.

5 Circular dichroism of biomolecules

5.1 Introduction

CD is commonly used to study biological macromolecules in one of two ways:

(1) to probe changes in the conformation of the macromolecule itself, and

(2) to probe its interaction with small molecules, especially achiral ones whose induced CD is due solely to their interaction with the macromolecule.

5.2 CD of polypeptides and proteins

Proteins are linear polymers of well-defined sequences (the primary structure) of amino acids that are folded in well-defined ways, with the overall shape of a protein molecule being crucial for its biological activity. Proteins form regular secondary structural units because the peptide O=C–N– link between amino acids is planar and rigid yet it has a large degree of rotational freedom about its bonds to the rest of the protein chain (*Figure 14*). This both constrains the possible relative orientations of neighbouring residues and allows a variety of intramolecular hydrogen bonding arrangements between the C–O of one peptide unit and the N–H of another unit. The resulting limited set of common chiral secondary structural units have more-or-less well defined CD signatures (*Figure 15*). It is this feature that makes CD so useful in the study of proteins.

5.3 Protein UV spectroscopy

UV spectra of proteins are usually divided into the 'near' and 'far' UV regions. The near UV in this context means 250–300 nm and is also described as the aromatic region, though transitions of disulphide bonds (cystines) also contribute to the total absorption intensity here. The far UV (<250 nm) is dominated by transitions of the peptide backbone of the protein, but transitions from some side chains also contribute in this region and, especially if the protein α-helical content is low, may give rise to erroneous protein structure determinations (see Section 5.4).

The lowest energy transition of the peptide chromophore is an $n{\rightarrow}\pi^*$ transition analogous to that in ketones, and the next transition is $\pi{\rightarrow}\pi^*$. The $n{\rightarrow}\pi^*$ transition has an extinction coefficient of ~100 mol^{-1} dm^3 cm^{-1}; it occurs at about 210–230 nm (depending mainly upon the extent of hydrogen bonding of the oxygen lone pairs) and its electric character is polarized more or less along the carbonyl bond. The $\pi{\rightarrow}\pi^*$ transition ($\varepsilon \sim$ 7000 mol^{-1} dm^3 cm^{-1}) is dominated

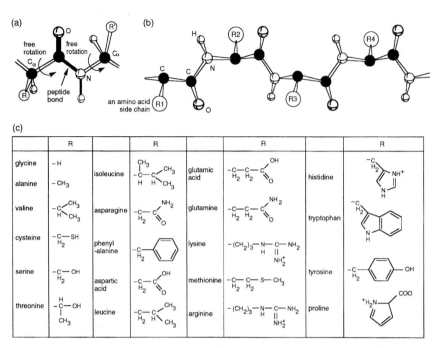

Figure 14 (a) The peptide bond, (b) the primary structure of a protein, and (c) the commonly occurring L-amino acids. (Source: *Circular dichroism and linear dichroism*, A. Rodger and B. Nordén, 1997, by permission of Oxford University Press.)

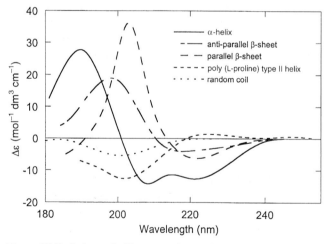

Figure 15 Typical protein CD spectra for particular secondary structural motifs used in CD protein structure fitting programs.

by the carbonyl π-bond and is also affected by the involvement of the amide nitrogen in the π-orbitals; its electric dipole transition moment is polarized somewhere near the line between oxygen and nitrogen and it is centred at 190 nm. In an α-helix (see below), the electric dipole coupling of the π→π* transitions on neighbouring residues results in a long wavelength component of this transition at ~208 nm.

The aromatic side chains, phenylalanine, tyrosine and tryptophan, all have transitions in the near UV region. The indole of tryptophan has two or more transitions in the 240–290 nm region with total maximum extinction co-efficient ε_{max}(279 nm) ~ 5600 mol^{-1} dm^3 cm^{-1}; tyrosine has one transition with ε_{max}(274 nm) ~ 1400 mol^{-1} dm^3 cm^{-1}; phenylalanine also has one transition with ε_{max}(258 nm) ~ 190 mol^{-1} dm^3 cm^{-1}; and a cystine disulphide bond absorbs from 250–270 nm with ε_{max} ~ 300 mol^{-1} dm^3 cm^{-1}. Although trypto-phans have by far the most intense transitions in this region, many proteins have few tryptophans compared with the other aromatic groups, so the region is not necessarily dominated by tryptophan transitions. Metallo-proteins have so-called prosthetic or extrinsic groups with additional chromophores that often have transitions in the visible region of the spectrum as well as the UV.

5.4 Protein structure determination from CD

At the present the main use of CD in the study of proteins is as an empirical gauge of protein structure and conformation using the CD induced into the backbone amide transitions from ~190–240 nm (the far UV or peptide region of the spectrum). In the absence of any contributions to the CD from side chain transitions, this has proved to be a very successful approach (12–22). Distinctive CD spectra (*Figure 15*) have been described for pure conformations such as the α-helix, β-sheets (with different ones sometimes being given for parallel and antiparallel sheets), β-turns and also the 'random' coil. At least in principle, the CD spectrum of a native protein is then the sum of the appropriate percentages of each component spectrum. However, relative arrangements of structural units and motifs such as disulphide bonds contribute to the observed CD spectrum.

5.4.1 α-Helix CD

The (right handed) α-helix is the dominant secondary structure in many pro-teins and on average accounts for about one third of the residues in globular proteins. It is a well-defined structural motif where the nth peptide unit forms hydrogen bonds between its C-O and the N-H of the $(n + 4)$th peptide and between its N-H and the $(n - 4)$th C-O; there is a 0.15 nm translation and 100° rotation between two consecutive peptide units, giving 3.6 amino acid residues per turn. Its helix pitch (number of residues times distance between α-Cs on neighbouring residues) is 0.54 nm.

The CD spectra of α-helices are characterized by a negative band with separate maxima of similar magnitude at 222 nm (the $n \to \pi^*$ transition) and 208 nm which is part of the $\pi \to \pi^*$ transition (see Section 5.3 and *Figure 15*). The α-helix is the only motif where the $\pi \to \pi^*$ transition has such a long wavelength component. The α-helix CD is also larger in magnitude than that due to other motifs so it is apparent upon the most casual inspection of a spectrum. It is no surprise therefore that the various empirical CD fitting programs that are available are fairly successful in determining the percentage of α-helical content of a protein. It should, however, be noted that the magnitude of the CD signal does vary with variations in the helix, and the helix length (the signal behaves as if it were for a helix about four residues shorter than it in fact is, which corresponds to the number of unanchored hydrogen-bonding groups) and the interactions with neighbouring structural units.

5.4.2 3_{10} helix CD

The 3_{10} helix is a right handed helix with 3.0 residues per turn and a helix pitch of 0.6 nm.

5.4.3 Proline like 3/1-helix CD

The polyproline type II helix is that adopted by the polypeptide chains of collagen. It has 3.0 residues per turn and a helix pitch of 0.94 nm.

5.4.4 β-sheet CD

An alternative efficient formation of hydrogen bonds occurs between a sheet of parallel or antiparallel runs of amino acids; these are known as a β-sheets. The runs of amino acids face in alternate directions so that alternate amino acids hydrogen bond to neighbouring runs on each side. The spectroscopic characterization of β-sheets has proved more difficult than that of α-helices due to the practical reason that they are less soluble in solvents with a good UV transmission, and due to the intrinsic reason that they are generally structurally less well-defined: they may be parallel or antiparallel and of varying lengths and widths. Furthermore, an extended β-sheet is usually found to show a marked twist, rather than to be planar. Such tertiary structure influences the overall CD spectrum.

The general characteristics of β-sheet CD may be taken to be a negative band at about 216 nm and a positive band of comparable magnitude near 195 nm (13). This level of characterization of the spectrum might be deemed to be not much worse than that of α-helices were it not for the fact that what is often called the 'random coil' but is more correctly referred to as 'other' structural features (see below) have their CD maxima at similar wavelengths and are often of opposite sign from those of the β-sheet. This means that an empirical fitting program may incorrectly weight these spectra (and the β-turn components) in an attempt to better account for the wavelength and magnitude variations that occur.

5.4.5 β-Turn CD

Typically the strands of an antiparallel β-sheet are linked by β-turns where the nth peptide unit forms hydrogen bonds with the $(n + 3)$rd peptide unit. However, the label β-turn is usually used to include all possible turns that occur, not simply the ones that enable a single strand to become an antiparallel β-sheet. About one quarter of the residues in globular proteins then fall into this structural group. Despite this range of structures, a 'typical' β-turn CD spectrum has been identified which has a weak red-shifted negative $n{\rightarrow}\pi^*$ band near 225 nm, a strong positive $\pi{\rightarrow}\pi^*$ transition between 200 nm and 205 nm, and a strong negative band between 180 nm and 190 nm.

5.4.6 'Random' coil CD

The label random coil describes the fully denatured state of the protein. However, the term is also often used collectively of the parts of the protein that are not α-helices, β-sheets or turns. A spectrum such as that illustrated in *Figure 15* results for a truly random coil. The other structural features in an average protein often have a strong negative CD signal just below 200 nm, a positive band at about 218 nm in many systems, and perhaps a very weak negative band at 235 nm. Folded proteins have no true random coil elements.

5.4.7 Aromatic region

The intensity of the CD induced into the achiral aromatic side chains is very dependent on their environment so, in contrast to fluorescence, it is not possible with CD to limit consideration specifically to the tryptophans. Groups which can be essentially ignored in the absorption spectrum may become significant in the CD. In particular, disulphide groups (covalent bonds formed between cysteine residues in different parts of a protein) have transitions in the aromatic region of the spectrum. Although these transitions are weak ($\varepsilon_{max} \sim 300 \, mol^{-1} \, dm^3 \, cm^{-1}$ at 250 and 270 nm), they are magnetic dipole allowed and so may contribute significantly to the protein CD in this region, often with tails stretching beyond 300 nm (which aids their identification). As the side chains are usually isolated from one another their CD usually arises from interactions with the neigh bouring amino acid residues, as would be the case for a ligand bound to the macromolecule.

5.5 Determining the percentage of different structural units in a protein from peptide region CD spectra

There are a range of different computer programs available for determining the percentage of different structural motifs just from the CD spectrum (14). The cost of these programs does not necessarily correlate with their reliability. Most programs will give an accurate α-helix content, if you accurately know the concentration of your protein (usually in terms of residues rather than molecules). We shall outline the use of a program written by W. C. Johnson, M. Parthasarathy and A. Toumadje for determining the structure of proteins from CD data (15). The program uses ideas developed by many other workers and is applicable only to protein structure determination. Small peptides require a different approach as discussed below.

The program, CDsstr, is currently available via ftp from alpha.als.orst.edu by logging in as 'anonymous' and using your email address as the password. Changing directory by typing 'CD/pub/wcjohnson/CDsstr', then 'bin' to specify binary file transfer, and then typing 'mget *.*' will retrieve the program files.

CDsstr computes the percentages of secondary structures of a protein from CD data using the method of Hennessey and Johnson (16) and the single value decomposition algorithm of Compton and Johnson (17). Rather than using typical spectra for each type of secondary structure motif, CDsstr uses a basis set of 22 proteins with known secondary structure and known CD spectra. The program self-consistently chooses a minimum sub-basis set of eight proteins for a given problem using methods given in (12−15). The program gives percentages of α-helix, 3/10-helix, extended-β-strand, β-turns, polyproline-like 3/1-helix and 'others' which are explicitly not random coils. The advantage of this approach over that of representing the CD spectrum as a sum of component parts is that the CD due to aromatic and sulphur-containing side chains etc. as well as helix length variations and interactions between neighbouring secondary structural units are all accounted for because they are contained in the basis sets.

The current version of the program runs on a PC under Windows 95/98 or NT4. The program consists of several files. The 22 basis set CD spectra are contained in the file basCD.dta and the corresponding secondary structure data resides in the file secstr.dta. The protein CD data to be analysed are read from the file proCD.dta. The proCD.dta file should consist of one title line followed by the CD data in units of $\Delta\varepsilon$, in 1 nm increments with each data value on a new line and beginning at 260 nm and ending at 178 nm. Numerical values are read in Fortran 1F10.2 format (i.e. one floating point number with a total of 10 characters, including any minus sign and decimal point, of which two lie after the decimal point). This program apparently works reasonably well on data truncated even at 200 nm, but more data give better answers.

Protocol 2

Application of CDsstr to compute protein secondary structure from CD spectra

Equipment and materials

- A PC running Windows 95/98 or NT4
- A copy of the CDsstr program
- CD data for the proteins to be analysed

A. Running CDsstr for the first time

1 Work within windows.

2 Create a folder named CDsstr in the root directory of the hard disk (assumed to be drive c: in the rest of this protocol).[a]

3 Copy all the CDsstr program files into the new CDsstr folder.

4 Delete any file with an .out filename extension from the CDsstr folder.

5 Delete the file named proCD.dta if it exists.

6 Verify the correct operation of the program by running the calculation on the test files provided. To do this, open a DOS window by selecting 'Command Prompt'

Protocol 2 continued

from the 'Programs' menu in the 'Start' menu. If 'Command Prompt' is not present in the 'Programs' menu, open a DOS window by selecting 'Run' from the 'Start' menu and typing 'c:\windows\command.com' in the box then clicking 'OK'.[b]

7   At the command prompt type 'c:' then 'CD\CDsstr'.

8   Type 'copy proCD.tst proCD.dta'.

9   Type 'CDsstr' to initiate the test.

10 Enter the value of each variable as prompted: NbasCD = 22; Nwave = 83; Npro = 3; ncomb = 100; icombf = 100 000.

11 When the command prompt reappears, compare the values in the files anal.out and reconCD.out with those in anal.tst and reconCD.tst. If CDsstr is functioning correctly they should exactly match.

B. Using CDsstr for protein CD analysis

1   Delete, rename, or move any file with an .out filename extension remaining in the CDsstr folder.

2   Delete any previously used file named proCD.dta unless you wish to use it in the current run.

3   If it is not already available, prepare an input file containing the CD data of the protein(s) to be analysed. Enter a title on the initial line then enter the CD data as $\Delta\varepsilon$ (for molar residue concentration) with each value on a new line. Begin at 260 nm and continue in 1 nm increments down to 178 nm. Ensure that the numerical values entered comply with the Fortran 1F10.2 format. If multiple protein data sets are being analysed, begin each new set with a title line. Save the file as c:\CDsstr\proCD.dta.[c]

4   If using truncated protein CD data (truncation above 200 nm is not recommended), the basis set protein CD spectra in the basCD.dta file must also be truncated by removing data points from each data set to match the experimental data set size.[c]

5   Determine appropriate values (see above) for the variables NbasCD, Nwave, Npro, Ncomb and icombf.

6   Begin the analysis by opening a DOS window (see Section A above). Type 'c:'. Then type 'CD\CDsstr' at the command prompt.

7   Enter values for the program variables as prompted.

8   When the command prompt reappears, view the results of the analysis by inspecting the output files anal.out and reconCD.out.

[a] Do not place this folder within any subdirectory if you wish to follow this protocol exactly.

[b] CDsstr must be run in MS-DOS from within Windows. Restarting the computer in MS-DOS mode will not permit the program to run.

[c] A commercial spreadsheet is well suited to editing CD data files. Most CD data analysis software permits a data file to be exported as text. The text file can then be imported into the spreadsheet, easily formatted for CDsstr input and saved as undelimited ASCII text. Delete, rather than cut, data cells to remove unwanted text/data.

Upon execution, CDsstr prompts the user for the value of several program variables. NbasCD is the number of basis set CD spectra and should be set to 22 if the default basCD.dta and secstr.dta files are being used. Nwave is the number of wavelengths contained in basCD.dta and is 83 for the default basis set (i.e. with data down to 178 nm). The number of protein CD spectra contained in the proCD.dta file, Npro, should then be entered and must lie between 1 and 100 (3 for the test file proCD.tst). The value for the number of successful combinations to be considered, ncomb, should usually be set to 100 except for alpha helical proteins where the program's maximum value of 400 should be used. The final requested parameter value is the maximum number of trial combinations allowed before the program stops the analysis, icombf. A value of 100 000 is usually appropriate for this parameter. The calculated secondary structure output is written to the file anal.out along with the input data. A second output file, reconCD.out contains the average reconstructed CD spectrum.

Peptide CD structure analysis differs from that of proteins as peptides are usually a combination of a particular secondary structure and random coil (i.e. unordered) residues. Proteins seldom if ever have random coils, so the CD of the proteins in the basis set of the protein analysis program has no random coil component. Further, the analysis of the CD of a peptide is not underdetermined, and so does not require a flexible method like Variable Selection. A Convex Constraint Analysis (22) that extracts the component spectra has been developed into a peptide structure analysis program by Greenfield (14).

5.6 Other applications of protein CD

Protein CD may also be used to probe whether the protein structure is perturbed in any way upon interaction with other molecules or as a function of a variable such as pH or solvent. For example, α-helices may be stabilized or induced by trifluoroethanol (23). Conversely α-helices may be destabilized by dissolving them in non-polar media such as lipids (24). Such conformational changes may be followed using CD. Either aromatic (near UV) or backbone (far UV) CD may be used for a qualitative or empirical study of this kind. CD has also been extensively used in protein folding studies, some of which are described in more detail in Chapter 10. CD can also be used to give quantitative estimates of binding constants when ligands bind to the protein using methods analogous to those outlined for DNA–ligand systems (see Sections 5.10 and 5.11).

5.7 Membrane proteins

The absorption and CD spectra of membrane proteins are often distorted by artefacts that arise as consequences of light scattering (25) (due to the large size of the membrane particles relative to the wavelength of light) and absorption flattening (26) (due to local high protein concentrations within the particles). The best solution to these problems is to avoid them by incorporating the membrane proteins into small particles, such as unilamellar vesicles, where the protein concentration is low. A useful system for this is described below. If no appropriate small particle can be found for a given system, then the artefacts must be removed in order to use the data.

The light scattering is only a problem because scattered light that does not reach the PMT is assumed to have been absorbed. As light scattering increases with decreasing wavelength, the effect of scattering is to cause a net increase in magnitude of signals and a shift of maxima. The light scattering problem disappears if the configuration of the instrument is such as to collect all the scattered light as well as the unscattered light. Mao and Wallace (27) suggest that a collection angle of 90° is sufficient. Our experiments (28) with a reflecting surface designed to collect all scattered light and the Jasco J-715 instrument equipped with the large sample compartment suggest that, in a J-715, essentially all the scattered light is collected by moving the sample holder into the recess before the PMT. This results in a collection angle of at least 72°.

The absorption flattening problem can only be treated at the sample preparation stage. Somehow local high concentrations must be avoided. Detergent solubilization, with detergents including Triton X-100, n-octyl glucoside, and sodium dodecyl sulphate, generally reduces membrane particle sizes. However, care should be taken with this approach given the use of the same reagents to denature proteins. It cannot be assumed that the structure (and hence CD) of the solubilized and original protein are the same. An alternative is to sonicate a sample of a membrane protein with, for example, dimyristoylphosphatidyl-choline (27) and thus to form small unilamellar vesicles (less than 100 nm inside). A range of lysophospholipids and phospholipids have been used for this purpose and for incorporating membrane proteins in micelles (29). The size of the particles should be checked before use in a CD experiment.

5.8 DNA geometry and CD spectra

To a first approximation double helical DNA (or RNA) can be viewed as a more-or-less vertical spiral staircase, where the steps are made of pairs of nitrogenous bases hydrogen bonded together and the support is the backbone of alternating phosphate groups and ribose sugars (*Figure 16*). In isolation from the chiral sugar units the phosphates and the bases are achiral; however, when they are joined together with the sugar units they become part of a chiral molecule, and their transitions are expected to exhibit CD spectra. The helix of the standard B-DNA (the common solution polymorph) staircase is right-handed and the steps are *approximately* perpendicular to the helix axis. Some geometry parameters for B-DNA and the other common polymorphs A-DNA and Z-DNA are given in *Table 1*. Single-stranded DNAs are structurally less well-defined than duplex DNAs and their CD signal is smaller. CD spectra of naturally occurring RNAs (*Figure 17*), which are often assumed to be single stranded, clearly show naturally occurring RNAs to be mainly well-structured with extensive A-form duplex regions (see Section 5.9).

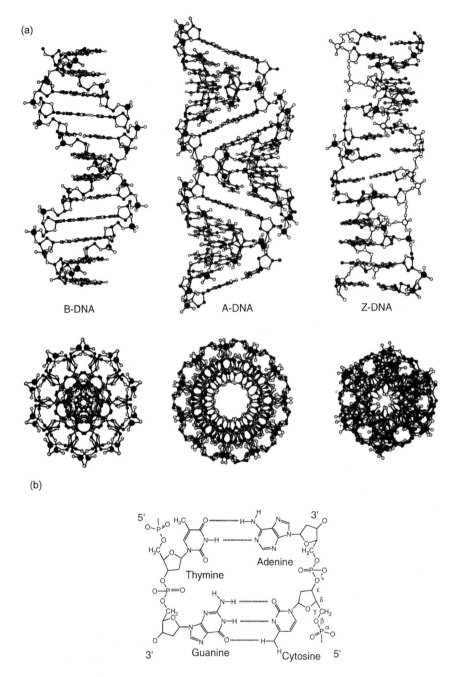

(a)

B-DNA   A-DNA   Z-DNA

(b)

Figure 16 (a) B-, A- and Z-DNA viewed perpendicular to the helix axis and down the helix axis. (b) Structural formula of DNA illustrating the definitions of torsion angles. (Source: *Circular dichroism and linear dichroism*, A. Rodger and B. Nordén, 1997, by permission of Oxford University Press.)

Table 1 Typical geometric parameters (see *Figure 8* for definitions) for standard A-, B- and Z- forms of DNA (30–34)

Parameter	A-DNA	B-DNA	Z-DNA
α	−85°	−47°	60°/160°
β	−152°	−146°	−175°/−135°
γ	46°	36°	178°/57°
δ	83°	156°	140°/95°
ε	178°	155°	−95°/−110°
ζ	−46°	−95°	−35°/−85°
Sugar conformation	C_3-*endo*	C_2-*endo*	C_3-*endo*/C_2-*endo*
Glycosidic bond	*anti*	*anti*	*anti* (C, T), *syn* (G)
Base roll	12°	0°	1°
Base tilt	20°	5°	9°
Base helical twist	32°	36°	11°/50°
Base slide	0.15 nm	0 nm	0.2 nm
Helix diameter	2.55 nm	2.37 nm	1.84 nm
Bases per helix turn	11	10.4	12
Base rise per base pair	0.23 nm	0.33 nm	0.38 nm
Major groove	Narrow, deep	Wide, deep	Flat
Minor groove	Broad, shallow	Narrow, deep	Narrow, deep

RNA differs from DNA in that every T is replaced by U, which has one less methyl group, and the backbone sugar is ribose instead of deoxyribose. Double helical RNA adopts the A-form geometry. The repeat unit of Z-DNA is a dinucleotide, necessitating two values of each angular parameter. ζ adopts a range of values in Z-DNA.

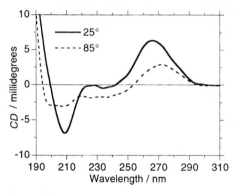

Figure 17 CD of a long (372 nucleotide) naturally occurring mRNA. At 25 °C the RNA is largely that of an A-form double helix; at 85 °C there is less chirality in the parts of the molecule being probed—in this case the RNA bases. This is due to there being less base pairing and base stacking at 85 °C. $A_{260\ nm}$ = 1.0 at 25 °C. (Source: *Circular dichroism and linear dichroism*, A. Rodger and B. Nordén, 1997, by permission of Oxford University Press.)

5.9 UV spectroscopy of the DNA bases

The UV absorbance of nucleic acids from 200–300 nm is due exclusively to transitions of the planar purine and pyrimidine bases. The backbone begins to contribute at about 190 nm. The accessible region of the spectrum is therefore

(a)

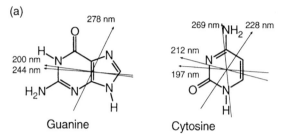

Guanine   Cytosine

Figure 18 Probable transition polarizations for UV transitions of guanine and cytosine. (Source: *Circular dichroism and linear dichroism*, A. Rodger and B. Nordén, 1997, by permission of Oxford University Press.)

dominated by $\pi \rightarrow \pi^*$ transitions whose exact assignment is still a matter of debate. All of them are polarized in the plane of the bases. Recent assignments for guanine and cytosine are illustrated in *Figure 18* (35–38). Adenine probably has five $\pi \rightarrow \pi^*$ transitions in the UV region (39). There may also be some $n \rightarrow \pi^*$ transitions present whose intensity (even in CD) is so small that they have not been definitively identified.

5.10 Nucleic acid CD

When we measure a CD spectrum of a polynucleotide with stacked bases, the magnitude is larger at 270 nm and significantly larger at 200 nm than that of the individual bases (4). The spectrum is dominated by the CD induced into transitions of the bases from their coupling with each other when the bases stack in a chiral (helical) fashion. When the CD is measured as a function of temperature, a melting curve is plotted that enables the double- to single-stranded transition to be followed. The resulting single-stranded, or approximately random coil, molecule still has significant CD at 270 nm due to the intrinsic base CD (*Figure 17*).

5.10.1 Empirical structural analysis of DNA CD

The simplest application of CD to DNA structure analysis is to identify the polymorph in a sample (4). Since the DNA CD from 200–300 nm is due to the skewed orientation of the bases, if the DNA is untwisted or the bases are tilted, a change from what is observed for B-DNA would be expected in the CD spectrum. Somewhat surprisingly, observed CD spectra often vary more with changes in base orientation (DNA polymorph) than as a function of the base composition of the DNA, though they are of course a sensitive function of DNA sequence as well.

The spectrum of calf thymus DNA shown in *Figure 19* is typical of B-form DNA. The CD signature of B-form DNA (*Figure 20*), as read from longer to shorter wavelength, is a positive band centred at 275 nm, a negative band at 240 nm, with the zero being around 258 nm. At 220 nm the CD signal is either positive or less negative; a small negative peak is then followed by a large positive peak

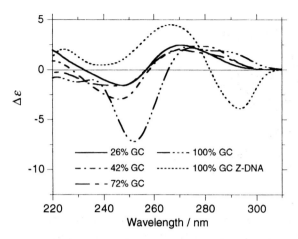

Figure 19 CD spectra of duplex DNAs (pH = 6.8, 5 mM NaCl, 1 mM cacodylate) of varying G–C base pair content: *Clostridium perfringens* (26% GC), calf thymus (42% GC), *Micrococcus lysodeikticus* (72% GC), poly[d(G–C)]$_2$ (100% GC). Also shown is Z-form poly[d(G–C)]$_2$ (100% G–C, pH = 6.8, 5 mM NaCl, 1 mM cacodylate, 50 μM [Co(NH$_3$)$_6$]$^{3+}$). The 195 nm negative signal of Z-DNA is obscured by the CD induced into the transitions of [Co(NH$_3$)]$^{3+}$ when this molecule is used to induce Z-form DNA. (Source: *Circular dichroism and linear dichroism*, A. Rodger and B. Nordén, 1997, by permission of Oxford University Press.)

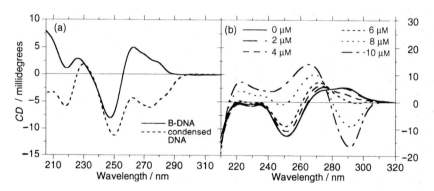

Figure 20 (a) Poly[d(A-T)]$_2$ (44 μM; 1 mM phosphate) in B-form (20 mM NaCl) and condensed (1 M NaCl and 45% v/v ethanol). (b) Poly[d(G–C)]$_2$ as a function of increasing spermine, [NH$_3$(CH$_2$)$_3$NH$_2$(CH$_2$)$_4$NH$_2$(CH$_2$)$_3$NH$_3$]$^{4+}$, concentration showing the B → Z transition. Spermine concentrations are indicated in the figure. (Source: *Circular dichroism and linear dichroism*, A. Rodger and B. Nordén, 1997, by permission of Oxford University Press.)

from 180–190 nm. When B-form DNA is stretched to have 10.2 bases per turn instead of 10.4, the 275 nm peak essentially disappears. This form of DNA can be induced by methanol or by wrapping the DNA around histone cores to form nucleosomes.

If B-DNA is compacted and the bases tilted and radially displaced from the centre of the helix (thus creating a hole when one looks down the helix axis) (*Figure 16* and *Table 1*) then A-form DNA results. A-DNA is characterized by a

133

positive CD band centred at 260 nm that is larger than the corresponding B-DNA band, a fairly intense, comparatively sharp negative band at 210 nm and a very intense positive band at 190 nm. The 250–230 nm region is also usually fairly flat though not necessarily zero. Naturally occurring RNAs adopt the A-form if they are duplex. A typical natural RNA spectrum is shown in *Figure 17*.

Z-form DNA (*Figure 16* and *Table 1*) does not readily form for all sequences. However, it is easily formed for poly[d(G–^{5}meC)]$_2$ in the presence of highly charged ions and also for poly[d(G–C)]$_2$ under appropriate conditions (40). Z-DNA is characterized by a negative CD band at 290 nm and a positive band at 260 nm (*Figure 19*). Care must be taken in using these signatures to identify Z-form DNA, since the same DNA in the A-form has a negative band at 295 nm and a positive band at 270 nm (41). Poly [d(I–C)]$_2$, I = inosine, is also negative at long wavelengths. A more definitive signal is the large negative CD signal in the 195–200 nm region for Z-DNA, whereas B-form DNA CD is near zero or positive in this region. For Z-DNA, the CD passes through zero between 180 and 185 nm. Condensed DNA also has a characteristic CD spectrum (*Figure 20*).

5.10.2 Quantitative structural analysis of DNA CD

With DNA, as discussed above, it is usually possible to identify qualitatively the overall polymorph that is present, but a more quantitative *empirical* analysis has defied extensive efforts. Somewhat perversely, however, theoretical analysis of DNA CD performed by calculating the spectrum for a proposed geometry is possible, whereas for proteins this approach has been less successful (41–43).

5.11 DNA/ligand interactions

Many DNA binding molecules (usually referred to as ligands or drugs even when they may have little or no therapeutic value) are themselves achiral. However, when they bind to DNA they acquire an induced CD (ICD) that is characteristic of their interaction (*Figure 21*). If the ligand transitions of interest occur in the DNA region of the spectrum we take the ICD to be the total CD of the system minus that of the DNA at the same concentration. This means any DNA conformational changes that alter the DNA CD or any ICD of the DNA itself are included in the net ICD of the interaction. As with application of CD to proteins and DNA the ligand ICD can be used on a number of levels. The simplest use is to note that it exists and therefore conclude that the molecule *does* bind to DNA. However, more information can usually be extracted.

5.11.1 Empirical analyses of ligand ICD

It is often very useful to measure a series of spectra where some variable such as the ionic strength, or mixing ratio (DNA:drug ratio), or temperature is changed. If the CD intensity changes but the shape of the spectrum remains the same during such an experiment, then it can be deduced that the ligand binding mode is unchanged though the amount of bound ligand may have changed (*Figure 21*). In such an experiment, if possible, it is best to keep the ligand concentration constant as then any changes are more easily apparent. If the

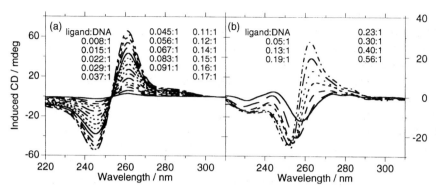

Figure 21 ICD of varying concentrations of anthracene-9-carbonyl-N^4-spermine upon interaction with (a) 160 μM poly[d(G–C)]$_2$, 5 mM NaCl and low ligand:DNA:phosphate mixing ratios, and (b) 130 mM poly[d(A–T)]$_2$, 5 mM NaCl and a wide range of ligand:DNA:phosphate mixing ratios. l = 1 cm. The smallest magnitude ICD signal corresponds to the lowest ligand concentration. (Source: *Circular dichroism and linear dichroism*, A. Rodger and B. Nordén, 1997, by permission of Oxford University Press.)

ligand binds in a single binding mode, or in a number of sites whose relative proportions are independent of DNA:drug ratio, then we may write

$$L_b = \alpha \times \text{ICD} = \alpha\rho \qquad\qquad 30$$

where L_b is the concentration of bound ligand, and α is a proportionality constant. If you can draw a plot of CD versus DNA:drug ratio that has a reasonably straight part, followed by a curve, and finally a levelling off part, then one of a number of the analysis methods may be used to determine equilibrium binding constants and site sizes (44–47).

Consider the equilibrium association constant K

$$K = \frac{L_b}{L_f S_f} \qquad\qquad 31$$

for the equilibrium

$$L_f + S_f \rightleftharpoons L_b \qquad\qquad 32$$

where L_f is the concentration of free ligands, and S_f is the free site concentration. The total site concentration, S_{tot} is given by

$$S_{tot} = \frac{C_M}{n} \qquad\qquad 33$$

where C_M is the macromolecule concentration. For DNA we usually use the concentration of bases, in which case n is the number of bases in a binding site. For a protein, C_M is usually taken to be the concentration of protein molecules in which case

$$n' = \frac{1}{n} \qquad\qquad 34$$

is the number of ligand binding sites on each protein molecule.

To perform a Scatchard plot (48) to analyse the data to give K directly, re-arrange *equation 31* as follows. If we know α (the proportionality constant, see *equation 30*) then from ICD spectra we determine L_b and hence L_f

$$\frac{r}{L_f} = \frac{KS_f}{C_M}$$

$$= \frac{K}{n} - rK$$

35

where

$$r = \frac{L_b}{C_M}$$

36

Thus, a plot of r/L_f versus r has slope $-K$ and y-intercept K/n. The x-intercept occurs where $r = 1/n$. The Scatchard plot is a better way of averaging over a data set than, for example, an arithmetic mean from direct application of *equation 31* for a number of data points. The problem with the Scatchard plot is that one needs to know the value of α in *equation 30* to determine L_b, and hence L_f from the ICD. Ensuring all the ligand in a solution is bound as required for a direct determination of α is often impossible with DNA. The 'Intrinsic method' was developed to solve this problem (*Figure 22*).

We write

$$K = \frac{\alpha\rho}{(S_{tot} - \alpha\rho)(L_{tot} - \alpha\rho)}$$

37

and rearrange this to give

$$L_{tot} = \frac{L_{tot}S_{tot}}{\alpha\rho} - S_{tot} + \alpha\rho - \frac{1}{K}$$

38

Thus, for two different total ligand concentrations, L_{tot}^j and L_{tot}^k, but the same macromolecule concentration, i.e. $S_{tot}^j = S_{tot}^k$,

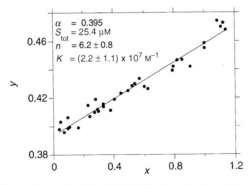

$\alpha = 0.395$
$S_{tot} = 25.4 \ \mu M$
$n = 6.2 \pm 0.8$
$K = (2.2 \pm 1.1) \times 10^7 \ M^{-1}$

Figure 22 Graph illustrating application of the intrinsic method for determining α and n using the data given in *Figure 21*. K has been determined using an average value from repeated application of *equation 37*. Alternatively, the data could have been used to perform a Scatchard plot. x and y are as defined in *equations 40* and *41*. (Source: *Circular dichroism and linear dichroism*, A. Rodger and B. Nordén, 1997, by permission of Oxford University Press.)

$$\frac{L_{tot}^{k} - L_{tot}^{j}}{\rho^{k} - \rho^{j}} = \frac{S_{tot}}{\alpha}\left(\frac{\dfrac{L_{tot}^{k}}{\rho^{k}} - \dfrac{L_{tot}^{j}}{\rho^{j}}}{\rho^{k} - \rho^{j}}\right) + \alpha \tag{39}$$

Thus a plot of

$$y = \frac{L_{tot}^{k} - L_{tot}^{j}}{\rho^{k} - \rho^{j}} \tag{40}$$

versus

$$x = \left(\frac{\dfrac{L_{tot}^{k}}{\rho^{k}} - \dfrac{L_{tot}^{j}}{\rho^{j}}}{\rho^{k} - \rho^{j}}\right) \tag{41}$$

(where any pair of data points considered have the same C_M) should be a straight line with slope $C_M(n\alpha)^{-1}$ and intercept α.

It is often convenient to perform experiments with constant ligand and varying macromolecule concentration. In this case C_M, and hence S_{tot}, are the variables and L_{tot} is fixed. Rather than *equations 40* and *41* we then use:

$$y = \frac{C_M^{k} - C_M^{j}}{\rho^{k} - \rho^{j}} \tag{42}$$

and

$$x = \left(\frac{\dfrac{C_M^{k}}{\rho^{k}} - \dfrac{C_M^{j}}{\rho^{j}}}{\rho^{k} - \rho^{j}}\right) \tag{43}$$

which gives a plot with slope L_{tot}/α and y-intercept $n\alpha$.

If there is a change in the shape of the spectrum as the mixing ratio or other experimental variable is changed, this implies that there is a change in the DNA–drug interaction as a function of the experimental variable. The change is usually due to occupancy of more than one binding site when the drug load on

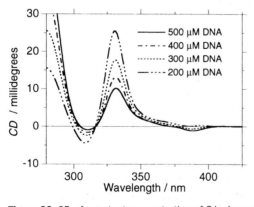

Figure 23 CD of constant concentration of 9-hydroxyellipticene (50 μM) in the presence of calf thymus DNA (concentrations indicated in figure, 5 mM NaCl, 1 mM phosphate buffer). (Source: *Circular dichroism and linear dichroism*, A. Rodger and B. Nordén, 1997, by permission of Oxford University Press.)

137

the DNA increases. Therefore the above methods for determining K cannot be used as *equation 30* does not hold. We may also be observing changes in the DNA conformation or to ligand–ligand interactions. A particularly dramatic change in the ligand ICD is observed if as the DNA concentration is *decreased* (so one would expect fewer ligands to bind and so a *decrease* in ICD), the ligands begin to stack together and the ICD *increases*. Under these circumstances, we generally observe a large excitonic CD. Thus spectra such as those of *Figure 23* immediately tell us that significant ligand–ligand interactions are occurring on the DNA.

Acknowledgement

This text has been adapted from Circular Dichroism and Linear Dichroism by Alison Rodger and Bengt Norden (1997) by permission of Oxford University Press.

References

1. Craig, D. P. and Thirunamachandran, T. (1984). *Molecular quantum electrodynamics: An introduction to radiation-molecule interaction.* Academic Press, London.
2. Barron, L. D. (1982). *Molecular light scattering and optical activity*, p. 1. Cambridge University Press, Cambridge
3. Grandison, A., McGinley, D., Shearer, T., Knight, L., Summers, E., Lyons, C. and Forde, C. (ed.) (1994). *Collins English dictionary* (3rd edn), p. 284. HarperCollins Publishers, Glasgow.
4. Rodger, A. and Nordén, B. In *Circular and linear dichroism* (ed. R. G. Compton, S. G. Davies, and J. Evans), p. 1. Oxford University Press, Oxford.
5. Bosnich, B. (1969) *Acc. Chem. Res.*, **2**, 266.
6. Moffitt, W., Woodward, R. B., Moscowitz, A., Klyne, W. and Djerassi, C. (1961) *J. Am. Chem. Soc.*, **83**, 4013.
7. Höhn, E. G., and Weigang, O. E. (1968) *J. Chem. Phys.* **48**, 1127.
8. Rodger, A. and Rodger, P. M. (1988). *J. Am. Chem. Soc.*, **110**, 2361.
9. Lightner, D. A., Bouman, T. D., Wijekoon, W. M. D. and Hansen, A. (1985) *J. Am. Chem. Soc.* **89**, 5805.
10. Fidler, J., Rodger, P. M., and Rodger, A. (1994) *J. Am. Chem. Soc.* **116**, 7266.
11. Schipper, P. E. and Rodger, A. (1986). *Chem. Phys.*, **109**, 173.
12. Fasman, G. D. (ed.) (1996). *Circular dichroism and the conformational analysis of biomolecules*, p. 1. Plenum Press, New York.
13. Greenfield, N. and Fasman, G. D. (1969). *Biochemistry*, **8**, 4108.
14. Greenfield, N. J. (1996). *Anal. Biochem.*, **235**, 1.
15. Johnson, W. C. Jr. (1999). *Proteins: Structure, Function and Genetics*, **35**, 307.
16. Hennessey, J. P. Jr. and Johnson, W. C. Jr. (1981). *Biochem.*, **20**, 1085.
17. Compton, L. A. and Johnson, W. C. Jr. (1986). *Anal. Biochem.*, **155**, 155.
18. Manavalan, P. and Johnson, W. C. Jr. (1987). *Anal. Biochem.*, **167**, 76.
19. Sreerama, N. and Woody, R. W. (1993). *Anal. Biochem.*, **209**, 32.
20. van Stokkum, I. H. M., Spoelder, H. J. W., Bloemendal, M., van Grondelle, R. and Groen, F. C. A. (1990). *Anal. Biochem.*, **191**, 110.
21. Dalmas, B. and Bannister, W. H. (1995). *Anal. Biochem.*, **225**, 39.
22. Perczel, A., Park, K. and Fasman, G. D. (1992). *Anal. Biochem.*, **203**, 83.
23. Shiraki, K., Nishikawa, K. and Goto, Y. (1995). *J. Mol. Biol.*, **245**, 180.

24. Aggell, A., Bell, M., Boden, N., Keen, J. N., Knowles, P. F., McLeish, T. C. B., Pitkeathly, M. and Radford, S. E. (1997). *Nature*, **386**, 259.

25. Bustamante, C., Tinoco, I., and Maestre, M. F. (1983). *Proc. Natl. Acad. Sci. USA*, **80**, 3568.

26. Gordon, D. J. and Holzworth, G. (1971). *Arch. Biochem. Biophys.*, **142**, 481.

27. Mao, D. and Wallace, B. A. (1984). *Biochemistry*, **23**, 2667.

28. Castiglione, E. and Rodger, A. *Unpublished data*

29. Rankin, S. E., Watts, A. and Pinheiro, T. J. (1998). *Biochemistry*, **37**, 12588.

30. Egli, M., Williams, L. D., Gao, Q. and Rich, A. (1991), *Biochemistry*, **30**, 11388.

31. Stryer, L. (1988). In *Biochemistry* (3rd edn), p. 1. W. H. Freeman and Company, New York.

32. Calladine, C. R. and Drew, H. R. (1992). *Understanding DNA: The molecule and how it works*, p. 1. Academic Press, Cambridge.

33. Beveridge, D. L. and Jørgensen, W. L. (1986). *Ann. N.Y. Acad. Sci.*, 482.

34. Gessner, R. V., Fredrick, C. A., Quigley, G. J., Rich, A. and Wang A. H.-J. (1989). *J. Biol. Chem.*, **264**, 7921.

35. Zaloudek, F., Novros, J. S. and Clark, L. B. (1985). *J. Am. Chem. Soc.*, **107**, 7344.

36. Matsuoka, Y. and Nordén, B. (1994). *J. Phys. Chem.*, **86**, 1378.

37. Clark, L. B. (1994). *J. Am. Chem. Soc.*, **116**, 5265.

38. Fülscher, M. P. and Roos, B. O. (1995). *J. Am. Chem. Soc.*, **117**, 2089.

39. Holmen, A., Broo, A., Albinsson, B. and Norden, B. (1997). *J. Am. Chem. Soc.*, **119**, 12240.

40. Parkinson, A. (1998) PhD Thesis, University of Warwick.

41. Williams, A. L. Jr., Cheong, C., Tinoco, I. Jr. and Clark L. B. (1986). *Nucleic Acids Res.*, **14**, 6649.

42. Rizzo, V. and Schellman, J. A. (1984) *Biopolymers,* **23**, 435

43. Lyng, R., Rodger, A. and Nordén, B. (1992) *Chem. Phys. Lett.,* **70**, 17

44. Nordén, B. and Tjerneld, F. (1976). *Biophys. Chem.*, **4**, 191.

45. Dieber, H., Secco, F. and Venturini, M. (1987). *Biophys. Chem.*, **26**, 193.

46. Fronaeus, S. (1950). *Acta Chem. Scand.*, **4**, 72.

47. Rodger, A. (1993). In *Methods in enzymology*, Vol. 226, p. 232. Academic Press, London.

48. Scatchard, G. (1949). *Ann. N. Y. Acad. Sci.*, **51**, 660.

General CD references

- Barron, L. D. (1982). *Molecular light scattering and optical activity*. Cambridge University Press, Cambridge.

- Craig, D. P. and Thirunamachandran, T. (1984). *Molecular quantum electrodynamics: An introduction to radiation-molecule interaction*. Academic Press, London.

- Harada, N. and Nakanishi, K. (1983). *Circular dichroic spectroscopy: exciton coupling in organic stereochemistry*. University Science Books, California.

- Mason, S. F. (1982). *Molecular optical activity and the chiral discrimination*. Cambridge University Press, Cambridge.

- Michl, J. and Thulstrup, E. W. (1986). *Spectroscopy with polarized light*. VCH, New York.

- Nakanishi, K., Berova, N. and Woody, R. W. (ed.) (1994). *Circular dichroism: Principles and applications*. VCH, New York.

- Richardson, F. S. (1979). Theory of optical activity in the ligand-field transitions of chiral transition metal complexes. *Chem. Rev.*, **79**, 17.

- Rodger, A. and Nordén, B. In *Circular and linear dichroism* (ed. R. G. Compton, S. G. Davies, and J. Evans). Oxford University Press, Oxford.

Chapter 5

Quantitative determination of equilibrium binding isotherms for multiple ligand–macromolecule interactions using spectroscopic methods

Wlodzimierz Bujalowski and Maria J. Jezewska
Dept. of Human Biological Chemistry & Genetics and The Sealy Centre for
Structural Biology, The University of Texas Medical Branch at Galveston,
Medical Research Building, 301 University Boulevard, Galveston,
TX 77555–1053, USA

1 Introduction

Thermodynamic studies provide information that is necessary in order to understand the forces that drive the formation of ligand–macromolecule complexes. Knowledge of the energetics of these interactions is also indispensable for characterization of functionally important structural changes that occur within the studied complexes. Quantitative examination of the equilibrium interactions are designed to provide the answers to the questions: What is the stoichiometry of the formed complexes? How strong or how specific are the interactions? Are there any cooperative interactions among the binding sites and/or the bound ligand molecules? Are the binding sites intrinsically heterogeneous? What are the molecular forces involved in the formation of the studied complexes, or, in other words, how do the equilibrium binding and kinetic parameters depend on solution variables (temperature, pressure, pH, salt concentration, etc.)?

Equilibrium isotherms for the binding of a ligand to a macromolecule represent the relationship between the degree of ligand binding (moles of ligands bound per mole of a macromolecule) and the free ligand concentration. *A true thermodynamic binding isotherm is model-independent and reflects only this relationship.* Only then, when such an isotherm is obtained, can one proceed to extract physically meaningful interaction parameters that characterize the free energies of interaction. This is accomplished by comparing the experimental isotherms to theoretical predictions based on specific binding models that incorporate known molecular aspects, such as intrinsic binding constants, cooperativity parameters, allosteric equilibrium constants, discrete character of the binding sites or overlap of potential binding sites, etc. (see below).

Any method used to quantitatively study ligand binding to a macromolecule must relate the extent of the complex formation to the free ligand concentration in solution. Numerous techniques have been developed to study equilibrium properties of specific and non-specific ligand–macromolecule interactions in which binding is directly monitored, including equilibrium dialysis, ultrafiltration, column chromatography, filter binding assay and gel electrophoresis (1–6). These direct methods are very straightforward; however, they are usually time consuming and some, like filter binding or gel shift assays, are non-equilibrium techniques which require many controls before the reliable equilibrium binding data can be obtained. Therefore, these direct methods are usually applied to systems where the indirect spectroscopic approaches cannot be used, due to the lack of suitable signal changes accompanying the formation of the complex.

Using indirect methods, the binding of the ligand is determined by measuring the physico-chemical parameter of the macromolecule–ligand mixture, most often a spectroscopic one, e.g. absorbance, circular dichroism or fluorescence (7–32). The change in the physico-chemical parameter is then correlated with the concentration of the free and bound species (11–12, 14–29). The advantages of using spectroscopic measurements are that these can be performed without perturbing the equilibrium and are relatively easy to apply.

In using a spectroscopic signal change, which accompanies the interactions, to calculate a binding isotherm, it is often assumed that a linear relationship exists between the fractional signal change and the fractional saturation of the ligand or the macromolecule. Although this is true when one deals with equilibrium binding of only a single ligand molecule at saturation, in the general case, however, in which multiple ligands bind to the macromolecule, the observed fractional signal change and the extent of binding may not have a simple linear relationship. For instance, this may occur if there are structurally or functionally different sites on the macromolecule, each possessing a different spectroscopic signal, and/or if there are cooperative interactions whose density changes with the extent of the degree of binding and affects the spectroscopic properties of the studied system (7, 25, 26, 29). The extent of any deviation from the ideal linear behaviour is usually unknown *a priori*, hence the degree of error introduced into the isotherm and the resulting binding parameters is also unknown. In other words, if a binding isotherm is obtained by indirect methods that involve assumptions about the relationship between the observed spectroscopic signal and the extent of ligand binding, then the isotherm and the interaction parameters that one obtains will be no more accurate than the assumptions. This may cause particular problems if the isotherm is being used to differentiate between alternative models for the interaction, since if one model does or does not 'fit' the isotherm, this may be due either to the failure of the model or to the failure of the assumptions on which the calculated isotherm is based. Therefore, it is crucial to obtain a thermodynamically rigorous binding isotherm, independent of any such assumptions (11, 12, 14–29). Determination of a model-independent isotherm constitutes the first step in a correct analysis of any ligand–macromolecule binding.

In this chapter, we discuss the use of spectroscopic approaches to study multiple-ligand binding phenomena and their rigorous thermodynamic analyses to obtain model-independent binding isotherms. In these approaches, one generally monitors a spectroscopic signal (absorbance, fluorescence, circular dichroism, NMR line width or chemical shift) from either the macromolecule or the ligand that changes upon formation of a ligand–macromolecule complex; however, we will limit our discussion to quantitative analysis as applied to the use of fluorescence. This spectroscopic technique is commonly available in a biochemical laboratory, moreover, the derived relationships are general and applicable to any physico-chemical signal used to monitor the ligand–macromolecule interactions.

The determination of the correct relationship between the spectroscopic signal change and the degree of saturation of the ligand or the macromolecule is a prerequisite for obtaining a thermodynamically rigorous binding isotherm. The methods which we will discuss allow one to use a spectroscopic signal to obtain a thermodynamically rigorous binding isotherm, even when direct proportionality does not exist (14, 25, 26). The only constraint on these methods is that they are not valid if the ligand or the macromolecule undergoes an aggregation or self-assembly process within the experimental concentration range used. Therefore, as in any case, knowledge of the self-assembly properties of the ligand and the macromolecule under study are essential before a rigorous analysis of a binding phenomenon can be undertaken.

2 Thermodynamic basis of quantitative spectroscopic titrations

Two types of titrations can be performed in order to examine a binding isotherm in studies of ligand–macromolecule interactions. In one case, the macromolecule is titrated with a ligand and will be referred to as a 'normal' titration, since the total average degree of binding $\Sigma \nu_i$, (average number of moles of ligand bound per total moles of macromolecule, L_B/M_T) increases as the titration progresses (11, 14–28). Notice, that we use symbol $\Sigma \nu_i$ instead of just ν to describe the total average degree of binding, because, in the general case of a multiple ligand binding system, the bound ligand molecules can be distributed over 'i' possible different bound states, all contributing to the total average degree of binding $\Sigma \nu_i$.

In the second case, the ligand is titrated with the macromolecule. This type of titration will be referred as a 'reverse' titration, since the binding density decreases throughout the titration (24). Generally, the type of titration that is performed will depend on whether or not the signal that is monitored is from the macromolecule (normal) or the ligand (reverse). As we pointed out, the first task in examining the ligand–macromolecule interactions is to convert a spectroscopic titration curve, i.e. a change in the monitored signal, as a function of the titrant concentration, into a thermodynamically rigorous, model-independent binding isotherm, which can then be analysed, using an appropriate binding model to extract binding parameters.

The thermodynamic basis for these methods is that the total average degree of binding, Σv_i, of the ligand on a macromolecule, including all different distributions of ligands bound in possible different states 'i', is a sole function of the free ligand concentration, L_F at equilibrium (12, 33). Therefore, at a given free ligand concentration the average value of Σv_i will be the same, independent of the concentration of the macromolecule, M_T. The unique values of L_F and Σv_i and the corresponding total ligand, L_T, and total macromolecule, M_T, concentrations must satisfy the mass conservation equation

$$L_T = (\Sigma v_i)M_T + L_F \qquad\qquad 1$$

Therefore, if the set of concentrations (L_{Ti}, M_{Ti}) can be found for which Σv_i and L_F are constant, then Σv_i and L_F can be determined from the slope and intercept, respectively, of a plot of L_T versus M_T, based on *equation 1*.

2.1 The signal used to monitor ligand–macromolecule interactions originates from the macromolecule

In this case, the signal monitors the progress of the saturation of a macromolecule and a 'normal' titration (addition of a ligand to a constant macromolecule concentration) is generally performed. Any physico-chemical intensive property of the macromolecule (e.g. fluorescence intensity, fluorescence anisotropy, absorbance, circular dichroism, viscosity, etc.) can be used to monitor the binding, if this property is affected by the state of ligation of the macromolecule.

As mentioned above, for a total macromolecule concentration, M_T, the equilibrium distribution of the macromolecule among its different ligation states, M_i, is determined solely by the free ligand concentration, L_F (12, 14, 33). Therefore, at each L_F, the observed spectroscopic signal, S_{obs}, is the algebraic sum of the concentrations of the macromolecule in each state, M_i, each weighted by the value of the physico-chemical property of that state, S_i. In general, a macromolecule will have the ability to bind n ligands, hence the general equation for the observed signal, S_{obs}, of a sample containing the ligand at a total concentration, L_T, and the macromolecule at a total concentration, M_T, is given by

$$S_{obs} = S_F M_F + \Sigma S_i M_i \qquad\qquad 2$$

where S_F is the molar signal of the free macromolecule and S_i is the molar signal of the complex, M_i, which represents the macromolecule with i bound ligands ($i = 1$ to n).

The mass conservation equation, which relates M_F and M_i to M_T, is given by

$$M_T = M_F + \Sigma M_i \qquad\qquad 3$$

The partial degree of binding, v_i ('i' moles of ligand bound per mole of macromolecule), corresponding to all complexes with a given number 'i' of bound ligand molecules is given by

$$v_i = \frac{iM_i}{M_T} \qquad\qquad 4$$

Therefore, the expression for M_i, the concentration of macromolecule with 'i' ligand molecules bound is

$$M_i = \left(\frac{v_i}{i}\right)M_T \qquad\qquad 5$$

Introducing *equations 4* and *5* into *equation 3* provides a general relationship for the monitored spectroscopic signal as

$$S_{obs} = S_F M_T + \Sigma M_T (S_i - S_F)\left(\frac{v_i}{i}\right) \qquad\qquad 6$$

Subsequently, we can define the quantity ΔS_{obs} as

$$\Delta S_{obs} = \frac{(S_{obs} - S_F M_T)}{S_F M_T} \qquad\qquad 7a$$

and

$$\Delta S_{obs} = \Sigma\left(\frac{\Delta S_i}{i}\right)(v_i) \qquad\qquad 7b$$

Notice, $\Delta S_{obs} = (S_{obs} - S_F M_T)/S_F M_T$ is the experimentally determined fractional signal change observed at the total ligand and macromolecule concentrations, L_T and M_T, and $(\Delta S_i/i) = (S_i - S_F)/i$ is the average molar signal change per bound ligand in the complex containing 'i' ligands. The quantity $\Delta S_{obs} = (S_{obs} - S_F M_T)/S_F M_T$ is referred to as the Macromolecular Binding Density Function (MBDF) (12, 14, 24–26).

Since $\Delta S_i/i$ is an intrinsic molecular property of the system, *equation 7* indicates that ΔS_{obs}, is only a function of the binding density distribution, Σv_i. Therefore, the total average degree of saturation of the macromolecule and the total average degree of binding of the ligand, Σv_i, must be the same for any value of L_T and M_T for which ΔS_{obs} is constant. Thus, when you perform a spectroscopic titration of a macromolecule with a ligand, at different macromolecule concentrations, M_T, the same value of ΔS_{obs} indicates the same physical state of the macromolecule, i.e. the same degree of macromolecule saturation with the ligand and the same Σv_i. Since Σv_i is a unique function of the free ligand concentration, then the value of L_F at the same degree of saturation must also be the same. The above derivation is rigorous and independent of any binding model and, as such, can be applied to any binding system, with or without cooperative interactions and with overlapping of binding sites (14–28).

2.1.1 A practical example of quantitative analysis of spectroscopic titration curves when the signal originates from the macromolecule

We will demonstrate the application of quantitative analysis to multiple ligand binding to a macromolecule using the data from studies of the binding of the ATP non-hydrolysable analogue, TNP-ATP, to the *E. coli* replicative helicase DnaB protein (14). The DnaB helicase forms a stable hexamer built of six chemically identical subunits (18, 21). The hexamer is specifically stabilized by the presence of magnesium ions in solution (14, 18, 21). The intrinsic DnaB tryptophan fluorescence is substantially quenched upon binding TNP-ATP, hence this signal was used to monitor the interaction. Therefore, in this case, the observed fractional protein fluorescence quenching, ΔS_{obs} (MBDF) at a given TNP-ATP concentration, is given by (see *equation 7*)

$$\Delta S_{obs} = \Sigma\left(\frac{\Delta S_i}{i}\right)(\nu_i) \qquad\qquad 8a$$

$$\Delta S_{obs} = \Sigma\Delta S_{i_{max}}\,\nu_i \qquad\qquad 8b$$

where $\Delta S_{i_{max}} = \Delta S_i/i$ is the average degree of protein fluorescence quenching per bound TNP-ATP, when 'i' nucleotide molecules are bound per DnaB hexamer.

Protocol 1

Fluorescence titration of the DnaB helicase with TNP-ADP

Equipment and reagents

- Stock solution of the DnaB helicase ($\sim 5 \times 10^{-5}$ M)
- Storage buffer (50 mM Tris/HCl, pH 8.1, 300 mM NaCl, 5 mM MgCl$_2$, 25% glycerol)
- Binding buffer (50 mM Tris/HCl, pH 8.1, 100 mM NaCl, 5 mM MgCl$_2$, 10% glycerol)

- TNP-ATP
- Fluorescence cuvette with an optical path of 0.5 cm
- Spectrofluorimeter

Method

1 Dialyse the stock solution of the DnaB helicase ($\sim 5 \times 10^{-5}$ M (hexamer)) from the storage buffer (50 mM Tris/HCl, pH 8.1, 300 mM NaCl, 5 mM MgCl$_2$, 25% glycerol) to the binding buffer (50 mM Tris/HCl, pH 8.1, 100 mM NaCl, 5 mM MgCl$_2$, 10% glycerol). The final concentration of the protein stock in the binding buffer will be approximately 3×10^{-5} M (hexamer). Measure the exact DnaB concentration using the extinction coefficient at 280 nm, $\varepsilon_{280} = 185\,000$ M^{-1} cm^{-1} (hexamer) (21, 22). After dialysis, always store protein solutions on ice.

2 Prepare two stock solutions of TNP-ATP. One solution having the nucleotide concentration 1×10^{-4} M and the second solution having the concentration 1×10^{-3} M. Using two (or more) solutions of the nucleotide allows you to add accurate volumes of TNP-ATP to the protein sample and to perform titrations over a large range of nucleotide concentrations.

3 Use a fluorescence cuvette with an optical path of 0.5 cm. This will allow you to use a small volume (450 μl) of the protein solution and decrease the corrections for the inner filter effect, due to the increasing absorption (at both excitation and emission wavelengths) as a result of the increasing concentration of TNP-ATP as the titration progresses (see below).

4 Set the excitation wavelength at 300 nm (the red edge of the tryptophan absorption spectrum). Excitation at 300 nm will additionally decrease the inner filter effect. Set

the emission wavelength at 345 nm (the maximum of the protein emission spectrum).

5 Set the spectrofluorimeter in the ratio mode. The signal coming from the emission PMT is now divided by the signal from the reference PMT. This mode allows you to eliminate any artefacts resulting from fluctuations, and/or a drift, of the lamp light intensity during the titration. Add 450 μl of the binding buffer to the cuvette and adjust the signal on the emission PMT to read approximately 0.01 in the ratio mode. This means that the light intensity coming from the emission PMT is approximately 1% of the intensity read by the reference PMT.

6 Take an aliquot of 7.5 μl of the buffer from the cuvette and add a 7.5-μl aliquot of the protein stock solution (3×10^{-5} M). Mix gently with a 200 μl automatic pipette. The concentration of the protein in the cuvette is now ~5×10^{-7} M (hexamer). Wait 15 min for the temperature of the sample to equilibrate to the required temperature of the measurement, e.g. 20 °C.

7 Titrate the protein solution with TNP-ATP using aliquots of 1 to 20 μl. Wait 2 min between subsequent additions of nucleotide aliquots to allow the sample to reach equilibrium. The final volume of the added nucleotide solution at the end of the titration should not exceed 15% of the initial total volume of the protein sample.

8 In an independent titration experiment, determine the effect of the presence of TNP-ATP on the fluorescence of the buffer alone for each titration point, using the same set-up for the instrument as used in the titration with the protein. This will be your background, B.

9 For each titration point, calculate the fractional fluorescence change using *equation 7a*. For each titration point 'i' and the titration point where absorption of the sample exceeds 0.01 at the excitation and emission wavelength correct the value of the fluorescence intensity, S_i, for the dilution and inner filter effect using the formula, $S_{icor} = (S_i - B_i)(V_i/V_o)10^{0.5l(A_{i\lambda ex} + A_{i\lambda em})}$ where S_{icor} is the corrected value of the fluorescence intensity at a given point of titration i, S_i is the experimentally measured fluorescence intensity, B_i is the background, V_i is the volume of the sample at a given titration point, V_o is the initial volume of the sample, l is the total length of the optical path in the cuvette expressed in cm, $A_{i\lambda ex}$ and $A_{i\lambda em}$ are the absorbances of the sample at excitation and emission wavelengths, respectively (14, 15, 33, 34). This formula provides an accurate correction for the inner filter effect up to the absorption value of ~0.3 at both excitation and emission wavelengths.

10 Plot the fractional fluorescence quenching ΔS_{obs} (macromolecular binding density function) as a function of the logarithm of the total TNP-ATP concentration.

2.1.2 Quantitative analysis of the fluorescence titrations

Fluorescence titrations of the DnaB protein (macromolecule) with TNP-ATP (ligand), at two different DnaB hexamer concentrations, in which the values of ΔS_{obs} have been plotted as a function of the total TNP-ATP concentration (L_T) are shown in *Figure 1*. Careful inspection of the plots already indicates that the bind-

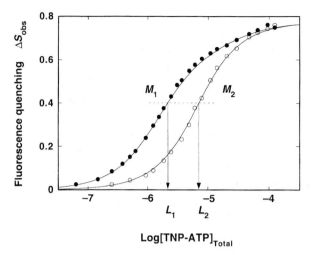

Figure 1 Fluorescence titration of the DnaB hexamer with TNP-ATP monitored by the fractional quenching of the protein fluorescence (macromolecular binding density function, MBDF) in 50 mM Tris/HCl (pH 8.1, 10 °C) containing 20 mM NaCl, 5 mM MgCl$_2$, and 10% glycerol at two different concentrations of the DnaB protein (●) $M_1 = 5.38 \times 10^{-7}$ M (hexamer); (○) $M_2 = 3.17 \times 10^{-6}$ M (hexamer). The solid lines are interpolations between data points added to separate the two data sets. The dashed horizontal line connects the points, at the same value of the MBDF, for each protein concentration. These points determine the set of total TNP-ATP concentrations, (L_1 and L_2), for which the average binding density, $\Sigma \nu_i$ and the free protein concentration, L_F, are constant and independent of the total protein concentrations, M_T. Reprinted with permission from Bujalowski, W. and Klonowska, M. M. (1993). *Biochemistry*, **32**, 5888–5900. Copyright (1993) American Chemical Society.

ing process is complex, i.e. these spectroscopic isotherms indicate the existence of two binding phases (see below). Also, depicted in *Figure 1* is the approach by which a single set of values of $\Sigma \nu_i$ and L_F can be obtained from these data. A horizontal line which intersects both curves and which defines one constant value of ΔS_{obs} is drawn. The point of intersection of this horizontal line with each titration curve defines the set of values of L_{Ti}, M_{Ti} for which L_F and $\Sigma \nu_i$ are constant, as discussed above. The example in *Figure 1* only has two titrations, hence only two sets of values of L_{Ti}, M_{Ti} are obtained ($i = 1$ and 2).

For the two titrations at macromolecular (DnaB hexamer) concentrations, M_{T1} and M_{T2} ($M_{T2} > M_{T1}$), two mass conservation equations can be written in the form of *equation 1* for each set of L_{Ti}, M_{Ti}. These two equations can then be solved for $\Sigma \nu_i$ and L_F, with the results given by

$$\Sigma \nu_i = \frac{(L_{T2} - L_{T1})}{(M_{T1} - M_{T2})} \qquad \text{9a}$$

$$L_F = L_T - \Sigma \nu_i (M_T) \qquad \text{9b}$$

In this manner, model-independent values of L_F and $\Sigma \nu_i$ can be obtained at any value of ΔS_{obs}, yielding a set of values for L_F and $\Sigma \nu_i$.

From a practical point of view, the accuracy of the determination will depend upon the used concentrations of the macromolecule and the ligand. Thus, the most accurate estimates are obtained in the region of the titration curves where the concentration of a bound ligand is comparable to its total concentration, L_T. In other words, the concentration of the bound ligand must constitute a significant fraction of the total ligand concentration in solution. In practice, this limits the accurate determination of the degree of binding, Σv_i, to the region of the titration curves where the concentration of the bound ligand is at least ~10% of the L_T. Therefore, a selection of proper concentrations of the macromolecule is crucial for obtaining L_F and Σv_i over the largest possible region of the titration curves, although the accuracy of the determination of Σv_i is mostly affected in the region of the high concentrations of the ligand approaching the maximum saturation. Such a selection of macromolecule concentrations is usually based on preliminary titrations which provide initial estimates of the expected affinity between the ligand and the macromolecule.

In the first step of the analysis of a multiple ligand binding system, the obtained values of Σv_i, corresponding to given values of ΔS_{obs}, are used to:

(1) determine the maximum stoichiometry of the macromolecule–ligand complex, and

(2) determine the relationship between the signal change, ΔS_{obs}, and the average number of bound ligand molecules, i.e. in the considered case, the average number of bound TNP-ATP molecules per the DnaB hexamer.

Both objectives can be achieved by plotting ΔS_{obs} as a function of Σv_i. The dependence of the quenching of the DnaB fluorescence upon the rigorously determined number of TNP-ATP molecules bound per hexamer is presented in *Figure 2*. The selected concentrations of the DnaB helicase allowed us to obtain an accurate (± 5 %) estimate of the average degree of binding up to 5.2 ± 0.3 TNP-ATP molecules bound. Although the maximum value of Σv_i cannot be directly determined, due to the discussed inaccuracy at the high ligand concentration region, the plateau of the fluorescence quenching ΔS_{max}, corresponding to the maximum saturation can be determined with the accuracy of ±5% (see *Figure 1*). Therefore, knowing the maximum extent of the protein fluorescence ($\Delta S_{max} = 0.78$) one can perform a short extrapolation of the plot (ΔS_{obs} versus Σv_i) to this maximum value. Such extrapolation of the data presented in *Figure 2* shows that, at saturation, the DnaB protein hexamer binds six molecules of TNP-ATP, thus, establishing the maximum stoichiometry of the DnaB–nucleotide complexes (14).

Notice, the plot of ΔS_{obs}, as a function of Σv_i, shown in *Figure 2* is strongly non-linear. The largest protein fluorescence quenching, up to ~0.21 ± 0.03, occurs upon binding the first TNP-ATP molecule. Average binding of the first three molecules causes quenching of ~0.55 ± 0.03, which corresponds to the first step in the binding isotherm (see *Figure 1*). Binding of the next three nucleotides, in the lower affinity phase, gives the maximum saturation of six TNP-ATP molecules per DnaB hexamer and is accompanied by an additional quenching of only ~0.21 ± 0.05. The dashed line represents a hypothetical

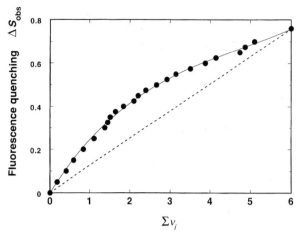

Figure 2 The dependence of the fractional quenching of the DnaB protein fluorescence upon the total average number of TNP-ATP molecules bound per DnaB hexamer in 50 mM Tris/HCl (pH 8.1, 10°C) containing 20 mM NaCl, 5 mM MgCl$_2$, and 10% glycerol (●). The average number of nucleotides bound, at a particular value of the fluorescence quenching, has been determined from the two titrations shown in *Figure 1*, using *equations 9a* and *9b*. The dashed line represents the theoretical situation where there is strict proportionality between the degree of nucleotide binding and the quenching of the DnaB protein fluorescence. The solid line is generated using the values of parameters *a*, *b* and *c* (see text) obtained from the non-linear least-square fit, based on the third degree polynomial described by *equation 15*, with intrinsic binding constant $K = 5.9 \times 10^5$ M^{-1} and cooperativity parameter, $\sigma = 0.55$. Reprinted with permission from Bujalowski, W. and Klonowska, M. M. (1993). *Biochemistry*, **32**, 5888–5900. Copyright (1993) American Chemical Society.

situation when strict proportionality between the degree of nucleotide binding and the quenching of the DnaB protein fluorescence exists. It is clear that very pronounced non-linearity exists between the extent of the protein fluorescence quenching and the degree of nucleotide binding, with the first three nucleotides binding with higher affinity and causing 70% of the total change of the protein fluorescence. Therefore, an incorrect isotherm would have been calculated if a linear relationship had been assumed for this system (14).

In principle, as shown in *Figures 1* and *2*, only two titration curves are needed to obtain values of Σv_i and L_F that cover the large range of the degree of binding. This is possible for the data in *Figure 1*, since the signal change (tryptophan fluorescence quenching) for the DnaB–TNP-ATP interaction is large and the affinity under the conditions used is fairly high. In this case, the data can be analysed using *equations 9a* and *9b*. However, if necessary, more accurate values of Σv_i and L_F can be obtained if more than two titrations are performed and the data are graphically analysed. For a case in which 'n' titrations are performed, then 'n' sets of values of L_{Ti}, M_{Ti} (i = 1 to n) will be obtained, one for each titration curve that is intersected by each horizontal line. A plot of L_T versus M_T can then be constructed, which will result in a straight line, according to *equation 1*, from which the values of L_F and Σv_i can be obtained from the intercept and the slope. If a plot of L_T versus M_T, determined in this manner, is not linear, this may indicate one or more inconsistent sets of titration data or aggregation phenomena associated with the ligand or the macromolecule, and these data should be viewed with caution.

2.1.3 Application of a binding model to analyse the binding isotherm

The analysis performed so far is completely model-independent, i.e. no specific mechanism of the nucleotide binding to the DnaB hexamer has been assumed. Once this analysis has been performed, one can postulate, on the bases of the physical properties of the system, a specific binding model which describes the binding isotherms and allows the experimenter to obtain intrinsic affinities and parameters characterizing the cooperative interactions (if any) within the system.

Briefly, the data show that binding of nucleotides to the DnaB hexamer occurs in two phases. Three nucleotides bind in the high affinity phase and the next three nucleotides bind in the low affinity step. The simplest explanation for this behaviour is that the DnaB hexamer, built of six chemically identical subunits, exhibits negative cooperativity between binding sites (14, 22). The reader should consult the original papers for a full discussion of this aspect of the nucleotide binding studies to the DnaB hexamer (14, 15, 22).

Binding of a ligand to a protein which forms a homohexamer in two steps, each including the same number of bound nucleotides (three), can be described in the simplest way by a hexagon model (14–16, 21). In this model, the ligand can bind to any of six initially equivalent binding sites with the same intrinsic binding constant K. However, the cooperative interactions, characterized by a parameter σ, are limited to only two neighbouring sites. If we assign the quantity $x = KL_F$, where L_F is the free ligand concentration, then the partition function for the hexagon model, Z, is described by (14, 16, 22)

$$Z = 1 + 6x + 3(3 + 2\sigma)x^2 + 2(1 + 6\sigma + 3\sigma^2)x^3 + 3(3\sigma^2 + 2\sigma^3)x^4 + 6\sigma^4 x^5 + \sigma^6 x^6 \qquad 10$$

The total average degree of binding $\Sigma\nu_i$ is defined by the standard statistical thermodynamic formula (14, 33)

$$\Sigma\nu_i = \frac{\partial \ln Z}{\partial \ln L_F} \qquad 11$$

which provides the expression for $\Sigma\nu_i$ as

$$\Sigma\nu_i = [6x + 6(3 + 2\sigma)x^2 + 6(1 + 6\sigma + 3\sigma^2)x^3 + 12(3\sigma^2 + 2\sigma^3)x^4 + 30\sigma^4 x^5 + 6\sigma^6 x^6]/Z \qquad 12$$

Because the rigorous thermodynamic approach described above allowed us to obtain both $\Sigma\nu_i$ and L_F, we can use *equation 12* to directly obtain K and σ, from the thermodynamic isotherm ($\Sigma\nu_i$ as a function of L_F), and to perform the optimization of the binding parameters, K and σ, using a non-linear least-square fit. *Figure 3(a)* shows the dependence of $\Sigma\nu_i$ upon the logarithm of $[TNP-ATP]_{Free}$.

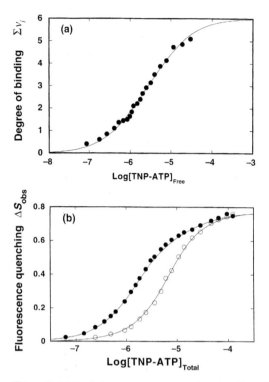

Figure 3. (a) The total average number of nucleotide molecules bound per DnaB hexamer, Σv_i, as a function of the free concentration of TNP-ATP in 50 mM Tris/HCl (pH 8.1, 10°C) containing 20 mM NaCl, 5 mM MgCl$_2$ and 10% glycerol (●). The solid line is a non-linear least-square fit, according to the hexagon model (*equations 10–12*) which provides the intrinsic binding constant $K = 5.9 \times 10^5$ M^{-1} and cooperativity parameter, $\sigma = 0.55$ (data from 14). (b) Fluorescence titrations of the DnaB hexamer with TNP-ATP monitored by the fractional quenching of the protein fluorescence (macromolecular binding density function, MBDF) in 50 mM Tris/HCl (pH 8.1, 10°C) containing 20 mM NaCl, 5 mM MgCl$_2$ and 10% glycerol at two different concentrations of the DnaB protein: (●) 5.38×10^{-7} M; (○) 3.17×10^{-6} M (hexamer). The solid lines are computer fits of the experimental binding isotherms, according to the hexagon model, obtained by the empirical function approach. First, the fractional protein fluorescence change (MBDF) has been calculated using *equation 15*, and using the values of $a = 0.286$, $b = -0.0484$ and $c = 3.609 \times 10^{-3}$. Subsequently, the fluorescence titration curves were subjected to the non-linear fit which provided the set of interaction parameters, intrinsic binding constant, $K = 5.9 \times 10^5$ M^{-1}, and cooperativity parameter, $\sigma = 0.55$. Reprinted with permission from Bujalowski, W. and Klonowska, M. M. (1993). *Biochemistry*, **32**, 5888–5900. Copyright (1993) American Chemical Society.

Notice, the isotherm is more scattered than the original titration curve, and the isotherm covers ~75% of the total binding curve due to the inaccuracy of determining the degree of binding in the high concentration range of the nucleotide. The solid line is a non-linear fit of the binding isotherm which provides $K = 5.9 \times 10^5$ M^{-1} and $\sigma = 0.55$.

In practice, higher accuracy in estimating the binding parameters could be obtained if one analyses the entire original fluorescence titration curves (*Figure 1*), without being limited to the region where the degree of binding can be accurately estimated. A general method which allows such an analysis of the entire fluorescence titration curves, for this very complex binding system, is discussed in the next section.

2.1.4 Empirical function approach

In order to analyse the original fluorescence titration curve, all ΔS_{imax} molecular parameters in *equation 8b* must be known. In the simplest case, all ΔS_{imax} parameters could be the same, i.e. binding to each site would be characterized by the same quenching parameter, ΔS_{max}. In such a case, the dependence of ΔS_{obs}, as a function of $\Sigma \nu_i$, would be strictly linear (dashed line in *Figure 2*) and *equation 8b* would reduce to

$$\Delta S_{obs} = \Delta S_{max} (\Sigma \nu_i) \qquad\qquad 13$$

If ΔS_{max} is known, *equation 13* can be directly used to fit the fluorescence titration curves (see *Figure 1*) and to extract K and σ. As we discussed above, obviously, this situation does not apply in this case. Sometimes, it is also possible to determine all optical parameters in the analytical formula relating ΔS_{obs} to the molecular parameters ΔS_{imax}, as defined by *equation 8b* (35, 36). However, in the general case, this is not necessary, and in the case of the DnaB protein, which has six nucleotide binding sites and possibly multiple ΔS_{imax} for a given number of bound nucleotide molecules, it is practically impossible. For instance, in the case of the hexagon model for the DnaB–nucleotide system, the protein, with a given number of nucleotide molecules bound, can exist in multiple configurations, e.g. the hexamer with three ligands bound has 20 possible configurations (see *equation 10*). Although some of these configurations are physically indistinguishable, resulting from simple statistical effects of binding three molecules to the initially independent six sites, there are configurations which differ by the number of cooperative interactions, and they may also differ by the values of the individual molecular quenching constants ΔS_{imax}. As an example, *equation 14* describes the minimum analytical relationship between the observed experimental quenching, ΔS_{obs}, the individual molar quenching parameters, ΔS_{imax}, and the interaction parameters, K and σ for the hexagon model (14)

$$\Delta S_{obs} = (6\Delta S_1 x + 6(3\Delta S_{21} + 2\sigma\Delta S_{22})x^2 + 6(\Delta S_{31} + 6\sigma\Delta S_{32} + 3\sigma^2\Delta S_{33})x^3 + \\ 12(3\sigma^2\Delta S_{41} + 2\sigma^3\Delta S_{42})x^4 + 30\sigma^4\Delta S_5 x^5 + 6\sigma^6\Delta S_6 x^6)/Z_H \qquad 14$$

For clarity, subscript 'max' has been omitted at individual molar quenching constants ΔS_{imax}.

Equation 14 contains 12 independent parameters, two interactions and 10 optical parameters. For instance, there are three physically distinguishable configurations of the hexamer with three ligands bound which have different densities of the cooperative interactions; therefore, there are three possible different molar quenching constants, ΔS_{31}, ΔS_{32} and ΔS_{33}, characterizing each of these configurations. As we mentioned above, in order to apply *equation 14* to obtain interaction parameters, K and σ from a single fluorescence titration curve, all ten optical constants, ΔS_i, must be known. It should be pointed out that, at the saturating concentration of the ligand, the observed experimental maximum quenching, is $\Delta S_{obs} = \Delta Q_6$.

The problem of finding all optical parameters can be avoided by using the following empirical function approach. This approach can be used for any ligand–macromolecule systems, where the determination of all optical constants in an analytical equation is practically impossible. An empirical function, usually polynomial, is found, which relates the experimentally determined dependence of the average quenching, ΔS_{obs} upon $\Sigma \nu_i$, as shown in *Figure 2*. This can be achieved by using a non-linear least-square fit of ΔS_{obs} as a function of $\Sigma \nu_i$. In the case of the DnaB–nucleotide interactions, a minimum third degree polynomial is necessary to describe this function as defined by (14)

$$\Delta S_{obs} = a(\Sigma \nu_i) + b(\Sigma \nu_i)^2 + c(\Sigma \nu_i)^3 \qquad \qquad 15$$

where a, b and c are fitting constants. This function is then used to generate a theoretical isotherm for a binding model and to extract the true binding parameters from the experimentally obtained single fluorescence isotherm. To generate the theoretical isotherm, the value of the degree of binding, $\Sigma \nu_i$, is calculated using *equation 12* for a given free nucleotide concentration and initial estimates of the intrinsic binding constant, K, and cooperativity parameter, σ. The obtained $\Sigma \nu_i$ is then introduced into *equation 15* and the fluorescence quenching, corresponding to this value of the degree of binding, is calculated. These calculations are performed for the entire fluorescence titration curve. Using *equation 15*, we performed the fit of the empirical plot of ΔS_{obs} as a function of $\Sigma \nu_i$ shown in *Figure 2* (solid line) which provided parameters $a = 0.286$, $b = -0.0484$ and $c = 3.609 \times 10^{-3}$. These parameters define the empirical dependence of ΔS_{obs} upon $\Sigma \nu_i$. Subsequently, *equation 15* with the obtained a, b and c, combined with *equation 12*, which defines the selected model of the nucleotide binding to the DnaB hexamer, is used in the non-linear fit of the original fluorescence titration curves with only two fitted parameters, K and σ. The solid lines in *Figure 3(b)* are the non-linear least-square fits of the entire fluorescence titration curves, according to the procedure described above.

2.2 Signal used to monitor the interactions originates from the ligand

In this case, some spectroscopic property (e.g. ligand fluorescence intensity), S, of the ligand changes upon binding to the macromolecule, hence the signal monitors the apparent degree of saturation of the ligand and a 'reverse' titration (addition of a macromolecule to a constant ligand concentration) is generally performed (12, 24).

Consider the general case of equilibrium, multiple-ligand binding (total concentration, L_T) to a macromolecule (total concentration, M_T) where there can be 'i' states of the bound ligand, with each state possessing a different molar signal, S_i. The observed signal, S_{obs}, from the ligand solution, in the presence of the macromolecule, has contributions from the free ligand and the ligand bound to the macromolecule in any of its 'i' possible bound states and can be expressed by

$$\Delta S_{obs} = S_F L_F + \Sigma S_i L_i \qquad 16$$

where S_F and L_F are the molar signal and concentration of the free ligand, respectively, and S_i and L_i are the molar signal and concentration of the ligand bound in state 'i', respectively.

Equation 16 is valid when the molar signal of each species is independent of concentration (i.e. in the absence of ligand and macromolecule aggregation). The concentrations of the free and bound ligand are related to the total ligand concentration by the conservation of mass equation

$$L_T = L_F + \Sigma L_i \qquad 17a$$

where

$$L_i = v_i M_T \qquad 17b$$

where v_i is the partial degree of ligand binding for the ith state. Substituting equations 17a and 17b into equation 16 and rearranging provides

$$\Delta S_{obs} - S_F L_T = M_T \Sigma (S_i - S_F) v_i \qquad 18$$

Notice, $S_F L_T$ is the initial signal from the ligand before titration with the macromolecule. Dividing both sides by $S_F L_T$ and next multiplying by (L_T/M_T) yields

$$\left[\frac{(S_{obs} - S_F L_T)}{S_F L_T}\right]\left(\frac{L_T}{M_T}\right) = \Sigma\left[\frac{(S_i - S_F)}{S_F}\right]v_i \qquad 19$$

which can be rewritten as

$$\frac{(S_{obs} - S_F L_T)}{S_F L_T}\left(\frac{L_T}{M_T}\right) = \Sigma(\Delta S)_i v_i \qquad 20a$$

and

$$\Delta S_{obs}\left(\frac{L_T}{M_T}\right) = \Sigma(\Delta S)_i v_i \qquad 20b$$

It should be pointed out that $\Delta S_{obs} = (S_{obs} - S_F L_T)/S_F L_T$ is the experimentally observed fractional change in the signal from the ligand (with respect to the signal of the total ligand free, $S_F L_T$) in the presence of the ligand and the macromolecule at total concentrations, L_T and M_T, and $(\Delta S)_i = (S_i - S_F)/S_F$ is the molar signal change from the ligand when it is bound in state 'i'. If the fluorescence intensity of the ligand is being monitored, then ΔS_{obs} corresponds to the fractional change of the ligand fluorescence. Equation 20 is general and independent of the spectroscopic method used to monitor the interactions.

Equation 20b indicates that the quantity $\Delta S_{obs}(L_T/M_T)$ is equal to $\Sigma(\Delta S)_i v_i$, the sum of the partial degrees of binding for all 'i' states of the ligand–macromolecule system, weighted by the intrinsic signal change for each bound state. The weighting factor, $(\Delta S)_i$, is a molecular quantity which is constant for a particular binding state 'i', under a given set of experimental conditions (temperature, buffer etc.). Therefore, the quantity, $\Sigma(\Delta S)_i v_i$, is constant for a given distribution of the degree of binding among different possible states, Σv_i.

At equilibrium, the values of L_F and $\Sigma \nu_i$ are constant for a given value of $\Delta S_{obs}(L_T/M_T)$, independent of the macromolecule concentration, M_T. Hence, under identical solution conditions, one can obtain thermodynamically rigorous measurements of $\Sigma \nu_i$ and L_F from plots of $\Delta S_{obs}(L_T/M_T)$ versus M_T for two or more titrations performed at different total ligand concentrations. This is accomplished by obtaining the set of concentrations (L_{Ti}, M_{Ti}), from each titration, for which the quantity $\Delta S_{obs}(L_T/M_T)$ is constant, and solving for $\Sigma \nu_i$ and L_F. The procedure is therefore analogous to the case in which a signal from the macromolecule is monitored during the titration. The quantity, $\Delta S_{obs}(L_T/M_T)$, is referred to as the Ligand Binding Density Function (LBDF) (11, 24).

2.2.1 A practical example of quantitative analysis of spectroscopic titration curves when the signal used to monitor the interactions originates from the ligand

As an example of the analysis when the signal from the ligand is monitored, we discuss a study of the interaction of a fluorescent nucleotide analogue, εADP (ligand), with the E. coli DnaB protein (macromolecule), where, instead of the protein fluorescence quenching, the fluorescence of the analogue has been used to monitor the binding (24). The analogue has its adenine modified with chloroacetaldehyde to provide a fluorescent etheno-derivative (37–39). As we discussed above, the DnaB hexamer binds six nucleotide molecules and the binding is characterized by the negative cooperativity. The fluorescence of the εADP is increased by ~21% upon binding to the DnaB helicase (24). To increase this signal change and, in turn, to obtain a higher resolution of the titration experiments, acrylamide has been added to the solution. Acrylamide is a very efficient dynamic quencher of the etheno-derivative of adenosine in solution (15, 24, 39). This extra dynamic quenching process (40, 41), which does not affect the thermodynamics of the nucleotide–enzyme interactions, is much less efficient for the nucleotide bound to the DnaB protein than for the free nucleotide, leading to a much larger change in the nucleotide fluorescence upon formation of the complex with the protein, thus, increasing the resolution of the titration curves. The application of the differential dynamic quenching of the ligand or the macromolecule fluorescence to increase the resolution of the binding experiments is thoroughly discussed in (24).

Protocol 2

Fluorescence titration of εADP with the DnaB helicase

Equipment and reagents

- Stock solution of the DnaB helicase (~7 × 10^{-5} M (hexamer))
- Storage buffer (50 mM Tris/HCl, pH 8.1, 300 mM NaCl, 5 mM MgCl$_2$, 25% glycerol
- Fluorescence cuvette with an optical path of 0.5 cm
- Binding buffer (50 mM Tris/HCl, pH 8.1, 100 mM NaCl, 5 mM MgCl$_2$, 100 mM acrylamide, 10% glycerol)
- εADP
- Spectrofluorimeter

Method

1 Dialyse the stock solution of the DnaB helicase ($\sim$7 $\times$ 10^{-5} M (hexamer)) from the storage buffer (50 mM Tris/HCl, pH 8.1, 300 mM NaCl, 5 mM MgCl$_2$, 25% glycerol) to the binding buffer (50 mM Tris/HCl, pH 8.1, 100 mM NaCl, 5 mM MgCl$_2$, 100 mM acrylamide, 10% glycerol). The final concentration of the protein stock in the binding buffer will be approximately 5 $\times$ 10^{-5} M (hexamer). Measure the exact DnaB concentration using the extinction coefficient at 280 nm, $\varepsilon_{280} = 185\ 000$ M^{-1} cm^{-1} (hexamer) (21). After dialysis, always store protein solutions on ice.

2 Prepare two stock solutions of the protein. One solution having the DnaB concentration 1 $\times$ 10^{-6} M and the second solution having the concentration $\sim$5 $\times$ 10^{-5} M. Using two solutions of the helicase allows you to add accurate volumes of the protein to the nucleotide sample and to perform titrations over a large range of the protein concentration.

3 Prepare stock solutions of εADP, from 3 $\times$ 10^{-5} to 3 $\times$ 10^{-4} M. Determine the nucleotide concentration using the absorption coefficient at 294 nm, $\varepsilon_{294} = 2900$ M^{-1} cm^{-1} (37, 38). Using different stock solutions will allow you to change the concentration of the εADP in the sample by adding the same volume of different nucleotide stock solutions.

4 Use a fluorescence cuvette with an optical path of 0.5 cm. This will allow you to use a small volume (450 μl) of the fluorescent nucleotide solution.

5 Set the excitation wavelength at 325 nm (the red edge of the εADP absorption spectrum). Excitation at 325 nm, far from the protein absorption maximum, will eliminate, to a great extent, the correction for the background of the protein fluorescence. Set the emission wavelength at 410 nm (the maximum of the εADP emission spectrum).

6 Set the spectrofluorimeter in the ratio mode. The signal coming from the emission PMT is now divided by the signal from the reference PMT. This mode allows you to eliminate the artefacts resulting from any fluctuations of the lamp and the possible, although slow, drift of the lamp intensity during the titration. Add 450 μl of the binding buffer to the cuvette and adjust the signal on the emission PMT to read approximately 0.01 in the ratio mode. This means that the intensity measured by the emission PMT is approximately 1% of the intensity read by the reference PMT.

7 Set the polarizer in the excitation channel at the vertical position and the polarizer in the emission channel at 54.7° (magic angle). Performing titrations using 'magic angle' conditions will eliminate the artefacts in the fluorescence intensity measurements, due to the possible changes in the fluorescence anisotropy of the sample. These artefacts, which commonly occur, result from the different sensitivities of the emission PMT for both the vertically and horizontally polarized light. Although this may not be a problem when the fluorescence intensity of a large DnaB hexamer is studied, it could induce artefacts when the emission of the relatively small εADP (affected by the binding to the large DnaB hexamer) is examined, as in this case (32, 42).

Protocol 2 continued

8 Take an aliquot of 20 μl of the buffer from the cuvette and add a 20-μl aliquot of the εADP stock solution (e.g. 1×10^{-4} M). Mix gently with a 200 μl automatic micropipette. The concentration of εADP in the cuvette is now $\sim 4.5 \times 10^{-6}$ M. Wait 15 min for the temperature to equilibrate to the temperature of the measurement, e.g. 20 °C.

9 Titrate the εADP solution with the DnaB protein using aliquots of 1 to 20 μl. The final volume of the added protein solution at the end of the titration should not exceed 15% of the total volume of the nucleotide sample.

10 In an independent titration, titrate the buffer with the DnaB protein solution and determine the contribution of the protein fluorescence to the observed fluorescence of the εADP sample for each titration point. At high protein concentrations such a contribution will become significant, in spite of the fact that the excitation is set at 325 nm (far from the protein absorption spectrum), due to the impurities in the protein sample and the discrete band pass of the excitation and emission monochromator. This is your background, B.

11 For each titration point calculate the fractional fluorescence increase, ΔS_{obs}, using the formula $\Delta S_{obs} = (S_{obs} - S_F L_T)/S_F L_T$. For each titration point 'i' correct the value of the fluorescence intensity, S_i, for the dilution and background using the formula, $S_{icor} = (S_i - B_i)(V_i/V_o)$ where S_{icor} is the corrected value of the fluorescence intensity at a given point of titration 'i', S_i is the experimentally measured fluorescence intensity, B_i is the background, V_i is the volume of the sample at a given titration point, V_o is the initial volume of the sample.

12 Plot the fractional ligand (εADP) fluorescence increase, ΔS_{obs}, as a function of the logarithm of the total DnaB protein concentration.

2.2.2 Quantitative analysis of the fluorescence titrations

A series of fluorescence titration curves of εADP with the DnaB protein, in 50 mM Tris/HCl (pH 8.1, 10 °C) containing 5 mM MgCl$_2$, 100 mM NaCl and 10% glycerol, at different nucleotide concentrations and in the presence of 100 mM acrylamide, is shown in *Figure 4*. At higher nucleotide concentrations, the curves are shifted toward higher concentrations of the DnaB helicase as more enzyme is necessary to saturate the increased amount of εADP in the sample. All curves reach the same plateau of the relative fluorescence increase, ΔF, at saturating concentrations of the DnaB helicase. As we pointed out above, in general, the fractional change of the ligand fluorescence upon the macromolecule concentration does not necessarily strictly correspond to the fractional ligand saturation. This is never *a priori* known for any multiple ligand binding system. However, the estimate of the degree of binding and the free ligand concentrations can be obtained by using the LBDF approach.

Figure 5 shows the plot of $\Delta S_{obs}(L_T/M_T)$ as a function of the DnaB protein concentration, obtained using the fluorescence titrations presented in *Figure 4*. For the different total concentrations of εADP, at the same value of the binding

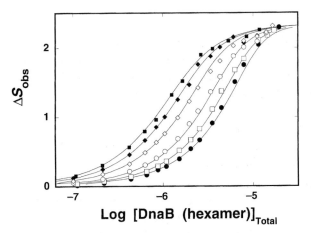

Figure 4 Fluorescence titrations ('reverse titrations') of εADP, at different concentrations of the nucleotide, with the DnaB helicase in 50 mM Tris/HCl (pH 8.1, 10 °C) containing 100 mM NaCl, 5 mM MgCl$_2$ and 100 mM acrylamide (λ_{ex} = 325 nm, λ_{em} = 410 nm). The concentrations of the nucleotide are: (■) 4 × 10^{-6} M; (◆) 6 × 10^{-6} M; (◇) 1 × 10^{-5} M; (○) 2 × 10^{-5} M; (□) 3 × 10^{-5}; (●) 4 × 10^{-5} M. The solid lines are computer fits of the experimental binding isotherms, according to the hexagon model, using a single set of binding parameters, K = 4 × 10^5 M^{-1}, σ = 0.4 and ΔS_{max} = 2.35. Reprinted from Jezewska, M. J. and Bujalowski, W. (1997) 'Quantitative analysis of ligand–macromolecule interactions using differential dynamic quenching of the ligand fluorescence to monitor the binding.' *Biophysical Chemistry*, **64**, 253–269. Copyright (1997) with permission from Elsevier Science.

density function, the total degree of binding, $\Sigma\nu_i$, and the free εADP concentrations, [εADP]$_{Free}$, must be the same, thus, allowing for their determination (see *equation 20*). As mentioned above, even though all of the 'reverse' titrations shown in *Figure 4* span the same full range of εADP fluorescence increase, they do not span the same range of the degree of binding as seen from the LBDF plot in *Figure 5*. As a result, multiple titrations at different values of L_T are required to span the full binding density range. A horizontal line, which intersects the LBDF curves is drawn, defining a constant value of the LBDF, $\Delta S_{obs}(L_T/M_T)$ (dashed line in *Figure 5*). The points of intersection of the horizontal line with each binding density function curve determine the set of values (M_{Ti}, L_{Ti}) for which L_F and $\Sigma\nu_i$ are constant, as shown in *Figure 5* for one constant value of the LBDF. Based on *equation 1*, the average degree of binding, $\Sigma\nu_i$, and L_F, can then be determined from the slope and intercept of a plot of L_T versus M_T at each constant value of $\Delta S_{obs}(L_T/M_T)$. By repeating this procedure for a series of horizontal lines that span the range of values of $\Delta S_{obs}(L_T/M_T)$ as a function of L_F, the values of $\Sigma\nu_i$ can be obtained and a binding isotherm can be constructed (11, 24).

In the construction of a series of LBDF plots, as shown in *Figure 5*, one should cover as wide a range of ligand concentrations as possible; however, care should be taken to avoid large changes in the total ligand concentration between two successive titrations, since this may bias the determination of L_F and $\Sigma\nu_i$. In our experience, six to eight titrations, using successive total ligand concentrations that differ by a factor of 1.5–2, will generate an accurate set of data.

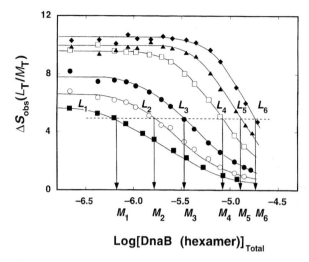

Figure 5 Dependence of the ligand binding density function (LBDF), $\Delta S_{obs}(L_T/M_T)$, on the logarithm of the total DnaB protein concentration, at different total concentrations of the εADP (L_T) in 50 mM Tris/HCl (pH 8.1, 10 °C) containing 100 mM NaCl, 5 mM MgCl$_2$ and 100 mM acrylamide: (■) L_1 = 4 × 10^{-6} M; (○) L_2 = 6 × 10^{-6} M; (●) L_3 = 1 × 10^{-5} M; (□) L_4 = 2 × 10^{-5} M; (▲) L_5 = 3 × 10^{-5} M; (◆) L_6 = 4 × 10^{-5} M. The solid lines are interpolations between the data points which separate different data sets and have no theoretical basis. The horizontal dashed line connects points at the same value of the LBDF, at different total εADP concentrations (L_1, L_2, L_3, L_4, L_5 and L_6) and total DnaB protein concentrations (M_1, M_2, M_3, M_4, M_5 and M_6), at which [εADP]$_{Free}$ and the total degree of nucleotide binding, $\Sigma\nu_i$, on the DnaB hexamer are the same. Reprinted from, Jezewska, M. J. and Bujalowski, W. (1997) 'Quantitative analysis of ligand–macromolecule interactions using differential dynamic quenching of the ligand fluorescence to monitor the binding.' *Biophysical Chemistry*, **64**, 253–269. Copyright (1997) with permission from Elsevier Science.

The model-independent binding isotherm constructed from the full analysis of the data in *Figures* 4 and 5, for the binding of εADP to the *E. coli* DnaB protein, is plotted in *Figure* 6. The solid line is a theoretical isotherm, according to the hexagon model, constructed using the intrinsic binding constant $K = 4 \times 10^5$ M^{-1} and the cooperativity parameter $\sigma = 0.4$ (*equation 12*). These values of K and σ are, within experimental accuracy, the same as the values which have been independently obtained using the rigorous fluorescence titration method in which the quenching of the protein fluorescence has been used to monitor the binding (14, 15, 22).

2.2.3 Correlation between the fractional signal change, ΔS_{obs}, and the average degree of binding, $\Sigma\nu_i$

Measurements of the average degree of binding, $\Sigma\nu_i$, as a function of the free ligand concentration, enable one to determine the relationship between the average signal change and the fraction of the bound ligand. This is not necessary in order to obtain a binding isotherm, as discussed above (*Figure* 6); however, if ΔS_{obs} is found to be directly proportional to the fraction of the

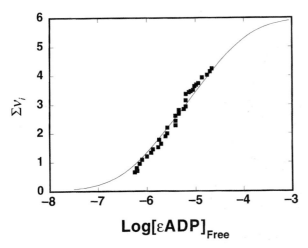

Figure 6 The dependence of the total average degree of binding of εADP on the DnaB helicase hexamer, as a function of the logarithm of the free nucleotide concentrations [εADP]$_{Free}$ in 50 mM Tris/HCl (pH 8.1, 10 °C) containing 100 mM NaCl, 5 mM MgCl$_2$ and 100 mM acrylamide. The solid line is the non-linear least-square fit of the binding isotherm, according to the hexagon model (*equations 10–12*), which provides intrinsic binding constant $K = 4 \times 10^5$ M^{-1} and cooperativity parameter $\sigma = 0.4$. Reprinted from Jezewska, M. J. and Bujalowski, W. (1997) 'Quantitative analysis of ligand–macromolecule interactions using differential dynamic quenching of the ligand fluorescence to monitor the binding.' *Biophysical Chemistry*, **64**, 253–269. Copyright (1997) with permission from Elsevier Science.

bound ligand (L_B/L_T), then binding isotherms can be constructed with much greater ease from a titration at a single ligand concentration (see below). The dependence of the fractional fluorescence increase of ΔADP as a function of the fraction of nucleotide molecules bound to the DnaB protein, $L_B/L_T = \Sigma \nu_i (M_T/L_T)$, is shown in *Figure 7*. The value of L_B/L_T has been determined using the binding density function plots shown in *Figure 5*.

It is clear that, in the studied binding of εADP to the DnaB hexamer using the fluorescence of εADP to monitor the binding, there is a linear correspondence between the relative increase of the nucleotide fluorescence, ΔS, and the fraction of the ligand bound, L_B/L_T, in the examined range of the degree of binding. However, it is important to check the relationship between the observed fluorescence change (or any signal used to monitor the binding) and L_B/L_T over a wide range of binding densities, since the signal change can, in general, be dependent upon the degree of binding (see *Figure 2*). In the case of the *E. coli* DnaB protein–εADP interactions, this direct proportionality holds for the degree of binding up to ~5.3 (24). Therefore, under these conditions, the fractional fluorescence increase of εADP is equal to the fraction of the bound nucleotide, i.e. $\Delta S_{obs}/\Delta S_{max} = L_B/L_T$. When ΔS_{obs} is directly proportional to L_B/L_T, one can determine the maximum extent of protein fluorescence quenching, ΔS_{max}, from a linear extrapolation of a plot of ΔS_{obs} versus L_B/L_T to $L_B/L_T = 1$, as shown in *Figure 7*. This short extrapolation (dashed line) to $L_B/L_T = 1$ gives the

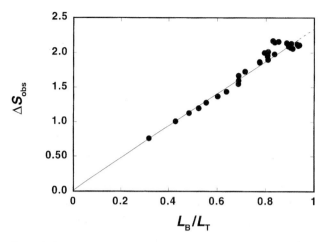

Figure 7 Relationship between the relative fluorescence increase and the fractional saturation of the nucleotide for εADP binding to the DnaB helicase in 50 mM Tris/HCl (pH 8.1, 10 °C) containing 100 mM NaCl, 5 mM MgCl$_2$ and 100 mM acrylamide. The selected concentration of εADP is 2×10^{-5} M. The concentration of the bound nucleotide has been calculated from [εADP]$_\text{Bound}$ = $(\Sigma \nu_i)$[DnaB]$_\text{Total}$, where the total average degree of binding of εADP on the DnaB protein, $\Sigma \nu_i$, has been determined by using the LBDF approach described in the text. Reprinted from Jezewska, M. J. and Bujalowski, W. (1997) 'Quantitative analysis of ligand–macromolecule interactions using differential dynamic quenching of the ligand fluorescence to monitor the binding.' *Biophysical Chemistry*, **64**, 253–269. Copyright (1997) with permission from Elsevier Science.

value of the maximum relative increase, $\Delta S_\text{max} = 2.35$, of the εADP fluorescence upon saturation with the DnaB protein in studied solution conditions (24).

2.2.4 Generation of binding isotherms from a single titration when $\Delta S_\text{obs}/\Delta S_\text{max} = L_B/L_T$

The LBDF analysis allows one to rigorously determine a model-independent binding isotherm and to determine the relationship between the ΔS_obs and the fraction of the bound ligand, L_B/L_T. The LBDF method is time consuming, since six to eight titrations are required to construct a single, precise binding isotherm over a wide range of binding densities, which is necessary if the relationship between ΔS_obs and L_B/L_T is not known *a priori*. However, if it is determined from the LBDF analysis that a linear relationship exists between ΔS_obs and L_B/L_T over a wide range of degree of binding (as in the case of the εADP–DnaB protein system discussed in the previous section), then one can use this relationship to determine the average degree of binding and the free ligand concentration from a single titration curve (17, 19, 22–24). For such a simple case, *equation 20* reduces to

$$\frac{\Delta S_\text{obs}}{\Delta S_\text{max}} = \frac{L_B}{L_T} \qquad \text{21a}$$

and it follows that

$$L_F = \left(\frac{\Delta S_{max} - \Delta S_{obs}}{\Delta S_{max}}\right) L_T \qquad\qquad 21b$$

and

$$\Sigma \, \nu_i = \left(\frac{\Delta S_{obs}}{\Delta S_{max}}\right)\left(\frac{L_T}{M_T}\right) \qquad\qquad 21c$$

Thus, a single titration can be used to obtain $\Sigma \nu_i$ as a function of L_F. However, we stress that one should not simply assume that the fractional signal change is equal to the fraction of the bound ligand, since, if it is not true, this could lead to significant errors. On the other hand, if direct proportionality does not exist between the signal change and the fraction of the bound ligand over a wide range of binding densities, the true binding isotherm can still be constructed without any assumptions by using the LBDF analysis (see *Figure 6*).

3 Summary

Understanding macromolecular interactions, such as those involving ligands and macromolecules, requires detailed knowledge of the energetics and kinetics of the formed complexes. Spectroscopic methods are widely used in characterizing the energetics (thermodynamics) and kinetics of ligand–macromolecule interactions in solution. These methods are very convenient to use, do not require large quantities of material, and, most importantly, do not perturb the studied processes. However, spectroscopic methods are indirect, i.e., the interactions are measured through monitoring changes of some physico-chemical parameter (e.g. fluorescence, absorbance, circular dichroism) accompanying the studied complex formation. In such studies, it is indispensable to determine the relationship between the observed signal and the degree of binding in order to obtain thermodynamically meaningful interaction parameters. The approaches presented in this chapter describe general quantitative methods of the analyses of macromolecular binding through spectroscopic measurements which allow an experimenter to determine the thermodynamically rigorous degree of binding or the degree of macromolecule saturation and the free ligand concentration. The method enables one to obtain the total degree of binding, $\Sigma \nu_i$, and the free ligand concentration, L_F, hence, to construct the entire *model-independent* binding isotherm. Only when the thermodynamically rigorous isotherm is obtained, can it be analysed by using the thermodynamic models which incorporate the known molecular aspects of the ligand–macromolecule interactions, like cooperativity, allosteric conformational changes, overlap of potential binding sites, etc.

We have illustrated the application of the methods to situations where the signal used to monitor the interactions originates from a macromolecule or a ligand, using as examples the *E. coli* replicative helicase DnaB protein interactions with nucleotide cofactors (14, 24).

The methods have been discussed as applied to studying ligand–macromolecule interactions using the fluorescence intensity to monitor the binding. However, these approaches can generally be applied to any ligand–macromolecule system by monitoring the binding and using any spectroscopic signal originating from a ligand or a macromolecule, thus, allowing the experimenter to construct model-independent binding isotherms.

Acknowledgements

We thank Gloria Drennan Davis for help in preparing the manuscript.

References

1. Riggs, A. D., Bourgeois, S., and Cohn, M. (1970). *J. Mol. Biol.*, **53,** 401.
2. Jensen, D. E. and von Hippel, P. H. (1977). *Anal. Biochem.*, **80,** 267.
3. Draper, D. E. and von Hippel, P. H. (1978) *J. Mol. Biol.*, **122,** 321.
4. Garner, M. M. and Revzin, A. (1981). *Nucl. Acids Res.*, **9,** 3047.
5. Fried, M. and Crothers, D. M. (1981). *Nucl. Acids Res.*, **9,** 6505.
6. Cassel, J. M. and Steinhardt, J. (1969). *Biochemistry*, **8,** 2603.
7. Holbrook, J. J. (1973). *Biochem. J.,* **128,** 921.
8. Boschelli, F. (1982). *J. Mol. Biol.*, **162,** 267.
9. Kelly, R. C., Jensen, D. E., and von Hippel, P. H. (1976). *J. Biol. Chem.*, **251,** 7240.
10. Porschke, D. and Rauh, H. (1983). *Biochemistry*, **22,** 4737.
11. Bujalowski, W. and Lohman, T. M. (1987). *Biochemistry*, **26,** 3099.
12. Lohman, T. M. and Bujalowski, W. (1991). *Methods in Enzymology*, **208,** 258.
13. Heyduk, T. and Lee, J. C. (1990). *Proc. Natl. Acad. Sci.*, **87,** 1744.
14. Bujalowski, W. and Klonowska, M. M. (1993). *Biochemistry*, **32,** 5888.
15. Bujalowski, W. and Klonowska, M. M. (1994). *Biochemistry*, **33,** 4682.
16. Bujalowski, W. and Klonowska, M. M. (1994). *J. Biol. Chem.*, **269,** 31359.
17. Bujalowski, W. and Jezewska, M. J. (1995). *Biochemistry*, **34,** 8513.
18. Jezewska, M. J. and Bujalowski, W. (1996). *J. Biol. Chem.*, **271,** 4261.
19. Jezewska, M. J., Kim, U.-S., and Bujalowski, W. (1996). *Biochemistry*, **35,** 2129.
20. Bujalowski, W. and Porschke, D. (1988). *Biophysical Chem.*, **30,** 151.
21. Bujalowski, W., Klonowska, M. M., and Jezewska, M. J. (1994). *J. Biol. Chem.*, **269,** 31350.
22. Jezewska, M. J., Kim, U.-S., and Bujalowski, W. (1996). *Biophysical Journal*, **71,** 2075.
23. Jezewska, M. J. and Bujalowski, W. (1996). *Biochemistry,* **35,** 2117.
24. Jezewska, M. J. and Bujalowski, W. (1997). *Biophysical Chem.*, **64,** 253.
25. Jezewska, M. J., Rajendran, S., and Bujalowski, W. (1997). *Biochemistry,* **36,** 10320.
26. Jezewska, M. J., Rajendran, S., and Bujalowski, W. (1998). *Biochemistry*, **37,** 3116.
27. Jezewska, M. J., Rajendran, S., and Bujalowski, W. (1998). *J. Biol. Chem.,* **273,** 9058.
28. Jezewska, M. J., Rajendran, S., Bujalowska, D., and Bujalowski, W. (1998). J. *Biol. Chem.,* **273,** 10515.
29. Holbrook, J. J. and Gutfreund, H. (1973) *FEBS Lett.*, **31,** 157.
30. Witting, P., Norden, B., Kim, S. K., and Takahashi, M., (1994). *J. Biol. Chem.*, **269,** 5700.
31. Halfman, C. J. and Nishida, T. (1972) *Biochemistry*, **11,** 3493
32. Lakowicz, J. R. (1983). *Principles of fluorescence spectroscopy*. Plenum Press, New York.
33. Hill, T. L. (1985). *Cooperativity theory in biochemistry*. Springer-Verlag, New York.
34. Parker, C. A. (1968). *Photoluminescence of solutions*. Elsevier, Amsterdam.
35. Bujalowski, W. and Lohman, T. M. (1989). *J. Mol. Biol.*, **207,** 249.

36. Bujalowski, W. and Lohman, T. M. (1989). *J. Mol. Biol.*, **207,** 268.
37. Secrist, J. A., Bario, J. R., Leonard, N. J., and Weber, G. (1972). *Biochemistry* **11,** 3499.
38. Leonard, N. J. (1984). *Crit. Rev. Biochem.,* **15,** 125.
39. Ando, T. and Asai, H. (1980). *J. Biochem.*, **88,** 255.
40. Eftink, M. R. and Ghiron, C. A. (1981). *Anal. Biochem.*, **114,** 199.
41. Eftink, M. R. (1991). *Biophysical and biochemical aspects of fluorescence spectroscopy*, Ch. 1 (ed. G. Dewey). Plenum Press, New York, London.
42. Azumi, T. and McGlynn, S. P. (1962). *J. Chem. Phys.*, **37,** 2413.

Chapter 6
Steady-state kinetics

Athel Cornish-Bowden

Institut Fédératif 'Biologie Structurale et Microbiologie', CNRS-BIP,
31 chemin Joseph-Aiguier, B.P. 71, 13402 Marseille Cedex 20, France

1 Introduction to rate equations, first-order, second-order reactions etc.

All of chemical kinetics is based on rate equations, but this is especially true of steady-state enzyme kinetics: in other applications a rate equation can be regarded as a differential equation that has to be integrated to give the function of real interest, whereas in steady-state enzyme kinetics it is used as it stands. Although the early enzymologists tried to follow the usual chemical practice of deriving equations that describe the state of reaction as a function of time there were too many complications, such as loss of enzyme activity, effects of accumulating product etc., for this to be a fruitful approach. Rapid progress only became possible when Michaelis and Menten (1) realized that most of the complications could be removed by extrapolating back to zero time and regarding the measured initial rate as the primary observation.

Since then, of course, accumulating knowledge has made it possible to study time courses directly, and this has led to two additional subdisciplines of enzyme kinetics, *transient-state kinetics*, which deals with the time regime before a steady state is established, and *progress-curve analysis*, which deals with the slow approach to equilibrium during the steady-state phase. The former of these has achieved great importance but is regarded as more specialized. It is dealt with in later chapters of this book. Progress-curve analysis has never recovered the importance that it had at the beginning of the twentieth century.

Nearly all steps that form parts of the mechanisms of enzyme-catalysed reactions involve reactions of a single molecule, in which case they typically follow *first-order kinetics*:

$$v = ka \qquad\qquad 1$$

or they involve two molecules (usually but not necessarily different from one another) and typically follow *second-order kinetics*:

$$v = kab \qquad\qquad 2$$

In both cases v represents the rate of reaction, and a and b are the concentrations of the molecules involved, and k is a *rate constant*. Because we shall be regarding the rate as a quantity in its own right it is not usual in steady-state kinetics to represent it as a derivative such as $-da/dt$.

A third case that is useful to consider in catalysed reactions is the *zero-order reaction*, in which v apparently does not depend on any concentration:

$$v = k \qquad\qquad 3$$

Although this behaviour is apparent rather than real it is useful to consider it because some of the constants in enzyme kinetics are zero-order rate constants.

In each of the equations above the k is an *overall rate constant*. However, it is often convenient to consider the rate with respect to one reactant, the concentrations of the others being maintained constant. For example, if the second-order reaction above is considered in relation to a only, b being regarded as a constant:

$$v = kb \cdot a \qquad\qquad 4$$

then it is evident that the rate is proportional to a, and it is said to be *first-order with respect to a*. The 'constant' kb, which is only constant of course if b is constant, is called a *pseudo-first-order rate constant*.

Although the individual steps in an enzyme mechanism may all be either first-order or second-order reactions, the mechanism normally includes several steps, with the result that the complete reaction rarely has a simple order. The fundamental problem, therefore, is to set up conditions in which the rate is easily measurable and then to determine how it varies with the conditions.

2 Units

All of the quantities required in steady-state kinetics are either concentrations, normally measured in mol litre^{-1} or a submultiple, rates, measured in mol litre^{-1} s^{-1}, or rate constants, with units that vary according to the type of rate constant: a first-order rate constant has dimensions of reciprocal time, and is typically measured therefore in s^{-1}, and a second-order rate constant has dimensions of reciprocal time multiplied by reciprocal concentration, and is typically measured therefore in mol^{-1} litre s^{-1}. It is obvious from elementary dimensional considerations that a pseudo-first-order rate constant has the same dimensions and units as a first-order rate constant, and that constants of different order cannot be meaningfully compared.

By far the most common practice is to regard the rate as an intensive quantity, i.e. to normalize it to a standard volume, so for any given conditions the rate is the same whether it refers to a 0.5 ml cuvette or to a 100 litre fermenter. However, in some specialized applications it is more usual to define the rate as a rate of conversion, i.e. as an extensive quantity measured in mol s^{-1}. If this is done, then the dimensions and units of all the rate constants must change correspondingly. If all the processes occur in a single compartment, there is not usually any reason to use extensive quantities, but it becomes much more convenient, if not essential, when one is concerned with transport between compartments of different volumes. These applications will not be considered in this chapter.

3 Basic assumptions in steady-state kinetics

The essential requirement in a steady-state experiment is to set up conditions such that the rate is essentially constant during a period convenient for measuring it (typically a period of several minutes), or that it decreases sufficiently slowly for an accurate estimate of its value at zero time to be possible. There are several experimental designs that will in principle produce such conditions, but only one of them is in widespread use, and that is to work with very low enzyme concentrations (typically much lower than those that exist in the cell), and in particular to use a concentration much smaller than (no more than 1% of) those of the substrate(s) and any other small molecules that may be present.

Although the need for low enzyme concentrations is primarily derived from the need to create a steady state, it has an obvious economic advantage that ensures that steady-state methods will continue to be widely used even though transient-state methods may in principle generate much more information. Transient-state methods typically involve direct observation of the enzyme or its complexes with the reactants, and they require enzyme concentrations high enough for direct observation, and consume large amounts of enzyme in each experiment. To avoid wasting materials, therefore, steady-state experiments are often used for preliminary exploration even when the ultimate aim is to study the transient state.

Initial-rate conditions also require that there is no reverse reaction (at least in the extrapolated case at zero time). This can be achieved without making any assumptions about the value of the equilibrium constant by studying conditions in which no products are present at zero time. This requirement does not exclude studying the effects of products on the reaction rate as long as there is more than one product (provided that at least one product is missing from the reaction mixture the rate of the reverse reaction will still be zero at zero time). It becomes more complicated for reactions with only a single product, but these are not common and will not be considered in this chapter.

If the essential experimental requirement of a very low enzyme concentration is fulfilled, a reaction can be studied over a period in which the reactant concentrations change so little that they can be regarded as constants, so there is no need to distinguish between their initial and instantaneous values. In the case of a reaction involving the conversion of a single substrate A, the equation that describes the steady-state behaviour is the Michaelis–Menten equation:

$$v = \frac{k_0 e_0 a}{K_m + a} = \frac{V a}{K_m + a} \qquad\qquad 5$$

Here e_0 is the total enzyme concentration, i.e. the total of all forms of enzyme including free enzyme and any intermediates in the reaction, and a is the substrate concentration, and the parameters are k_0, the *catalytic constant* (or *turn-*

over number), and K_m, the *Michaelis constant*. The form shown in the middle is more fundamental than that on the right, but the second form, in which $k_0 e_0$ is written as the *limiting rate V*, is often used because the enzyme concentration is not always known in meaningful units. V is the limit that v approaches as the enzyme becomes *saturated*, i.e. when a becomes very large, and K_m is the value of a at which $v = 0.5V$, i.e. at which the rate is *half-maximal*. Notice that k_0 is a first-order rate constant, K_m is a concentration and V is a zero-order rate constant, or in other words a rate. The ratio k_0/K_m is a second-order rate constant, and is called the *specificity constant* and given the symbol k_A: it is a more fundamental constant than K_m in the analysis of enzyme mechanisms (i.e. it has a simpler mechanistic meaning), but the equation is usually written in terms of V (or k_0) and K_m nonetheless.

One may derive *equation 5* from the following model,

$$A + E \underset{k_{-1}}{\overset{k_1}{\rightleftharpoons}} EA \underset{k_{-2}}{\overset{k_2}{\rightleftharpoons}} EP \overset{k_3}{\rightarrow} E + P \qquad 6$$

which assumes that the reaction passes through an enzyme–substrate complex, EA, that undergoes catalytic transformation to an enzyme–product complex, EP, which then breaks down to form products. A simpler mechanism containing only one intermediate also generates the Michaelis–Menten equation, but this is too simple to be realistic. Conversely, there are infinitely many more complex mechanisms that also generate it, so although the three-step mechanism provides a useful basis for discussion one can never be certain from steady-state measurements alone that the true mechanism is not more complex.

The Michaelis–Menten parameters can be defined as follows in terms of the rate constants shown in the above reaction:

$$k_0 = \frac{k_2 k_3}{k_{-2} + k_2 + k_3} \qquad 7$$

$$k_A = \frac{k_1 k_2 k_3}{k_{-1} k_{-2} + k_{-1} k_3 + k_2 k_3} \qquad 8$$

$$K_m = \frac{k_{-1} k_{-2} + k_{-1} k_3 + k_2 k_3}{k_1 (k_{-2} + k_2 + k_3)} \qquad 9$$

Note that none of the three parameters has a simple transparent meaning. The interpretations commonly attributed to them depend on additional simplifying assumptions that are not always correct. For example, K_m is often said to be equal to the equilibrium constant k_{-1}/k_1 for dissociation of A from EA, but the expression in *equation 9* does not take this form unless k_2 is very small, and it is not safe to regard K_m as a measure of the equilibrium dissociation constant.

4 Measurement of specific activity

According to *equation 5* the measured initial rate of reaction is proportional to the total concentration of enzyme. However, in practice, this is something that needs to be established by experiment rather than just assumed to be true, because if the enzyme becomes inactivated when it is diluted, or if it exists in multiple states of aggregation with different activities, or if the enzyme solution is contaminated with an inhibitor, then the simple expectation may not be realized.

Let us consider the last of these to illustrate the sort of problems that can arise. If a concentration e_0 of enzyme implies a concentration αe_0 of a competitive inhibitor then the rate equation is not *equation 5* but *equation 15* (below), with the inhibitor concentration i written as αe_0. Notice that this means that e_0 appears not only as a factor of the numerator of the rate expression, but also as a term in the denominator, so the whole expression is no longer proportional to e_0. This makes the rate expression as a function of e_0 have the same form that it has as a function of the substrate concentration a, so it approaches a limit of $k_0 K_{ic}/K_m \alpha$ instead of increasing indefinitely as e_0 increases. Clearly this limit depends on all four of the quantities that it contains, but only one of them, α, is (in principle) under the control of the experimenter. If α cannot be made small enough to be negligible then use of the assay may need to be restricted to a low range of enzyme concentrations where the effect of the inhibitor is imperceptible.

Once conditions have been determined that give a good proportionality over a useful range, these can be used as the basis of an activity assay for the enzyme. As noted above, the rate is an intensive quantity, i.e. its magnitude does not vary with the total size of the system considered. The quantity required from an assay, however, is the amount of enzyme, which is an extensive quantity because it does vary with the size of the system. It can be obtained by multiplying the rate obtained by the volume of the compartment in which the reaction takes place. The unit for measuring enzyme activity is mol s^{-1}, as the rate of conversion of a particular substrate under specified conditions. For example, a reaction rate of 5 μM s^{-1}, or 5 μmol litre^{-1} s^{-1}, obtained in a 0.5 ml reaction volume would correspond to an enzyme activity of $5 \times 0.5 \times 10^{-3} = 0.0025$ μmol s^{-1}.

Notice that the definition of enzyme activity implies that all of the conditions, including the identity of the substrate, are specified. Thus a change of rate observed when a different substrate is used in the assay does not imply a change in enzyme activity (though if the change of substrate is adopted as part of the definition of the standard assay, it will imply a change in the definition of enzyme activity).

The SI unit of enzyme activity, 1 mol s^{-1}, may also be called the katal, with symbol kat, but this term has not been widely adopted. It is a very large unit for most purposes, but its submultiples the nanokatal (nkat) and microkatal (μkat) are of a convenient size for measuring enzyme activities in the laboratory. In the past it was usual to define a 'unit' of enzyme activity measured in μmol min^{-1} (= 16.67 nkat), and this is still used, though it is not very satisfactory because the name is uninformative and fails to convey the incorporation of a factor of 10^{-6} and a non-SI unit (the minute) in the definition.

5 Graphical determination of K_m and V

Equation 5 defines a hyperbolic dependence of rate on substrate concentration, and in principle one could plot v against a, determine V from the asymptotic rate at high substrate concentrations, and once this is known K_m could be found as the value of a at which $v = 0.5V$. In practice, this is a bad method (and is little used) because it is very difficult to decide accurately where the asymptote ought to be drawn. It is nearly always drawn too close to the curve, with under-estimation of V. This is because the curve in fact approaches the asymptote very slowly: even at a concentration of $10K_m$ the rate is still only 91% of the limit. The usual practice, therefore, is to transform *equation 5* into one of the following three forms, which underlie the three straight-line plots illustrated in *Figures 1–3*:

$$\frac{1}{v} = \frac{1}{V} + \frac{K_m}{V}\cdot\frac{1}{a} \qquad\qquad 10$$

$$\frac{a}{v} = \frac{K_m}{V} + \frac{1}{V}a \qquad\qquad 11$$

$$v = V - K_m \cdot \frac{v}{a} \qquad\qquad 12$$

The *double-reciprocal plot*, often attributed to Lineweaver and Burk (2), is illustrated in *Figure 1* and is based on *equation 10*; it is the mostly widely used straight-line plot, but it is also the least satisfactory, because it distorts the effect of experimental error to such an extent that it is difficult to form any visual judgement of where the best line should be drawn. Either of the other two plots is better. The plot of a/v against a (*equation 11*) is associated with Woolf (3) or Hanes (4) and is illustrated in *Figure 2*; notice that the definitions of the slope and intercept on the ordinate are the opposite way round from those that apply to the double-reciprocal plot. The plot of v against v/a (*equation 12*) is variously attributed to Eadie (5) or to Hofstee (6) and is illustrated in *Figure 3*. It has the particular advantage that the entire observable range of v values, from 0 to V, is

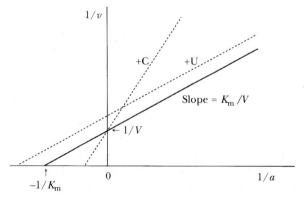

Figure 1 The double-reciprocal plot. The dashed lines labelled +C and +U illustrate how the line is displaced by the presence of a competitive or uncompetitive inhibitor, respectively.

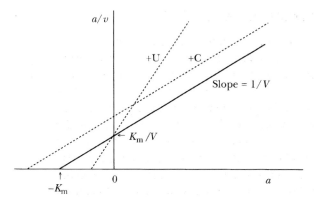

Figure 2 The plot of a/v against a. Notice that the definitions of the slope and the intercept on the ordinate are the opposite way round from those in *Figure 1*. The dashed lines labelled +C and +U illustrate how the line is displaced by the presence of a competitive or uncompetitive inhibitor, respectively.

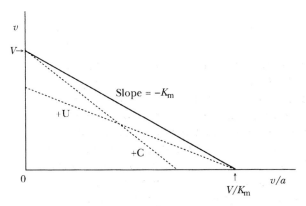

Figure 3 The plot of v against v/a. Notice that this plot maps the entire range of observable rates (from 0 to V) onto a finite range of paper. The dashed lines labelled +C and +U illustrate how the line is displaced by the presence of a competitive or uncompetitive inhibitor, respectively.

plotted in a finite range; this makes it easy to judge by eye if an experiment has been well designed. On the other hand, it has the disadvantage that v, normally the less reliable measurement, contributes to both coordinates, and errors in v cause deviations along lines through the origin rather than parallel with one or other axis.

These plots are valuable for illustration purposes, but for actual estimation of kinetic parameters it is more objective to use appropriate computer programs. For this purpose it is not sufficient just to apply unweighted linear regression to the straight-line plots, as this suffers from the same statistical distortions as the plots themselves. Full treatment would require more space than is available here, but may be found elsewhere (7). The following two equations for calculating best-fit values of K_m and V give satisfactory results if the standard deviation of v is approximately constant when expressed as a percentage of the true value, as is usually at least approximately correct:

$$K_m = \frac{\Sigma v^2 \Sigma(v/a) - \Sigma(v^2/a)\Sigma v}{\Sigma(v^2/a^2)\Sigma v - \Sigma(v^2/a)\Sigma(v/a)} \qquad 13$$

$$V = \frac{\Sigma(v^2/a^2)\Sigma v^2 - [\Sigma(v^2/a)]^2}{\Sigma(v^2/a^2)\Sigma v - \Sigma(v^2/a)\Sigma(v/a)} \qquad 14$$

6 Inhibition of enzyme activity

The simplest kind of enzyme inhibition is *competitive inhibition,* characterized by a rate equation of the following form:

$$v = \frac{k_0 e_0 a}{K_m(1 + i/K_{ic}) + a} \qquad 15$$

in which i is the concentration of the inhibitor and K_{ic} is the *competitive inhibition constant.* (The qualification 'competitive' and the second subscript c are usually omitted if only this simplest kind of inhibition is being considered.) Competitive inhibition arises when a molecule binds reversibly to the free enzyme in such a way as to prevent the substrate from binding. Although this occurs most obviously when the inhibitor is similar enough to the substrate in structure to replace it in the substrate binding site, but not similar enough to undergo reaction, competitive inhibition can also arise in more complex ways.

The opposite extreme from competitive inhibition is *uncompetitive inhibition,* and is defined by the presence of a factor that affects the variable term in the denominator of the Michaelis–Menten equation instead of the constant term:

$$v = \frac{k_0 e_0 a}{K_m + a(1 + i/K_{iu})} \qquad 16$$

The inhibition constant K_{iu} is an *uncompetitive inhibition constant.* This type of inhibition is not common, but it is important as a component of *mixed inhibition,* when both competitive and uncompetitive effects occur simultaneously:

$$v = \frac{k_0 e_0 a}{K_m (1 + i/K_{ic}) + a(1 + i/K_{iu})} \qquad 17$$

There is no particular reason for the two inhibition constants K_{ic} and K_{iu} to be equal, and most of the mechanisms for mixed inhibition suggest that they ought to be different. However, the case where $K_{ic} = K_{iu}$ is often given an undeserved prominence in discussions of inhibition, largely because experiments done many years ago suggested that it was a more common phenomenon than it is. It is called *non-competitive inhibition* and its rate equation is the same as *equation 17,* but with both K_{ic} and K_{iu} written simply as K_i.

All of these kinds of inhibition are conveniently discussed in terms of *apparent Michaelis–Menten parameters,* i.e. the parameters that replace the ordinary parameters in the Michaelis–Menten equation when an inhibitor is present. In the general case, *equation 17*, these are as follows:

$$k_0^{\mathrm{app}} = \frac{k_0}{1 + i/K_{\mathrm{iu}}} \qquad\qquad 18$$

$$k_A^{\mathrm{app}} = \frac{k_A}{1 + i/K_{\mathrm{ic}}} \qquad\qquad 19$$

$$K_m^{\mathrm{app}} = \frac{K_m \,(1 + i/K_{\mathrm{ic}})}{1 + i/K_{\mathrm{iu}}} \qquad\qquad 20$$

Note that the first two expressions have the same form, and both simplify to independence of i in the event that one or other inhibition term is negligible. The expression for the apparent value of K_m is more complicated, especially when one considers how it varies with the different types of inhibition: it increases with the concentration of a competitive inhibitor, it decreases as the concentration of an uncompetitive inhibitor increases, it may change in either direction as the concentration of a mixed inhibitor increases, or it is independent of inhibitor concentration if the inhibition is non-competitive. In general it is simplest to regard k_A as the parameter affected by competitive inhibition, negligibly so when the competitive component is negligible; k_0 as the parameter affected by uncompetitive inhibition, negligibly so when the uncompetitive component is negligible; and K_m just as the ratio of the two, i.e. $K_m = k_0/k_A$.

The effects of the different kinds of inhibition on the common plots are illustrated in *Figures 1–3* and follow naturally from *equations 18–20*. Any competitive effect alters the apparent value of k_A: hence it increases the slope of the plot of $1/v$ against $1/a$ (*Figure 1*), it increases the ordinate intercept of the plot of a/v against a (*Figure 2*), and it decreases the abscissa intercept of the plot of v against v/a (*Figure 3*). Conversely, any uncompetitive effect increases the ordinate intercept of the plot of $1/v$ against $1/a$, increases the slope of the plot of a/v against a, and decreases the ordinate intercept of the plot of v against v/a. When both components of the inhibition are present both kinds of effects occur.

7 Specificity

Specificity is a fundamental property of enzymes, but it is often assessed in an unsatisfactory way, by comparing the kinetic parameters for different reactions measured in isolation from one another. In the cell, specificity must clearly refer to the capacity of an enzyme to react selectively with one substrate when others are present simultaneously, and any satisfactory measure of specificity should take account of this (8). The equation for reaction of one substrate A in the presence of a competing substrate A' follows a form similar to that for competitive inhibition, *equation 15*:

$$v = \frac{k_0 e_0 a}{K_m(1 + a'/K'_m) + a} \qquad\qquad 21$$

with the inhibitor concentration replaced by a', the concentration of A', and the inhibition constant by K'_m, the Michaelis constant for the reaction of A' considered in isolation. The rate of reaction of A' is given by the same equation with an obvious transposition of symbols:

$$v = \frac{k'_0 e_0 a'}{K'_m(1 + a/K_m) + a'} \qquad 22$$

Dividing one equation by the other, the ratio of rates is the ratio of substrate concentrations multiplied by the ratio of specificity constants:

$$\frac{v}{v'} = \frac{k_0/K_m}{k'_0/K'_m} \cdot \frac{a}{a'} = \frac{k_A a}{k'_A a'} \qquad 23$$

This result explains the choice of the name specificity constant. *Equation 23* applies at all concentrations, and the specificity constant measures specificity at all concentrations. This point is worth emphasizing, because at very low substrate concentrations the rate of an enzyme-catalysed reaction follows second-order kinetics approximately, and the specificity constant is the second-order rate constant in these conditions. As this property was well known long before the relation to specificity was generally recognized it has sometimes generated an incorrect idea that there is a similar concentration restriction for specificity.

8 Activators

Activation is the opposite effect to inhibition, and an activator is, naturally enough, a species that increases the rate of an enzyme-catalysed reaction. In practice, however, activation has been studied much less than inhibition, and this section will be correspondingly brief. The most important point to make is that various other phenomena are also sometimes called activation, and these should not be confused with true activation. For example, certain enzymes, such as the digestive enzyme pepsin, are initially synthesized as inactive 'zymogens', pepsinogen in this case, which are transformed into active enzyme by a post-translational partial proteolysis, a process that is sometimes called zymogen activation. Another case is really a confusion: for some enzymes, most notably ATP-dependent kinases, the true substrate is a complex ion containing a metal, typically $MgATP^{2-}$, and such enzymes are sometimes said to be activated by Mg^{2+} ions.

True activation is, as noted, the converse of inhibition, and can be classified and analysed in similar ways. In general, whenever a term of the form i/K_i occurs in an inhibition equation one can conceive of a corresponding activation type where a term K_x/x defines the effect of an activator X. It is important to realize, however, that the activation analogue of competitive inhibition cannot be explained by any mechanism in which it makes sense to talk about competition between substrate and activator, and thus a term like 'competitive activation' makes no sense even though it is the inverse of competitive inhibition. Such activation may occur, for example, if the substrate binds only to the enzyme–

activator complex and not to the free enzyme, so a more meaningful term might be compulsory activation.

9 Environmental effects on enzyme activity

As well as inhibitors and activators, enzymes are affected by various environmental effects, of which the most important are pH and temperature.

9.1 pH

Mechanistically, most pH effects are inhibition and activation by one particular inhibitor and activator, the proton, and the equations that describe them are the same as those for inhibition and activation. There are, however, some important differences. First, the fact that the proton concentration can be varied and controlled over a very wide range makes it possible to observe and measure properties that would be difficult to observe if they occurred with other species. Second, unlike most inhibitors and activators the proton usually displays both properties, acting simultaneously as an inhibitor and an activator. This gives rise to the familiar bell-shaped curve illustrated in *Figure 4*, which describes the response of an enzyme to variation in the pH:

$$k = \frac{\tilde{k}}{1 + \dfrac{h}{K_1} + \dfrac{K_2}{h}} \qquad 24$$

In this equation k represents some kinetic constant for the reaction, such as the catalytic constant or the specificity constant, $\tilde{k}$ is a notional pH-independent

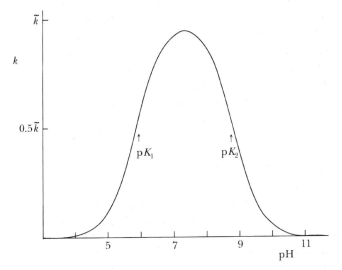

Figure 4 Bell-shaped pH profile. The parameter k plotted in the ordinate represents either the catalytic constant k_0 or the specificity constant k_0/K_m, and the limit $\tilde{k}$ is the value it would have if all of the enzyme existed in the catalytically active state of protonation.

form of the same constant that defines the value it would have if all the enzyme could be fixed in the active state of protonation, h is the hydrogen-ion concentration and K_1 and K_2 are the acid dissociation constants for two successive deprotonation steps. This is, of course, the simplest possible simultaneous activation and inhibition, and much more complex cases also occur. It should be obvious from inspection of the equation that k approaches zero at both low pH, where h/K_1 is large, and at high pH, where K_2/h is large, but is larger at intermediate pH values. If K_1 and K_2 are well separated it may approximate to $\tilde{k}$ in this intermediate region, but may be much less.

We may expect both the free enzyme and all intermediate complexes that occur in an enzyme-catalysed reaction to exist in multiple states of protonation, and so we can expect any parameter of the reaction to vary with pH. In general the pH dependence of the specificity constant k_A or k_0/K_m provides information about the protonation states of the free enzyme, and that of the catalytic constant k_0, provides information about the protonation states of the enzyme–substrate complex(es). The pH dependence of K_m is usually more complicated, as it involves contributions from all of the different forms of enzyme; it is most easily analysed by regarding K_m as $k_0/(k_0/K_m)$.

9.2 Temperature

Temperature effects on enzymes follow the same principles as those that affect simple chemical reactions, based on the Arrhenius treatment (9), but in practice they are very complicated. This is because all enzyme mechanisms involve multiple steps, and all of the rate constants describing the different steps vary separately with temperature. It is unlikely, therefore, that useful information will emerge from studies of temperature dependence unless the experimenters are willing to go rather deeply into the theory. If one is working with an enzyme for which the mechanism is known in considerable detail, and for which the rate constants of individual steps can be measured, then the temperature dependence of any of these steps can be analysed just as if it were obtained in a chemical kinetic experiment. Apart from this, temperature dependencies are most likely to be useful when a break (abrupt change in slope) in an Arrhenius plot coincides with a known physical phenomenon; for example, if one is studying a lipid substrate that is known to undergo a phase transition at a particular temperature, then a break in the Arrhenius plot at that temperature may be helpful for identifying which form of the lipid reacts with the enzyme.

It is important to realize, however, that most of the abrupt changes in slope that are reported are not real. The reasons for this are primarily psychological: once a line or set of lines have been drawn over the experimental points in a graph it is easy to be convinced that they fit the points rather better than they do in reality, and because of this it is not difficult to 'fit' the data in an Arrhenius plot so that the change in slope occurs in a place that is easy to interpret. Consider the data in *Figure 5*. Is the break at (a) 9 °C or (b) 17.5 °C? Or is there in reality no break at all but a gradual change of slope over the whole range plotted, as in (c)?

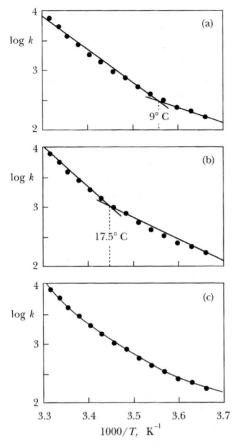

Figure 5 Dependence of a rate constant k on temperature (Arrhenius plot). In all three cases the same points are shown, but in (a) and (b) they are interpreted as lying on pairs of straight lines and in (c) as lying on a smooth curve.

10 Cooperativity

All of the cases considered so far can be regarded as generalizations of the Michaelis–Menten equation, (*equation* 5). However, although many enzymes do behave in this way, at least as a first approximation, there are some important exceptions. It is simple to calculate from *equation* 5 that if $a = K_m/9$ then $v = 0.1V$ and if $a = 9K_m$ then $v = 0.9V$; in other words, spanning the 10–90% range of available rates requires an 81-fold range of substrate concentrations, almost two orders of magnitude. Similar calculations may be done with any of the equations of the Michaelis–Menten type, whether for additional substrates, inhibitors or activators, and they imply that as long as enzymes follow Michaelis–Menten kinetics, their rates vary very little with the sort of changes in substrate and other concentrations that are likely to occur in the cell. As effective regulation of metabolism requires that at least some enzymes respond sensitively to changes in the concentrations of effectors, it is not surprising that Michaelis-Menten kinetics is not universal and that some enzymes display *cooperativity*.

The most obvious indication of cooperativity is that the plots expected to give straight lines give curves, and in principle one can determine the degree of cooperativity by analysing the degree of curvature, but it is easier and more usual to use the *Hill plot* (10). This exists in various forms, depending on whether we are studying cooperativity with respect to a substrate, an inhibitor or an activator. As cooperative inhibition is one of the most important cases in practice, it is used in *Figure 6* to illustrate the plot, which is then a plot of $\log[v_i/(v_0 - v_i)]$ against $\log i$, where v_0 is the uninhibited rate and v_i is the rate at a concentration i of inhibitor. If *equation 15* applies, i.e. if there is no cooperativity, then it is a simple matter to show that

$$\log [v_i/(v_0 - v_i)] = \log [K_{ic}(1 + a/K_m)] - \log i \qquad 25$$

i.e. that the slope of the Hill plot ought to be -1. The slope is negative because this is an example of inhibition; the equivalent plots for effects of substrates or activators are positive, and in all cases the slope, ignoring the sign, is called the *Hill coefficient*.

As *equation 25* illustrates, a Hill coefficient of 1 is characteristic of Michaelis-Menten kinetics, or *non-cooperative kinetics*. A Hill coefficient greater than 1 indicates that the rate is more sensitive to changes in concentration than expected from the simple model. Values in the range 1.5 to 4 are quite commonly reported. Values greater than 4 are very rare, probably because of the difficulty in evolving an enzyme structure with sufficiently tight coupling between different sites to generate a higher degree of cooperativity. Values between 1 and 1.5 are also uncommon, either because the relatively weak

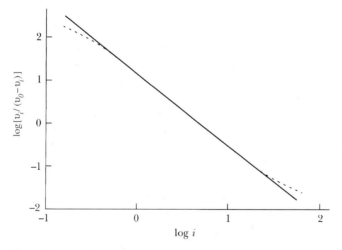

Figure 6 Hill plot for inhibition. As noted by the dashed lines at the two extremes, any Hill plot should in principle approach a slope of unity at the extremes, even though in practice the curvature may often be hard to detect and the points may appear to fit a straight line over the whole experimental range.

degree of cooperativity often passes unnoticed by the experimenter or possibly because it offers no particular physiological function. Hill coefficients less than 1 also occur, indicating a rate that is less sensitive to changes in concentration than expected from the simpler model, and this is called *negative cooperativity*.

The Hill coefficient is best regarded as a purely empirical value to allow the degree of cooperativity to be measured and compared with other cases. Contrary to what is sometimes claimed, there is no physically plausible model that generates straight Hill plots with slopes other than 1 (or −1), and the expectation is that any cooperative Hill plot should be curved at the extremes, because the extreme slopes should approach 1 or −1. That this curvature is not often reported simply reflects the difficulty of carrying out sufficiently accurate experiments over a wide enough range of conditions.

11 Experimental conditions for kinetic studies

In deciding the conditions for any steady-state experiment the primary object-ive is to obtain accurate initial rates over a sufficient range of conditions to define the kinetic parameters of interest. Ideally the rate should be constant, i.e. the progress curve should be straight, during the period of measurement. If this is impossible, then it is essential to realize that unless integrated rate equations are used (for a full progress-curve analysis) all of the theory will assume *initial* rates, and that therefore the slope to be used is the initial slope, not the average slope over the period of measurement. Many experimenters report a 'linear' phase followed by a tailing-off, but this is nearly always an illusion (cf. *Figure 5*, which illustrates the same psychological problem in a different context), because the tailing-off normally begins at zero time. What this implies in practice is that the slope required is that of the tangent drawn to touch the progress curve at zero time, not that of a chord drawn to fit the first few data points.

When the curvature is appreciable it may be necessary to use curve-fitting to estimate the initial slope accurately. Some methods based on integrated rate equations have been described, but for most purposes it is sufficient to use a 'generic' equation such as a quadratic function to describe the progress curve. If this is done there are several pitfalls to be avoided.

In a few cases the rate may initially increase rather than decrease, and when this occurs one should try to determine the cause, or at least conditions in which it does not happen. No simple model predicts an initial acceleration, and so if it occurs it implies either that more complex analysis will be required or that there is an experimental artefact to be eliminated. An example of the first type of explanation might be a slow conversion of the enzyme into a different and more active state when the assay is initiated: this may well be interesting and important and full understanding of the enzyme mechanism may require it to be properly analysed, but this is a detailed topic that is beyond the scope of this chapter. A more trivial explanation might be an assay in which the temper-ature is inadequately controlled and allowed to increase during the first part of the assay: this just needs to be eliminated by more careful experimentation.

12 Concluding remarks

Steady-state analysis of initial rates has formed the basis of most kinetic studies of enzymes for almost a century, and as such it includes much more detail than one can hope to cover in a single chapter. Some important topics, such as the analysis of reactions with more than one substrate, have not been discussed at all, and nearly all have been dealt with very briefly. A considerably more complete account of all of the material in this chapter may be found elsewhere (11).

References

1. Michaelis, L. and Menten, M. L. (1913) *Biochem. Z.* **49**, 333–369.
2. Lineweaver, H. A. and Burk, D. (1934) *J. Am. Chem. Soc.* **56**, 658–666.
3. Woolf, B. (1932) cited by J. B. S. Haldane and K. G. Stern in *Allgemeine chemie der enzyme,* pp. 119–120. Steinkopff, Dresden and Leipzig.
4. Hanes, C. S. (1932) *Biochem. J.* **26**, 1406–1421.
5. Eadie, G. S. (1942) *J. Biol. Chem.* **146**, 85–93.
6. Hofstee, B. H. J. (1952) *J. Biol. Chem.* **199**, 357–364.
7. Cornish-Bowden, A. (1995) *Analysis of enzyme kinetic data.* Oxford University Press, Oxford.
8. Fersht, A. (1985) *Enzyme structure and mechanism,* pp. 150–154. Freeman, New York.
9. Arrhenius, S. (1889) *Z. Physik. Chem.* **4**, 226–248.
10. Hill, A. V. (1910) *J. Physiol.* **40**, iv–vii.
11. Cornish-Bowden, A. (1995) *Fundamentals of enzyme kinetics* (2nd edn). Portland Press, London.

Chapter 7
Spectrophotometric assays

T. J. Mantle* and D. A. Harris[†]
*Department of Biochemistry, Trinity College, Dublin 2, Ireland
[†] Department of Biochemistry, University of Oxford, South Parks Road, Oxford OX1 3QU

1 Introduction

Chapter 1 describes the principles and practice of spectrophotometry, and in this chapter we will consider how this technique can be used to measure the amount of analyte in solution, when the product of a reaction between the analyte and a 'very useful reagent' produces a measurable change in absorbance. Some good examples of this type of reaction are given together with various protocols for routine applications such as the measurement of protein concentration.

If there are no contaminating species absorbing at the wavelength of interest, it may be possible to measure the analyte directly simply by taking absorbance measurements. Spectrophotometry has also proved to be fundamental for rate measurements using a wide variety of enzyme-catalysed and non-catalysed reactions. We will also see that it is often the case that we will use an enzyme reaction that has gone to completion to measure the amount of an analyte that is limiting in the reaction. Furthermore, in some cases we can get the reaction to cycle so that the amplified rates can provide a highly sensitive assay, see, for example, the Tietze method (1) for measuring glutathione (Section 3.3). We will also briefly cover the use of so-called 'plate readers' as these are being used increasingly as 'multi-cuvette spectrophotometers'. Finally, 'plate readers' will be compared with 'centrifugal analysers', much beloved in clinical chemistry labs. The reader should note that some methods are dealt with in detail here to allow selected protocols to be used directly. Also, a number of useful reference works for enzyme assays are given, including (2–4), which should be consulted. With a little thought novel assays can often be developed based on the various 'coupling' methodologies described below.

1.1 Spectrophotometers

The design of standard spectrophotometers has been covered in Chapter 1 and we should simply mention that most modern 'plate readers' work in essentially the same way. The major difference is in their beam geometry. In spectrophotometers, samples are read through cuvettes with a horizontal light path which is normally 1 cm in length allowing the Beer–Lambert Law (see Section

1.2) to be applied in its simplest form (constant light path). In 'plate readers' the vertical light beam results in a pathlength that depends on the volume of fluid in each well. The major drawback to this configuration is that the pathlength is not maintained by the dimensions of the cuvette (as in a conventional spectrophotometer) and small variations in the volume pipetted into the well are a potential source of error. For most ELISA uses these errors are acceptable; however, to use these instruments to read absorbances it is advisable to use a series of dilutions of a standard analyte to set up a calibration curve using a conventional spectrophotometer, and then to compare the readout for the same solutions in a 'plate reader'. It is then possible to calculate the pathlength in the 'plate reader' and to make the appropriate corrections. This problem has recently been addressed by Molecular Devices (*www.moldev.com*) where the pathlength is automatically corrected using near infrared measurements. The same problem does not apply to 'centrifugal analysers' as the pathlength in this case is determined by the dimensions of the cuvette. Another downside with earlier 'plate readers' was the restriction placed on the wavelengths that could be used due to the limited range of filters supplied; however, the introduction of interference filters has effectively overcome this problem.

1.2 Beer–Lambert Law

It will not hurt to remind ourselves that the use of spectrophotometry in assays rests on the operation of the Beer-Lambert Law (described in Chapter 1) which may be written as:

$$A = \varepsilon\,c\,l \qquad\qquad 1$$

where A is the absorbance, and is equal to the log of the ratio of the incident (I_o) to transmitted (I) light, ε is the extinction coefficient and l is the length of the path cell (usually, but not always 1 cm). It is important to emphasize that A has a logarithmic relationship to light transmittance and that as A increases from 1 to 2 only 10% and 1%, respectively, of the incident light is transmitted. In earlier spectrophotometers, this low transmission could result in inaccurate readings although this is not normally a problem with the current generation of instruments. However, deviations from the Beer–Lambert law may be observed above an absorbance of 1 for non-instrumental reasons and calibration of the system to be used is a wise precaution.

1.3 The nature of the sample

One of the joys of spectrophotometric assays is that very often the analyte in the sample does not have to removed from other contaminating compounds, as long as it is the only component of the mixture that is absorbing light at the wavelength of interest. For example, many dehydrogenases use NAD^+ as cofactor. When reducing equivalents are transferred from the oxidizable substrate, NAD^+ is reduced to NADH which has a characteristic absorption at 340 nm. As NAD^+ does not absorb at this wavelength (it has a major absorption peak at 260 nm) the reaction can be conveniently monitored simply by measuring the increase in absorbance at 340 nm. Obviously if the reverse reaction is being followed, i.e. the oxidation of NADH, then this is measured as a decrease in A_{340}. This provides a good example of the experimental problems caused by the non-linear response of a spectrophotometer to a change in concentration of a chromophoric solute (see Section 1.2 above and Chapter 1). If an assay contains 200 μM NADH, then the absorbance at 340 nm in a cuvette with a 1-cm pathlength will be approximately 1.24 and this may exceed the linear range of the response. Thus under these conditions a reaction in which the NADH is converted to NAD^+ may have a progress curve which appears to have a lag period followed by an accelerating rate of decrease in absorbance. Such situations benefit from the use of a cell with a shorter pathlength. This problem commonly arises in experiments in which the K_m of NADH (or any other chromophoric substrate or coenzyme) is being determined where high concentrations are desired to allow approach to V (see Chapter 6).

Although it is sometimes not necessary to separate the analyte from other components of the mixture, the sample often requires some 'work up' to facilitate absorbance measurements. The most common type of interference in spectrophotometric assays is light scattering due to the presence of particulate material (e.g. mitochondria, red cells etc.). This leads to a decrease in transmission and therefore an artefactual apparent increase in absorbance. This type of particulate interference can often be removed by introducing a suitable centrifugation step.

2 Some general comments on, and practical aspects of, assay design

It is important that assays are specific, sensitive, accurate and precise. It should go without saying that it is also important that they should also be as rapid and as inexpensive as possible. There is no point in having a simple assay which gives imprecise results; however, it is equally true that very sensitive assays are commonly less robust and more expensive and should be avoided unless required.

The specificity of biochemical assays often depends not on the chromophores measured (these are fairly restricted e.g. p-nitrophenol, NADH, phosphomolybdate, chloronaphthol etc.) but on the specificity of the enzyme or antibody employed. For example, the use of LDH to measure lactate will also give an increase at 340 nm with a small number of related compounds. However, this is not usually a problem as, of these structures, only lactate is usually encountered in most biological systems.

As described above, it is often the case that an enzyme reaction is used to monitor an analyte and the reaction will be allowed go to completion. If the signal is not amplified in some way (see Section 3.3), then the sensitivity is limited by the magnitude of the extinction coefficient. For example, the extinction coefficient for NADH at 340 nm is 6220 M^{-1} cm^{-1} which means that a solution of 1 μM has an absorbance of 0.006. This reflects the lowest amount of NADH that can be measured with confidence using spectrophotometry. If greater sensitivity is required for measuring NADH, then fluorimetry (excitation wavelength 340 nm, emission wavelength 460 nm) can be used.

2.1 Accuracy and precision

A repeated assay on the same sample yields a range of values that can be expressed as the mean, $\bar{x}$, +/– the standard deviation σ. For a precise assay σ is small compared with the mean ($\bar{x}$) and it is desirable that the $\sigma/\bar{x}$ (the % error) should be less that 5%. If the scatter is small then the reproducibility is good and the assay has good resolution (the ability to distinguish between two close values). Elements that affect precision are random errors such as pipetting, weighing etc.

An accurate assay is one where the measured value of x ($\bar{x}$) is close to the true value, X, i.e. $X - \bar{x}$ is small. If we are measuring a reaction spectrophotometrically to an end point there are a number of possible systematic errors. For example, the reaction may not go to completion if a concentration of reagent is too low, or the time allowed, while appropriate for a high concentration of analyte, is too short at lower concentrations. A trivial, but occasionally important source of inaccuracy can occur by working away from the wavelength of maximum absorbance (λ_{max}). This is particularly applicable to compounds with 'narrow' absorption bands and is further exacerbated by use of wide slit widths on the spectrophotometer (see Chapter 1).

In several experiments accuracy may be less important than precision. For example, if during the purification of an enzyme a protein assay is used based on albumin as a standard, the colour change observed for 1 mg of albumin may be different to that observed with 1 mg of the enzyme under study. However, as long as the error is systematic and not random, the method can still be used to monitor the increasing specific activity of the enzyme preparation at different stages of the purification.

In most cases, however, accuracy is critical. In measuring the stoichiometries of ligand binding to proteins it is essential to know the true molecular weight and concentration of the protein. We shall therefore now discuss a number of practical aspects that should allow the absolute calibration of assays for macromolecules, particularly proteins and DNA

2.1.1 Volume

Assay mixtures contain, in principle, three components, i.e. the sample containing the material to be assayed, a diluent and a 'colour' producing reagent (which of course may mediate an absorbance change in the UV region of the spectrum). If the volume of the sample is changed to obtain a larger or smaller colorimetric response, this change is balanced by using a correspondingly adjusted volume of diluent to keep the sum of the two components constant. It is, however, often convenient, particularly in automated systems, to pre-dilute a range of different concentrations of a standard for a calibration curve and then to add a constant volume of different concentrations of sample.

The total volume in the assay is often chosen to be between 2 and 2.5 ml which will ensure that the entire light beam passes through the solution in most 1-cm pathlength cells. With any new instrument it is often useful to make up a solution and add decreasing volumes (e.g. 3, 2.9, 2.8 ml etc.) until a deviation in absorption (or fluorescence) is recorded which will indicate that not all of the light is passing through the solution. For visualization, it may also be useful to place a piece of paper into the cuvette holder and by applying a wavelength in the *visible* portion of the spectrum to locate the position of the beam.

Assay volumes can be scaled down to a final volume of 1 ml or below if semi-microcells (0.2–0.4 cm wide; 1-cm pathlength) are available. However, it is important in this case to use a narrow slit width for the light beam, and in some cases it may be necessary to blacken the sides of the cuvette by insulation tape or even by using a permanent marker pen. Semi-microcuvettes with blackened sides can be purchased directly from manufacturers (e.g. Hellma).

2.1.2 Diluent

The diluent contains buffers ions and other reagents to allow colour development. For example, the Biuret and Lowry assays for protein consists of alkali, Cu^{2+} ions (to chelate to the protein) and tartrate to keep the Cu^{2+} in solution under alkaline conditions. The sample volume is rarely allowed to exceed 25% of the volume of diluent. It is especially important to consider that there may be compounds in the sample that interact with key components of the diluent. Common interactions that need to be watched for are chelators in the sample that may lower the free concentration of metals required in the diluent. Useful lists of compounds that interfere with the Lowry assay for proteins (see Section 4.3) can be found in (5) and (6).

2.1.3 Order of addition

Although the order of mixing of reagents does not often affect final colour development it is common sense to add the more labile/sensitive reagents last. If instability in a reagent produces some colour, then high blanks will be observed. It is always useful to check that the order of addition does not affect colour development.

3 End point and rate assays

3.1 End point assays

In these assays all of the metabolite/analyte is converted into some chromophoric compound. These have an advantage in that any variation in factors that may affect the time to completion (e.g. temperature, amount of coupling enzyme etc.) will not affect the final measurement, since this is made when the reaction has ended. It follows therefore, that these reactions must be essentially irreversible, i.e. $\Delta G°$ must be large and negative. If the equilibrium is poised it may be possible to pull the equilibrium over to the product side by including a trapping reagent in the assay. For example, in the estimation of ethanol by alcohol dehydrogenase, the inclusion of semicarbazide in the assay traps the aldehyde produced as the stable semicarbazone and so drives the reaction to completion. Even if the reaction is complete, the appropriate absorbance may not be measured if, as is often the case, the colour reagent produces a product that is stable for a limited period of time (and hence the colour is only stable for a limited time). In these cases, it is important to establish a maximum period for the assay so that decay of the coloured product is not a significant problem. An additional problem can occur with an enzyme-generated signal if the enzyme-catalysed reaction is 'leaky'. For example, the assay for ATP involves a 'double couple' of hexokinase and glucose 6-phosphate dehydrogenase (plus $NADP^+$, see also Section 6.3.2) in the presence of a high concentration of glucose. Under these conditions, the reaction is expected to be limited by ATP; however, when the enzymatically produced glucose 6-phosphate has been consumed there is a slow increase in A_{340} (due to NADH) as glucose itself is a poor substrate for G6PDH. In this case, it is important to extrapolate the progress curve to zero time.

3.2 Rate assays

If the substrate concentration [S] is well below the K_m, then the initial rate (v) exhibits an approximately linear relationship with [S], where the line goes through the origin with a slope that approximates to V/K_m. In contrast to end point assays, it is vitally important that all variables, such as temperature, pH, salt concentration etc. are strictly controlled to ensure that the only variable affecting the rate is [S]. Rate assays can be facilitated by use of *continuous flow analysers*, common in hospital pathology laboratories for the routine screening of metabolites in serum/urine samples from patients. In these, a sample is pumped for a short period of time into a continuous stream of reagents through thermostatted delay coils which 'incubate' the reaction mixture at a set temperature and for a set time before delivering the mixture to a flow-through cell in a colorimeter or fluorimeter. The injection of sample is followed by an injection of water or buffer as a wash period and then the next sample is injected to the system. The period of the incubation can be adjusted by changing the flow rate of the reagents or the length of the delay coil. The sensitivity of the system may be adjusted either electronically at the detector stage or by changing the ratio of the flow rates of the sample to that of the reagent. In theory, the system should record a series of rectangular peaks as the samples, separated by washes, flow through the detector. However, lateral diffusion causes these peaks to become rounded to some extent. It is essential to ensure that the rate of production of the product is linear for the total period of the incubation.

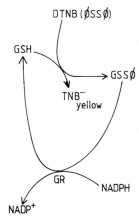

Figure 1 A hypothetical scheme to illustrate the Tietze reaction. DTNB is represented as φSSφ and reacts with GSH to form the mixed disulphide GSSφ. This appears to be a substrate for glutathione reductase and is reduced to regenerate GSH, which then enters another round with DTNB, continuing to produce the yellow anion TNB⁻ at a rate that is dependent on the initial concentration of GSH.

3.3 Rate assays involving amplification

There are a number of cases where 'rate assays' involve amplification. A classic example is the method of Tietze (1) to measure glutathione (GSH). This employs the enzyme glutathione reductase which catalyses the NADPH-dependent reduction of oxidized glutathione (GSSG) to form GSH. Gluathione reductase also appears to reduce the mixed disulphide formed when GSH reacts with Ellman's reagent (5,5′-dithiobis-2-nitrobenzoic acid; DTNB). In this assay, GSH initially reacts with Ellman's reagent (5,5′-dithiobis-2-nitrobenzoic acid; DTNB) liberating the yellow 5-thio-2-nitrobenzoate anion (TNB⁻) (which is stabilized as the quinonoid structure) and the mixed disulphide between glutathione and DTNB. The assay appears to work because the mixed disulphide, a substrate for glutathione reductase, is reduced back to GSH producing more of the yellow TNB⁻. The reaction cycles as shown in *Figure 1*, yielding an amplified signal (at 412 nm) due to the release of the yellow TNB⁻ ion.

4 Spectrophotometric assays for proteins

The measurement of protein concentration is an everyday occurrence in most laboratories working in the bimolecular sciences. An accurately determined protein concentration is an essential piece of information for many types of experiments that require a term for molarity of a protein, to monitor the various stages of a protein purification, to determine the amount of material to be added to an ELISA or coupled assay, or to normalize a number of different biological samples (e.g. number of cells or amount of cell extract). This type of assay must be straightforward and robust. It is also important that the assay can be used for a range of physically distinct protein preparations. For example, one may wish to measure the amount of protein in an electrophoretically homogenous preparation, in a subcellular fraction (e.g. mitochondrial suspension), whole cells or crude tissue homogenate.

4.1 A_{280}

For rapid, and generally fairly accurate estimations of protein concentration in soluble preparations the absorbance at 280 nm gives a good guide. This is based on the fact that most proteins have approximately the same fraction of tryptophan residues (which contributes the most to the A_{280} for most proteins, see Section 6.3, Chapter 2). It is generally found that an A_{280} of 1 is equivalent to a protein concentration of about 1 mg/ml. It is important not to conduct such measurements in solutions that have a low M_r component that absorbs strongly at this wavelength (e.g. ATP or ADP) or with pure proteins with abnormally low fractions of tryptophan. Additionally, the A_{215} can be utilized since this measures absorbance due to peptide bonds. In this case an A_{215} of 0.066 is equivalent to a protein concentration of 1 mg/ml. However, at this wavelength additional contaminants absorb, limiting its general usefulness.

If the material is contaminated with a UV-absorbing material whose spectrum is known, it may be possible to correct for the contaminant by measuring the absorbance at two wavelengths. A well-known example is protein contaminated by nucleic acids (or nucleotides), when the following equation can be used:

$$\text{mg/ml protein} = 1.55A_{280} - 0.76A_{260} \qquad\qquad 2$$

Protocol 1

A Biuret assay to determine protein concentration

Reagents

- Dissolve 1.5 g of $CuSO_4.5H_2O$ and 6 g sodium potassium tartrate (containing four waters of crystallization) in 500 ml boiled and then cooled distilled water.

- To this solution add 300 ml 10% (w/v) sodium hydroxide, mixing during the addition.
- Finally add 200 ml distilled water.

This solution is stable for several months at room temperature.

Method

NB. The sample should contain 0.5–4 mg protein in 0.2 ml water or buffer.

1 Construct a standard curve for the assay by adding 2.3 ml of the reagent (see above) to each of a number of samples (0.2 ml) containing between 0.5 and 5 mg BSA.

2 Add 2.3 ml of the reagent to 0.2 ml of the protein solution being assayed.

3 Leave the mixtures at room temperature for 2 h (alternatively heat to 80°C for 5 min, cool rapidly and stand at room temperature for 10 min).

4 Read the A_{540} of each sample against a blank of water or buffer plus reagent. Note that 1 mg of protein gives an A_{540} of approximately 0.1.

4.2 The Biuret method

This assay is not affected by the variation of the amino acid content of proteins as it is based on the complex of peptide bonds with Cu^{2+} in alkaline solution and is robust, albeit the least sensitive method, with a useful range of 1 to 5 mg.

The colour is stable for several days if the solutions are kept dark. A number of substances interfere with the assay including high concentrations of ammonia (which complexes with the Cu^{2+}) and reducing agents (including reducing sugars) which reduce the Cu^{2+} to Cu^{+}. It may be necessary to remove soluble materials that interfere with the assay. One of the easiest methods to use is to precipitate the protein using ice-cold trichloroacetic acid (TCA). It is usual to add an equal volume of 20% (w/v) TCA to the sample, leave the mixture on ice for 10 min and then pellet the precipitate by centrifugation. The pellet may be dried in air or washed twice with ice-cold acetone to remove residual TCA and assist the drying. The pellet from this procedure (or any other insoluble protein) may be solubilized by the addition of 0.2 ml 0.25% deoxycholate in 0.1 M NaOH, heating to 80°C for 5 min, allowing to cool and then proceeding at step 2 above.

4.3 The Lowry method

This method (7) has far greater sensitivity than the Biuret and has a working range of 5 to 40 μg, which probably explains why this reference is a citation classic. It suffers from greater interference (5, 6) than the Biuret method and also greater variability between proteins. The method uses two 'reagents', an alkaline copper reagent (similar to the Biuret reagent) and the Folin reagent (phosphomolybdate and phosphotungstate). It appears that after the formation of an initial copper–protein complex that there is some amino acid side chain mediated reduction of the Folin reagent (8).

Protocol 2

A Lowry assay to determine protein concentration

Reagents

- 1% $CuSO_4.5H_2O$ (A)
- 1% sodium potassium tartrate.$4H_2O$.(B)
- 2% sodium carbonate in 0.1 M NaOH (C)
- Folin–Ciocalteau reagent (D)

These reagents are stable for several months at room temperature with the exception of the Folin–Ciocalteau reagent which must be stored at 4°C. The working solutions must be freshly prepared by mixing equal volumes of reagents (A) and (B) and then diluting this mixture with 98 volumes of reagent (C) to give the alkaline copper solution. Reagent (D) must be diluted 1:1 v/v with distilled water before use.

Method

NB. The sample should contain 5–40 μg protein in 0.2 ml water or buffer.

1   Construct a standard curve by adding 2.1 ml alkaline copper solution to each of several samples (0.2 ml) of solution containing between 1 and 40 μg of BSA.

2   Add 2.1 ml of the alkaline copper solution to 0.2 ml of the protein solution to be assayed.

3   Leave the mixtures for 10 min.

4   Add 0.2 ml of the diluted Folin–Ciocalteau reagent to each sample and vortex immediately.

5   Leave the mixtures at room temperature for 1 h.

6   Read the A_{750} of each sample against a blank of water or buffer plus reagents. Note that 10 μg of protein gives an A_{750} of approximately 0.1.

As with the Biuret assays it may be necessary to remove soluble materials that interfere with the assay. Common interfering agents include a number of common buffers (e.g. HEPES) and neutral detergents (e.g. Triton X100). Indeed the Folin method is notorious for the number of compounds that interfere (5, 6). One of the easiest approaches is to precipitate the protein using ice-cold trichloroacetic acid (TCA) as described above under the Biuret assay. A number of variations on the original method have been developed to overcome interference, particularly by detergents in membrane-bound proteins (9).

4.4 The bicinchoninic assay

This is a modification of the Folin method, without the Folin reagent! In this case the Cu^+ formed does not mediate reduction of the Folin reagent but is chelated by bicinchoninic acid.

Protocol 3

A bicinchoninic assay to determine protein concentration

Reagents

- 1% bicinchoninic acid, 0.16% sodium tartrate, 2% sodium carbonate, 0.95% sodium bicarbonate in 0.1 M NaOH (pH 11.25) (B)
- A solution of 4% $CuSO_4.5H_2O$ in distilled / deionized water (A)

Solutions (A) and (B) are stable for several months at room temperature. To use, mix reagents (A) and (B) in a ratio of 1:50 to form the bicinchoninic reagent.

Method

NB. The sample should contain 5–40 μg protein in 0.2 ml water or buffer.

1   Construct a standard curve by adding 2.3 ml bicinchoninic reagent to samples of 0.2 ml of BSA (1–40 μg) in water.

2   Add 2.3 ml of the reagent to 0.2 ml of the protein solution to be assayed.

3   Leave the mixtures at room temperature for 2 h (or heat at 60 °C for 30 min).

4   Read the A_{562} of each sample against a blank of water or buffer plus reagents. Note that 10 μg of protein gives an A_{562} of approximately 0.1.

4.5 Dye-binding assay

On deprotonation, Coomasie Brilliant Blue G250 exhibits a blue colour that can be measured at 595 nm. This forms the basis of a quick, convenient and sensitive assay, as deprotonation is stabilized when the dye is bound to protein (10).

Protocol 4

A dye-binding assay for the estimation of protein concentrations

Reagents

- 100 mg of Serva Blue G dissolved in 50 ml ethanol, then diluted to 500 ml with water (A)

- Concentrated phosphoric acid (85%) diluted 1:5 v/v with water (B)

Solutions (A) and (B) are mixed in a 1:1 ratio on the day of use to produce the working reagent.

Method

NB. The sample should contain 2–20 μg protein in 0.2 ml water or buffer.

1 Construct a standard curve by adding 2.3 ml of the working reagent to each of several samples (0.2 ml) of BSA containing between 1 and 20 μg of protein.

2 Add 2.3 ml of the reagent to 0.2 ml of the protein solution being assayed.

3 Mix the solutions and read the A_{595} of each one immediately. Do not leave the mixture for periods longer than 30 min. Note that 5 μg protein gives an A_{595} of approximately 0.1.

The dye-binding assay does not accommodate SDS or deoxycholate and is therefore not useful for proteins solubilized in these detergents. However, the assay will operate with 0.2% Triton X-100.

4.6 Fluorimetric assay

The sensitivity of protein assays can be increased by several orders of magnitude (sensitive down to 10 ng protein) by using fluorescence detection methods. A number of compounds have been developed. One, the proprietary NanoOrange™ (available from Molecular Probes), is a dye that binds non-covalently to proteins with a large enhancement of fluorescence at 590 nm, and this can be used in an assay very similar to the colorimetric dye-binding assay described above.

There are also a number of reagents (originally developed for detection in liquid chromatography) which react with organic amines with a large increase in fluorescence, and which can thus be used to quantitate proteins in appropriate buffers. These include o-phthaldialdehyde (11), 3-(4-carboxybenzoyl)-quinoline-2-carbodialdehyde (CBQA) (12) and fluorescamine (a heterocyclic dione) (13). All are readily adaptable for use in microplate format.

Protocol 5

A fluorescence assay for the estimation of protein concentrations

Reagents

- 7.5 mg of fluorescamine dissolved in 25 ml dry acetone (A)
- Borate buffer pH 8.5 (3 g boric acid + 4.8 g sodium tetraborate in 200 ml water) (B)

Solution A should be prepared on the day of use, in a dry test tube, and contact with moisture (e.g. wet pipette tips) should be avoided.

Method

NB. The sample should contain 0.1–2 μg protein in 0.2 ml water or buffer.

1 Construct a standard curve by adding 2.1 ml borate buffer to each of several samples (0.2 ml) of BSA containing between 0.1 and 2 μg of protein.

2 Add 2.1 ml of buffer to 0.2 ml of the protein solution being assayed.

3 Mix on a vortex mixer and, while still mixing, add 0.2 ml fluorescamine solution.

4 Read the fluorescence of each sample after 2 min, using $\lambda_{ex} = 390$ nm and $\lambda_{em} = 465$ nm

Amines interfere with this assay; aliphatic amines yield fluorescent products with these compounds, while ammonia, although not giving a fluorescent product, will combine with (and hence use up) the reagent. Unlike the other methods for protein determination, these reagents will give responses with peptides in addition to proteins. Whether or not this is advantageous depends on the requirements of the investigator.

5 Spectrophotometric assays for nucleic acids

DNA and RNA can be prepared from cells or tissue extracts using phenol as a protein precipitant and then precipitating the nucleic acid using cold ethanol. The nucleic acid can be dissolved in dilute salt normally 10 mM Tris/EDTA, pH 7.5. The bases of nucleic acids absorb at 260 nm and this gives a useful method (akin to measuring protein by monitoring the A_{280}) for estimating the concentration of nucleic acids. A solution of nucleic acid at 1 mg/ml gives an A_{260} of approximately 20. It is useful to remember that a 1 mg/ml solution of protein gives an A_{280} of approximately 1.

More sensitive methods are based on measuring the fluorescence enhancement of nucleic acid–dye complexes such as those used for staining DNA in electrophoresis gels. The dye Hoeschst 33258 (14) can detect double-stranded DNA down to 100 ng, and YOYO-1 and the proprietary reagent PicoGreen™ dsDNA are sensitive down to 50 pg. All are available from Molecular Probes. OliGreen™ ssDNA has a similar high sensitivity for single-stranded DNA. Unfortunately, there are no dyes whose fluorescence is specifically enhanced by RNA.

6 Enzyme-based spectrophotometric assays

6.1 Some general points on assay design

Most enzyme-based assays involve an analyte that is of low molecular weight, so that large molecular weight material (particularly proteins, some of which may exhibit an activity that uses the analyte) is generally removed with a precipitating step. This often involves acid treatment (TCA or perchloric acid proving particularly effective), metal ions (e.g. $ZnCl_2$) or organic solvents (e.g. ethanol). The addition of acid is an effective precipitant of proteins but will not discriminate between the unwanted proteins and the coupling enzymes which will then be used, so it is vital that the acid is removed. TCA can be conveniently removed by extraction into ether or acetone, while perchloric acid can be neutralized with KOH which provides two functions: the first is neutralization and the second the precipitation of insoluble potassium perchlorate.

The components of the assay will normally be in a final volume of 2.5 ml although scaling down can normally be achieved for smaller cells or sample holders. Under these conditions it is usually desirable to dilute the sample extensively into the assay mixture so that any small effects on buffering capacity, ionic strength or potential inhibitors are minimized.

It is important to drive the reaction essentially to completion. For many coupling enzymes that are linked to ATP (e.g. hexokinase) this is not a problem. Some reactions, however, are at thermodynamic equilibrium and need some thermodynamic assistance to be dragged over to completion. A classic example of this type of strategy is the addition of semicarbazide to trap ketones in the thermodynamic sink of the semicarbazone (see above). An alternative is to link a reversible reaction with an enzyme linked reaction that will provide an alternative thermodynamic sink.

In addition to the above, many coupling enzymes require additional factors, such as Mg^{2+} for most kinases and K^+ for pyruvate kinase.

6.2 Amount of enzyme required

Various authors have considered the amount of enzyme required for an end point assay (15–17). While it is advisable not to be too wasteful, most common coupling enzymes are not too expensive and a few preliminary experiments will soon determine the minimum amount needed to give reliable data in a reasonable time. Putting the problem quantitatively, our ideal amount of enzyme should lead to the reaction being 'virtually' complete (from 100% to 1% of substrate remaining) in a reasonable time (say, 10 min). Since the enzyme will slow down as the substrate is used up, we cannot just calculate the amount of enzyme needed from its given activity (V) divided into the substrate available; instead we need to consider the integrated rate equation for the process. Remember that V is a measure of the amount of enzyme added ($V = k_{cat}e_o$) and so when we choose the amount of enzyme to add, we choose a V value.

For any enzyme reaction

$$v = -\frac{ds}{dt} = \frac{Vs}{K_m + s} \qquad\qquad 3$$

in which v is the rate, and s the substrate concentration. As the substrate is used up, and the reaction approaches completion, $s \ll K_m$, so that

$$-\frac{ds}{dt} = \frac{Vs}{K_m} \qquad\qquad 4$$

Integration gives us the total time for given change in s

$$-\int_{s0}^{s1}\frac{ds}{s} = -\frac{V}{K_m}\int_0^t dt \qquad\qquad 5$$

or

$$\ln\frac{s_0}{s_1} = \frac{V}{K_m}t \qquad\qquad 6$$

For a decrease in s from 100% to 1% , in a time of 10 min (i.e. values we considered reasonable above), the following expression must hold

$$2.3 \log\frac{100}{1} = \frac{V}{K_m}10 \qquad\qquad 7$$

or

$$V/K_m = 0.46 \text{ min}^{-1} \qquad\qquad 8$$

or

$$V = 0.46K_m \text{ min}^{-1} \qquad\qquad 9$$

This equation tells us that, for a reaction to go to 'completion' in 10 min, the amount of enzyme added should be an amount able to change the concentration of substrate in the assay by $0.46K_m$ per minute. If $K_m = 1$ mM, for example, for 99% completion in 10 min

$$V \geq 0.46 \text{ mM min}^{-1} (\geq 0.46 \text{ mmol litre}^{-1} \text{ min}^{-1}) \qquad\qquad 10$$

The concentration of enzyme is conventionally given in U/ml, where 1 U is the amount of enzyme that will convert 1 µmol substrate/min at saturation. If we add 1 U of enzyme to an assay volume of 3 ml, for example, then $V = 1/3$ µmol/ml/min = 0.33 mmol litre^{-1} min^{-1}. Thus, if $K_m = 1$ mM, for $V \geq 0.46$ mmol litre^{-1} min^{-1}, we need to add $\geq 0.46/0.33$ Units of enzyme if the assay volume is 3 ml.

These calculations are useful 'rangefinders', but a number of points should be borne in mind. It is important to determine experimentally the activity of the enzyme in the assay medium used. The units of activity that accompany commercial preparations are those measured under optimal conditions (as described by the manufacturer), and these conditions may not be those used in the assay medium being used. Also note (as set out above) that the amount of enzyme required will increase with increased assay volume. The reaction will, of course, proceed faster if you add more than the calculated amount and this may be convenient if the cost is not too great.

Commercial preparations of many enzymes are supplied as suspensions in high concentrations of ammonium sulphate or may be stabilized with glycerol. It is often, though not always, necessary to remove these 'protectants' before using the enzyme. This can be achieved by dialysis or gel filtration (this latter method will not work with ammonium sulphate suspensions). It is important to check the stability of the resultant enzyme stock solution. Some enzymes can be stored frozen as aliquots that may be used once and then discarded. Some may stand repeated freezing and thawing (this is unusual), while others may retain considerable activity at 4°C. It is worth investing some time to establish the best conditions for storing stock enzyme solutions and it is also advisable to check the activity of stocks on a regular basis. This is as useful a check on the substrates and buffers as on the stability of the enzyme solutions.

Remember, more enzyme, more activity, so that if the activity of a stock solution is declining on storage it may still be possible to add sufficient units of activity simply by adding more from the stock solution. This, of course, rather depends on the use of small volumes of concentrated stock enzyme solutions in the final assay mix.

6.3 Determination of glucose—a comparison of two methods

6.3.1 Peroxidase coupled method

Peroxidase catalyses a number of reactions coupling hydrogen peroxide breakdown with dye formation. This reaction has found favour in immunoblotting where chloronaphthol reacts with hydrogen peroxide to form a purple–blue product. To use the peroxidase reaction to measure glucose an additional enzyme is required that couples glucose oxidation to hydrogen peroxide production. This is the flavoprotein glucose oxidase, and the hydrogen peroxide produced is coupled to the formation of a green dye using the colourless leuco-dye 2,2'-azino-di-(3-ethylbenzthiazoline) 6-sulphonate [ABTS]. This simple format can be used with other enzymes that are specific for the substrate they oxidize and produce hydrogen peroxide as a product (for example, galactose oxidase, D-amino acid oxidase, monoamine oxidase etc.).

It is important to note that the order of addition of assay components may or may not be important, but should always be considered (see Section 2.1.3). In this case hydrogen peroxide is not stable so that the enzyme that produces this (glucose oxidase) is added last, to start the reaction.

6.3.2 Glucose 6-phosphate dehydrogenase (G6PDH) coupled method

This protocol couples hexokinase to G6PDH. The hexokinase catalyses the ATP-dependent formation of glucose 6-phosphate from the glucose added, while the G6PDH takes the phosphorylated product and catalyses the $NADP^+$-dependent formation of 6-phosphogluconolactone.

Protocol 6

An assay for glucose using glucose oxidase

Reagents

- 1 mM ABTS in 0.1 M sodium phosphate pH 7
- Glucose oxidase (100 U/ml) in 0.1 M sodium phosphate pH 7
- Peroxidase (20 U/ml) in 0.1 M sodium phosphate pH 7

These reagents are stable for one month at 4°C.

Method

NB. The sample should contain 1–10 nmol glucose in 0.2 ml.

1 Construct a standard curve by adding 2.1 ml ABTS to each of 0.2 ml samples of known concentrations of glucose (1–10 nmol glucose in 0.2 ml water).

2 Add 2.1 ml of ABTS reagent to 0.2 ml of the sample of glucose to be assayed.

3 Add 0.1 ml peroxidase to all of the above samples.

4 Preincubate for 5 min at 37°C.

5 Start the reaction by the addition of glucose oxidase (0.1 ml).

6 Read the A_{420} of each sample after 10 min against a blank of reagents with no glucose added.

Protocol 7

An assay for glucose using glucose 6-phosphate dehydrogenase

Reagents

- 200 mM Tris/HCl pH 8, containing 2 mM $MgCl_2$ and 0.2 mM EDTA
- 200 mM MgATP. This is prepared by dissolving ATP (1 mmol) in 4 ml $MgCl_2$ (250 mM) and titrating this solution to pH 7 using 1 M KOH. The final volume should be brought to 5 ml.
- 100 mM $NADP^+$. In contrast to NADH and NADPH which should not be kept in acid solutions, $NADP^+$ should be stored in buffer below pH 7. As phosphate inhibits G6PDH Tris/HCl pH 6.5 is suitable.
- G6PDH (20 U/ml)
- Hexokinase (100 U/ml)

Method

NB. The sample should contain 40–400 nmol glucose in 0.2 ml.

1 Add 2.1 ml Tris buffer to 0.2 ml samples of glucose (either containing known amounts between 40–400 nmol of glucose for setting up a standard curve or the unknown sample being assayed).

2 Add 0.05 ml MgATP and 0.025 ml $NADP^+$ to each sample.

3 Add 0.1 ml G6PDH enzyme, mix the components and preincubate at 37°C for 5 min.

4 Read the initial A_{340} and then add 0.025 ml hexokinase.

5 Monitor the reaction until there is no further change in A_{340}.

6 Take the difference between the two A_{340} values (ΔA_{340}).

7 Read the ΔA_{340} against a blank of reagents with no glucose added.

8 To calculate the amount of glucose in μmol:

$$\mu\text{mol} = \Delta A_{340} \times \frac{2.5}{6.22} \qquad\qquad 11$$

If non-trivial amounts of glucose 6-phosphate are present in the sample there will be a significant initial A_{340} due to the presence of this material. If the amount of glucose present in the same sample is relatively low, then the ΔA_{340} may be difficult to measure against the high initial reading. In this case the peroxidase-coupled reaction may be more suitable. Indeed the economics (ABTS is far cheaper than $NADP^+$) and sensitivity of the peroxidase-coupled reaction means that this is often the method of choice for glucose. However, as elaborated by Harris (18), the hexokinase/G6PDH coupled method has advantages for measuring ATP and $NADP^+$, and can be made more sensitive by using fluorescence measurements to detect NADPH. The cost may also be reduced by using a bacterial G6PDH which uses NAD^+ as the coenzyme instead of the more expensive $NADP^+$ (enzyme available from Boehringer Mannheim).

7 Luminescence-based assays

Measurement of luminescence provides the basis of highly sensitive assays for ATP, NADPH and H_2O_2. The production or disappearance of ATP can be measured by following light emission in the presence of fire-fly luciferase which catalyses the following reaction:

$$\text{ATP} + \text{luciferin} + O_2 \rightarrow \text{oxyluciferin PPi} + CO_2 + \text{AMP} + \text{light} \qquad 12$$

Likewise, NADPH can be measured using the bacterial luciferase which catalyses the following reactions:

$$\text{NADPH} + H^+ + \text{FMN} \rightarrow \text{NADPH} + FMNH_2 \qquad\qquad 13$$

$$FMNH_2 + \text{RCHO} + O_2 \rightarrow H_2O + \text{RCOOH} + \text{light} \qquad\qquad 14$$

In which RCHO is a long chain aldehyde (C8 to C12). Finally, earthworm luciferase can be used for determining H_2O_2. Further details can be found in (18).

Conventional spectrophotometers (which are configured to measure the difference between incident and transmitted light) do not normally show sufficient sensitivity for determination of bioluminescence. Digital fluorimeters and scintillation counters can often be adapted for measurement of luminescence but, most conveniently, a dedicated luminometer (which requires neither a mono-chromator or light source) should be used.

8 Spectrophotometric assays of enzymes

8.1 Some elementary enzyme kinetics

Having established a spectrophotometric assay for enzyme activity (see below), it is vital before setting off to do any detailed measurements of enzyme activity to check two factors. The first of these is that true initial rates are being measured. A plot of product (P) appearance (in this case some spectral change) against time (t) *must* be linear over the period of measurement to measure true initial rates (the slope of the [P]/t curve is, by definition the initial rate v). There are a number of reasons that such [P]/t curves may deviate from linearity:

1. The substrate has simply been used up. This can be checked by calculating the amount of product produced and determining whether a significant fraction of the substrate has been used. Furthermore, some reactions rapidly approach equilibrium (perhaps when only a small amount of substrate has been used up) and, if this is the case, the reaction will not proceed without some additional system for removing the product.

2. A build up of inhibitory product may cause deviation from linearity. There are numerous examples of this phenomenon, which is hardly surprising as by definition most products have dissociated from the enzyme surface and are therefore capable of binding to the enzyme or one of the enzyme-substrate or enzyme–product complexes.

3. The enzyme may be unstable under the conditions of assay. Selwyn (19) has devised a method to check this possibility, where a plot of product con-centration against the product of time and the enzyme concentration (et) is made, and should give the same curve whatever initial concentration of enzyme is used. If, however, the enzyme is unstable, the concentration of active enzyme will be time dependent and in such cases the graphs of [P] versus et will give different curves for each initial concentration of enzyme.

4. Another component of the assay other than the enzyme may be unstable. This may be checked by incubating the assay mixture without each of its components in turn for an appropriate period of time, and then to start the reaction by the addition of the 'missing component'. If the initial rate is identical, irrespective of which component is added last, then instability can be ruled out. Conversely, if the initial rate is lower when one component is included in the pre-incubation but not when it is added last, this indicates an unstable component.

Having established that genuine initial rates are being measured it is im-portant to establish that the initial rate (v) varies in a linear way with the enzyme concentration, e. Doubling the enzyme concentration should double the initial rate (see Chapter 6). With enzymes that exhibit concentration-dependent polymerization/depolymerization where the polymer is more or less active than the monomer, one may see v versus e plots show significant deviation from linearity curving upwards or downwards, respectively (20). This is not often a problem but should be checked, particularly with enzymes that have not been the subject of earlier detailed kinetic study. For a more detailed review on the principles of enzyme assays and kinetic studies see Chapter 6 and (15–17).

8.2 Continuous assays

These are the most straightforward of enzyme rate assays and are applicable to all of those enzymes that utilize NADH or NADPH. The spectral change (ΔA_{340}) is easily monitored by connecting the output from the spectrophotometer to a recorder to give a continuous measure of the reaction rate. Although the slopes on the recorder traces can be used directly as a measure of the initial rate (albeit in arbitrary units) it is best to immediately translate the change in A_{340} per unit time into μmol product/min/mg protein (the generally accepted units for initial rates). Knowing the molar extinction coefficient for NADH (6.22×10^3 M^{-1} cm^{-1}) the ΔA_{340} is easily converted into a concentration (μM) and, knowing the volume of the reaction mixture in the cuvette, into an amount in nmol or μmol.

For some reactions where protons are produced or consumed it is possible, by working at very low concentrations of buffer, to use an indicator as the buffer and to monitor spectrophotometric changes in this way. To optimize the sensitivity of the assay the pH should be close to the pK of the indicator and the assay should be conducted at the pH optimum for the enzyme, where small changes in pH have no significant effect on the activity of the enzyme. This may entail an extensive screening for indicators of the appropriate pK_a *which do not inhibit the enzyme*. Details of this approach with phosphofructokinase can be found in (21).

8.3 Stopped assays

It is not always possible to obtain such obliging enzymes and then one is forced to devise alternative assays. One approach that has been particularly fruitful is the design of non-physiological (chromogenic) assays which produce coloured products that are stable and convenient to measure spectrophotometrically. *p*-Nitrophenol derivatives of organic acids, phosphoric acid and sugars have produced excellent substrates for a variety of esterases, phosphatases and glycosidases where the yellow *p*-nitrophenolate anion liberated is easily measured at 410 nm. The pK_a of *p*-nitrophenol is 6.8 so that even at pH 7 this reaction can be monitored continuously (albeit not with the maximum sensitivity). However, for 'acid' phosphatase measurements, where the pH optimum is well below the pK_a of the *p*-nitrophenol, the reaction is terminated by alkali and the fully ionized *p*-nitrophenolate anion is then measured at 410 nm. In the case of the arylsulphatases A and B, nitrocatechol sulphate has been found to be a particularly good substrate, and the nitrocatechol liberated is seen as an orange–red anion when the reaction is stopped by the addition of alkali. In these cases the addition of alkali performs two functions; one to terminate the reaction at the required time and the second to produce the chromophore.

Another common procedure for stopped assays is the use of molybdate under acid conditions to complex and spectrophotometrically assay inorganic phosphate. This provides a useful alternative to *p*-nitrophenyl phosphate for phosphatase assays, as physiologically relevant phosphate esters can be used. For example, this assay also provides a useful method for measuring ATPase activity. Details of the phosphomolybdate assay are given here (however, see Section 8.5 for methodologies that utilize plate readers).

Protocol 8

A phosphomolybdate assay for inorganic phosphate released by ATPase activity

Reagents

- 200 mM MgATP. This is prepared by dissolving ATP (1 mmol) in 4 ml $MgCl_2$ (250 mM) and titrating this solution to pH 7 using 1 M KOH. The final volume should be brought to 5 ml. (B)

- 100 mM Tris/HCl pH 8, containing 50 mM KCl, 2 mM $MgCl_2$ and 0.2 mM EDTA (A)
- 40% trichloroacetic acid (TCA) (C)
- 1% ammonium molybdate in 1 M sulphuric acid (D)

Solutions (C) and (D) are stable for months at room temperature. The MgATP (B) can be stored at $-20\,°C$ in aliquots, that should be used once and discarded. The colour reagent is prepared on the day of use by dissolving 1 g $FeSO_4.7H_2O$ in solution (D).

Method

1   Mix 0.2 ml of sample with 0.8 ml of buffer (A).

2   Pre-incubate the mixture at 30 °C for 5 min.

3   Add 0.02 ml MgATP (B) to initiate the reaction.

4   Terminate the reaction after 10 min by the addition of 0.1 ml TCA (C).

5   Add 1.9 ml of colour reagent and read the A_{700} after 10 min.

NB. The A_{700} of a blank solution made up of reagents with no enzyme added should be measured and subtracted from the reading given in step 5 above. A standard curve may be constructed by using known amounts of inorganic phosphate in 0.2 ml of water and treated as above. 1 μmol phosphate released should give an A_{700} of approximately 1.

It should be stressed for any stopped assay that, after demonstrating linear product–time curves (i.e. to show genuine initial rates), in a preliminary experiment, half-time points should be incorporated into the protocol for the highest and lowest substrate concentration used, to demonstrate initial rate conditions are maintained. For example, if the rate is linear up to 20 min then 'half-time' points of 10 min should be included routinely.

8.4 Coupled assays

In these assays, the initial reaction product is colourless, but a second enzyme is included to convert this into a product measurable by spectrophotometry. For example, for an ATPase assay, the product ADP can be used to drive NADH oxidation (measurable by A_{340}) in the presence of phosphoenolpyruvate + pyruvate kinase (which generate pyruvate in the presence of ADP) and lactate dehydrogenase (which converts NADH to NAD^+ in the presence of pyruvate). The procedure is similar to that adopted for metabolite assays (see Section 6.). In these assays it is vital that the activity of the coupling enzyme does not limit the activity of the enzyme to be measured. Similar calculations to those described in Section 6 have been suggested (15–17), but it is generally advisable to monitor the reaction with a range of concentrations of the coupling enzyme, and to use an experimentally determined amount that is clearly not rate limiting under any condition measured and which shows no significant lag. It is important to note that the conditions employed will generally be those that are optimal for the enzyme to be assayed and that under these conditions the coupling enzyme may not be operating under optimal conditions. Thus it is important to demonstrate experimentally that the amount of coupling enzyme added is sufficient.

8.5 Plate readers

Several established protocols have been adapted for 96-well plate readers including catalase, hyaluronidase, acetylcholinesterase, protein phosphatases and membrane-bound ATPases (22–26). In several instances these have involved novel protocols that are well suited to the ELISA format. For example, a sensitive, rapid microtitre-based assay for hyaluronidase activity was described by Frost and Stern (23). The free carboxyl groups of hyaluronan are biotinylated in a one-step reaction using biotin-hydrazide. This substrate is then covalently coupled to a 96-well microtitre plate. At the completion of the enzyme reaction, residual substrate is detected with an avidin–peroxidase reaction that can be read in a standard ELISA plate reader. Because the substrate is covalently bound to the microtitre plate, artefacts such as pH-dependent displacement of the biotinylated substrate do not occur. The sensitivity permits rapid measurement of hyaluronidase activity from cultured cells and biological samples, with an interassay variation of less than 5%.

The standard protocol for measuring inorganic phosphate using variations of the Fiske and Subbarow method were described in Section 8.3. There have been a number of applications described recently that have adopted a plate-reader format (see, for example, 25–27). With careful planning most assays that use colorimetric reagents (e.g. those for measuring protein concentrations described in Section 4) can be modified to facilitate the use of a plate reader.

8.6 Centrifugal analysers

These analysers use centrifugal forces to mix samples and reagents that are held in wells arranged in concentric circles about the axis of rotation of a spinning disc. The contents of an inner well (e.g. sample) are moved radially outwards passing into other wells holding reagents. After mixing, the sample/reagent solutions continue to move outwards to cuvettes located towards the perimeter of the disc. The disc rotates rapidly presenting each cuvette sequentially to an optical detector. The reactions develop with time and the absorption (or fluorescence) of each cuvette is measured each time it passes through the detector to allow the rate of reaction or final value to be determined.

Since the earlier reviews on centrifugal analysers (28, 29) there has been significant progress in design and formatting. Artuch *et al* (30) have recently described an adaptation of some previously reported methods for the Cobas Fara II centrifugal analyser, using the supernatant of a unique deproteinized blood sample for the determination of four key analytes (lactate, pyruvate, β-hydroxybutyrate and acetoacetate) in bioenergetic studies.

The advent of molecular biology and biotechnology have realized and facilitated analyses for DNA or RNA sequences and these are now being married to spectrophotometric assays using centrifugal analysers. New technological advances have led to the automation of major parts of the assay process. Automated systems have been developed for amplification and detection of nucleic acid sequence for infectious agents, using the polymerase chain reaction, ligase chain reaction, strand displacement amplification, transcription-associated amplification, and nucleic acid sequence-based amplification. Development of such automated systems are based on accumulation and integration of new molecular biotechnology. There has appeared a fully automated PCR system (COBAS AMPLICOR), which amplifies target nucleic acid sequences, captures the biotinylated and amplified products on oligonucleotide-coated paramagnetic microparticles, and detects the products with an avidin–horseradish peroxidase conjugate. Automated systems provide improvement not only in terms of labour efficiency but also assay accuracy. Recently, the extraction of a specific sequence for hepatitis C virus RNA has been automated, and the RNA can be specifically extracted from serum by hybridization with probe-coated paramagnetic microparticles, and then subjected to *in vitro* amplification (31).

9 Spectrophotometric assays for protein amino acid side chains

There are a number of useful reagents that allow the stoichiometry and reactivity of amino acid side chains to be monitored spectrophotometrically. Although these are not always absolutely specific for the amino acid side chain chosen, the experimental protocol is straightforward and the inferences can normally be checked using other methodologies. *Note that these reagents are hazardous and users should follow the appropriate laboratory safety regulations for their use.*

9.1 Cysteine

The classic reagent is 5,5'-dithiobis-(2-nitrobenzoic acid) or DTNB, sometimes known as Ellmans reagent (32), which is absolutely specific for the thiol side chain of cysteine residues. By using a molar excess of DTNB over the protein thiol concentration in the presence of 1 to 10 mM EDTA at pH 8, the thiol content of a protein can be conveniently monitored spectrophotometrically by the release of a thionitrobenzoate ion (TNB⁻) which absorbs strongly at 412 nm (ε_{412} = 13 600 M⁻¹ cm⁻¹). If all of the cysteine residues are accessible to the reagent and all have a normal pK_a (approximately pH 8.7), then the reaction is rapid as long as the concentration of reagent is in excess. From the extinction coefficient of the TNB⁻ ion released, it is possible to calculate the concentration of thiol modified. Often a fraction of the protein's thiol groups are not available for reaction because of steric hindrance to the reagent by side chains of neighbouring residues. For monitoring total content, therefore, the protein is commonly unfolded before the reaction by the addition of SDS. However, useful information on protein conformation/folding can be achieved in the absence of SDS, where kinetic sets of 'fast' and 'slow' reacting thiols (reflecting cysteine side chains in distinct environments) can be monitored. Other reagents include 4,4′-dithiopyridine that reacts with thiol groups to release 4-thiopyridone that absorbs light at 324 nm with an ε_{324} = 19 800 M⁻¹ cm⁻¹.

9.2 Lysine

Another useful reagent is trinitrobenzene sulphonic acid (TNBS) which is almost totally specific for amino groups and hence reacts well with the ε-amino groups of reactive lysine residues. It will, however, also react with the thiol group of cysteine residues, but the S–TNP conjugate is not stable in mildly alkaline solutions. Typically, 1 to 10 mM TNBS in buffer (pH 7.5 to 9) is used to modify proteins. Reaction times are normally 1 to 4 h and the reaction is conveniently monitored at 340 nm (ε_{340} = 14 000 M⁻¹ cm⁻¹). The large absorbance change coupled with the fairly high degree of specificity means that this is a useful reagent for kinetic analysis of reactive amine groups (33).

An alternative reagent for lysine residues is pyridoxal phosphate which reacts with unprotonated amine groups to form a Schiff base. The reagent therefore modifies 'reactive' lysyl amino groups at pH values around 7.5. The unstable imine can be reduced using sodium borohydride (added in aliquots to a final concentration of 1 to 10 mM after having adjusted the pH of the medium to 4.5 to 6) to form a stable adduct which absorbs at 325 nm (ε_{325} = 9710 M⁻¹ cm⁻¹) and emits fluorescence at 425 nm. Further details can be found in (34).

9.3 Tyrosine

Tetranitromethane is a useful (*but extremely hazardous*) reagent for modification of tyrosine side chains to give a coloured product that is easily monitored by spectrophotometry (35). There are a number of side reactions, including the oxidation of cysteine and the modification of histidine, tryptophan and methionine residues at higher pH, coupled with an inherent problem of crosslinking via tyrosine residues. However, if a protein concentration of 2 to 10 mg/ml can be achieved then this reaction can be carried out at room temperature adding the reagent to a final concentration of 1 to 3 mM (from a stock solution in ethanol). The reaction is terminated by gel filtration to remove nitroformate (a co-product), and the nitrotyrosine is measured at 428 nm (ε_{428} 4100 M^{-1} cm^{-1}) at pH values greater than 10. Alternatively, pH 6.0 can be used where the absorption spectrum has a maximum at 360 nm and an isosbestic point at 381 nm with $\varepsilon_{381} = 2200$ M^{-1} cm^{-1}. It should be noted that tyrosines can sometimes be disubstituted by this reagent and this may lead to errors in calculations of the stoichiometry of the reaction.

9.4 Histidine

A similar spectrophotometric determination of histidine residues is based upon the reactivity of the imidazole side chain with diethylpyrocarbonate (sometimes termed ethoxyformic anhydride) at pH 6.0 to yield an N-carbethoxyhistidinyl derivative (36). The reagent will also react to varying degrees (depending upon pH) with other residues such as tyrosine, lysine or cysteine. The addition of hydroxylamine to the modified protein will release the carbethoxy group from modified histidines and tyrosines but not from cysteines or lysines. The reaction with the imidazole side chain of histidine may be monitored and quantified by an increase in absorbance at 240 nm ($\varepsilon_{240} = 3200$ M^{-1} cm^{-1} (36)). Again caution must be used as the use of excess reagent can give rise to disubstituted derivatives of histidine which have higher molar extinctions than the mono-substituted group and this may lead to over-estimation of the equivalents of histidine in a protein (37). The carbethoxy-imidazole derivative is unstable and is readily hydrolysed by water. The rate of the reaction increases with pH above pH 6.0: at pH 7.0 and 25 °C the half-life of the group is 55 h and at pH 10.0 it is 18 min (38). Furthermore, the diethylpyrocarbonate reagent itself is also hydrolysed by water and the type of buffer used affects the rate of hydrolysis of the reagent. The half-time for the breakdown of the reagent in 60 mM sodium phosphate buffer, pH 6.0 at 25 °C, is 24 min (36). Therefore, the rate of modification of histidines in a protein will decrease with time due to the loss of the functional reagent. The concentration of the reagent will need to be 'topped up' during a modification experiment to ensure that all available residues are reacted. Care must be taken to keep the stock reagent anhydrous and stock solutions should be made up in dry ethanol.

9.5 Tryptophan

There are a number of reagents (e.g. 2-nitrophenyl-sulphenyl chloride and 2-hydroxy-5-nitrobenzylbromide) which react with tryptophan side chains to give products which can be measured spectrophotometrically. However, these require an acid pH to allow the reaction to proceed with any degree of specificity (39, 40).

10 Concluding remarks

The continuing improvement of both the available chemical reagents and instrumentation for analytical purposes is leading to increases in sensitivity and decreases in the amount of sample required. Micro-colorimeters are commercially available that require only a few μl of sample to fill liquid wave guide capillary cells with pathlengths of several cm (see World Precision Instruments *http://www.wpiinc.com*). The use of spectroscopic methods for the routine determination of concentrations of small molecules or macromolecules or rates of reaction is therefore certain to flourish for many years to come.

References

1. Tietze, F. (1969) *Anal. Biochem.* **27**, 502.

2. Bergmeyer, H. U. (1974) *Methods of enzymatic analysis.* Academic Press, New York, London.

3. Barman, T. E. (1969 and 1974) *Enzyme handbook.* Springer Verlag, Heidelberg, New York.

4. Guilbault, G. G. (1970) *Enzymatic methods of analysis.* Pergamon, Oxford, London, New York.

5. Ji, T. H. (1973) *Anal. Biochem.* **52**, 517.

6. Peters, M. A. and Fouts, J. R. (1969) *Anal. Biochem.* **30**, 299.

7. Lowry, O. H., Rosebrough, N. J., Farr, A. L. and Randall, R. J. (1951) *J. Biol. Chem.* **193**, 265.

8. Legler, G., Muller-Platz, C. M., Mentges-Hettkamp, M., Pflieger, G. and Julich, E. (1985) *Anal. Biochem.* **150**, 278.

9. Rodriguez-Vico, F., Martinez-Cayuela, M., Garcia-Peregrin, E. and Ramirez , H. (1989) *Anal. Biochem.* **183**, 275.

10. Bradford, M. M. (1976) *Anal. Biochem.* **82**, 327.

11. Hernandez, L, Marquina, R, Escalona, J. and Guzman, N. A. (1990) *J. Chromatog.* **502**, 247.

12. Liu, J. P., Hsieh, Y. Z., Wiesler, D, and Novotny, M. (1991) *Analyt. Chem.* **63**, 408.

13. Bridges, M. A., McErlane, K. M., Kwong, E., Katz, S. and Applegarth, D. A. (1986) *Clin. Chem. Acta* , **157**, 73.

14. Labarca, C. and Paigen, K. (1979) *Anal. Biochem.* **102**, 344.

15. McClure, W. R. (1969) *Biochemistry* **8**, 2782.

16. Tipton, K. F. (1985) In *Techniques in the life sciences* (ed. K. F. Tipton) Vol. B1/II, Supplement BS113, p. 1. Elsevier, Limerick.

17. Tipton, K. F. (1992) Principles of enzyme assay and kinetic studies, in *Enzyme assays: A practical approach* (ed. R. Eisenthal and M. J. Danson). IRL Press, Oxford.

18. Harris, D. A. (1987) In *Spectrophotometry and spectrofluorimetry: A practical approach* (ed. C. L. Basford and D. A. Harris), pp. 49–90. IRL Press, Oxford, Washington.

19. Selwyn, M. J. (1965) *Biochim. Biophys. Acta* **105**, 193.

20. Hulme, E. C. and Tipton, K. F. (1971) *FEBS Lett* **12**, 197.

21. Hoffer, H. W. (1971) *J. Physiol. Chem.* **352**, 997.

22. Cohen, G., Kim, M. and Ogwu, V. (1996) *J. Neurosci. Methods* **67**, 53.

23. Frost, G. I. and Stern, R. (1997) *Anal. Biochem.* **251**, 263.

24. Doctor, B. P., Toker, L., Roth, E. and Silman, I. (1987) *Anal. Biochem.* **166**, 399.

25. Fisher, D. K. and Higgins, T. J. (1994) *Pharm. Res.* **11**, 759.

26. Sadrzadeh, S. M., Vincenzi, F. F. and Hinds, T. R. (1993) *J. Pharmacol. Toxicol. Methods* **30**, 103.

27. Drueckes, P., Schinzel, R. and Palm, D. (1995) *Anal. Biochem.* **230**, 173.

28. Burtis, C. A., Tiffany, T. O. and Scott, T. D. (1976) *Methods Biochem. Anal.* **23**, 189.

29. Savory, J. (1977) *Ann. Biol. Clin. (Paris)* **35**, 261.

30. Artuch, R., Vilaseca, M. A., Farre, C. and Ramon, F. (1995) *Eur. J. Clin. Chem. Clin. Biochem.* **33**, 529.

31. Albadalejo, J., Alonso, R., Antinozzi, R., Bogard, M., Bourgault, A. M., Colucci, G., Fenner, T., Petersen, H., Sala, E., Vincelette, J. and Young, C. (1998) *J. Clin. Microbiol.* **36**, 862.

32. Ellman, G. L. (1959) *Arch. Biochem. Biophys.* **82**, 70.

33. Freedman, R. B. and Radda, G. K. (1968) *Biochem. J.* **108**, 383.

34. Lilley, K. S. and Engel, P. C. (1992) *Eur. J. Biochem.* **207**, 533.

35. Sokolovsky, M., Riordan, J. F. and Vallee, B. L. (1966) *Biochemistry* **5**, 3582.

36. Miles, E. W. (1977) *Methods in Enzymology,* **47,** 431.

37. Avaeva, S. M. and Krasnova, V. I. (1975) *Bioorg. Khim.* **1**, 1600.

38. Means, G. E. and Feeney, R. E. (1971) *Chemical modification of proteins*. Holden-Day, San Francisco.

39. Fontana, A. and Scoffone, E. (1972) *Methods in Enzymology* **25**, 419.

40. Horton, H. R. and Koshland, D. E. (1965) *J. Am. Chem. Soc.* **87**, 1126.

Chapter 8
Stopped-flow spectroscopy

M. T. Wilson* and J. Torres[†]
* Department of Biological Sciences, Central Campus, University of
Essex, Wivenhoe Park, Colchester, Essex CO4 3SQ
[†] Department of Biochemistry, University of Cambridge,
Tennis Court Road, Cambridge CB2 1QT

1 Introduction

There was a time, fortunately some years ago now, when to undertake rapid
kinetic measurements using a stopped-flow spectrophotometer verged on the
heroic. One needed to be armed with knowledge of amplifiers, light sources,
oscilloscopes etc. and ideally one's credibility was greatly enhanced were one to
build one's own instrument. Analysis of the data was similarly difficult. To obtain
a single rate constant might involve a wide range of skills in addition to those
required for the chemical/biochemical manipulation of the system and could
easily include photography, developing prints and considerable mathematical
agility.

Now all this has changed and, from the point of view of the scientist attempting to solve problems through transient kinetic studies, a good thing too! Very
high quality data can readily be obtained by anyone with a few hours training
and the ability to use a mouse and 'point and click' programs. Excellent stopped
-flow spectrophotometers can be bought which are reliable, stable, sensitive and
which are controlled by computers able to signal-average and to analyse, in
seconds, kinetic progress curves in a number of ways yielding rate constants,
amplitudes, residuals and statistics. Because it is now so easy, from the technical
point of view, to make measurement and to do so without an apprenticeship in
kinetic methods, it becomes important to make sure that one collects data that
are meaningful and open to sensible interpretation. There are a number of
pitfalls to avoid. The emphasis of this article is, therefore, somewhat different
to that written by Eccleston (1) in an earlier volume of this series. Less time will
be spent on consideration of the hardware, although the general principles
are given, but the focus will be on making sure that the data collected means
what one thinks it means and then how to be sure one is extracting kinetic
parameters from this in a sensible way.

With the advent of powerful, fast computers it has now become possible to
process very large data sets quickly and this has paved the way for the application of 'rapid scan' devices (usually, but not exclusively, diode arrays), which
allow complete spectra to be collected at very short time intervals during a
reaction. In these circumstances there is a danger of being swamped by data.

Whereas some years ago a single mixing experiment in a stopped-flow apparatus would deliver some tens of absorbance values collected at a given wavelength, the same experiment may now yield thousands of absorbance values at hundreds of wavelengths and hundreds of time intervals. Analysis of these large data sets now uses matrix algebra methods (Singular Value Decomposition, SVD) to filter signal from noise and to assess, objectively, the number of linearly independent spectral components which contribute to the total spectral change accompanying the reaction, and to extract the pseudo-spectra (see later) and their time courses. Inherent to these methods is great data compression, thus permitting economic and efficient data storage. Once the number of components and their time courses are known global analysis of the data allows one to fit the kinetic processes to a chemical model by solving, iteratively, the differential equations which describe it. By so doing, not only can the rate constants be extracted but also the true (model-dependent) spectra of any intermediates on the reaction pathway. These analytical techniques are very powerful and easy to implement; however, they can mislead and it is important to retain one's chemical/biochemical common sense when examining the output. It is on these new techniques of rapid scan spectroscopy and the analysis of data that we also spend some time.

2 Features of the basic instrument

The stopped-flow spectrophotometer is in essence a simple device designed to mix two reactants rapidly (<1 ms in conventional instruments) and efficiently and to transfer the resulting mixture, approx. 100–150 μl in standard instruments, to an optical cell as rapidly as possible (generally in about 1 ms). This is achieved by driving together solutions of the reactants from the syringes where they are stored, by pushing simultaneously their plungers (*Figure 1*). The reactants flow through the optical cell and into a further syringe, the stop syringe. Flow stops when the plunger of this syringe meets an immovable block. The sol-

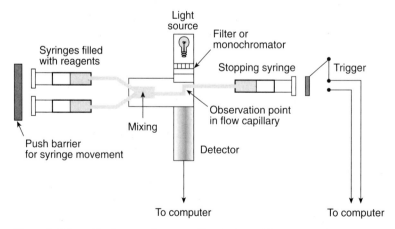

Figure 1 Schematic diagram of a conventional stopped-flow spectrophotometer.

ution within the optical cell (pathlength, l) now ages and the time course of the reaction may be followed by monitoring the absorbance at a suitable wavelength (λ). The advantage in following the change in absorbance (ΔA) is that it is directly related to the concentration (c) of the absorbing chromophore(s) through the appropriate molar *difference* extinction coefficient $\Delta\varepsilon_\lambda$, i.e. $\varepsilon_{\lambda\ product} - \varepsilon_{\lambda\ reactant}$

$$\Delta A_\lambda = \Delta\varepsilon_\lambda\, c\, l \qquad\qquad 1$$

Absorbance changes report directly concentration changes and thus may be fitted to the rate equations governing the reaction. Data capture is initiated just prior to the plunger stopping, thus recording some of the 'flow-period', see below. To achieve this rapid mixing and transfer the syringes are driven pneumatically at a relatively high pressure, usually about 10 bar, and it is essential that the system is hydraulically solid, i.e. no air bubbles should be present as these lead to optical artefacts.

2.1 Instruments available

There are now some excellent modern instruments available that are sensitive, stable and equipped with good, user-friendly, software; for example:

- Applied Photophysics Ltd
- Hi Tech Scientific
- Olis

(For further information see the List of Suppliers.) These instruments are provided with extensive instructions on how they should be used and the practical protocols provided below should be read in conjunction with these. The suppliers have informative websites that are worth visiting. These instruments, with suitable attachments, can all record spectra rapidly during the time course of a reaction, the Olis uses what amounts to a rapid scan monochromator which has the advantage that intense white light is not passed through the reaction mixture. The first two companies also manufacture small independent units that can be used in conjunction with any conventional spectrophotometer that has a reasonably fast A/D converter. These instruments can capture reaction time-courses taking less than ~1 s to complete.

When one considers the mechanical operations involved in accelerating stationary solutions to high velocity and the abrupt stopping of the mixture, they all possess remarkable stability. For single wavelength measurements using a PMT for detection these instruments are capable of determining rate constants from absorbance changes of 0.005A. This means that mechanical 'ringing' of the system induced by mechanical vibration is minimal.

The examples we have chosen to illustrate have been obtained using the Applied Photophysics instrument SX.18MV, which is provided with a PMT detection system and supplemented with a diode array for collecting spectra during a reaction time course.

3 Measurement at a single wavelength

3.1 Setting up

It is good practice to check from time to time that the stability of the instrument is within the manufacturer's specification. This is easily done by collecting what in a conventional spectrophotometer would be termed a 'baseline' by mixing buffer with buffer. In this, and all subsequent experiments described, the data are collected without using any electronic filters to remove noise. It is almost always best to reduce experimental noise by signal averaging a number of data sets rather than using damping circuits. Like most 'rules of thumb' there are exceptions and for very slow reactions with very small amplitude, damping may be necessary, but in these cases why use a rapid kinetic device?

Protocol 1

To test the stability of the detection system

Method

1   Fill working syringes with water or buffer.

2   Select a wavelength where the lamp output is high and the PMT is most sensitive, usually somewhere between 500–600 nm.

3   Set the sensitivity on the recording apparatus to a medium range, say full scale 0.2A units and set the PMT voltage (usually done automatically).

4   Wash the optical cell through up to 10 times by operating the pneumatic drive system. This should ensure that the system is bubble-free.

5   Once a reasonably flat baseline is obtained set the sensitivity to maximum, collect five traces and average these with the software provided.

6   Finally collect and average traces using the *internal* trigger, i.e. traces which do not involve mechanical mixing, flow and the arrest of flow. These traces allow one to monitor lamp fluctuations and other electronic noise.

Figure 2 illustrates the results of such a stability test. The electronic/lamp noise is essentially zero while the noise generated by the mechanical mixing amounts to ~001A. If the noise is above this, then one should clean the apparatus by purging with detergent solutions or as recommended by the manufacturer. Bubbles that can occasionally prove difficult to flush out may be removed by passing degassed water through the instrument. Lamp 'ripples' (~2 Hz) can sometimes be improved by slight lamp realignment.

3.2 Selecting the wavelength for a real experiment

If one is to use the stopped-flow apparatus in the absorbance mode it is important to know the spectra of the starting reactants and of the final products in order to select the best wavelength at which to collect kinetic data. This wavelength is

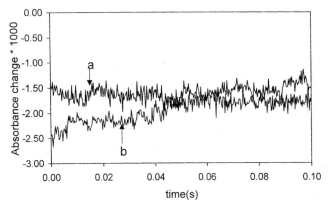

Figure 2 Baseline measurement to determine the stability of the instrument. Trace a is the average of three individual time courses, with buffer (0.1 M phosphate) in the optical cell, recorded by triggering data capture using the internal trigger. This shows the electronic and lamp noise inherent in the system and demonstrates that this is essentially random noise equivalent to << 0.001A. Trace b shows a similar average captured on mixing buffer with buffer. This demonstrates that mechanical/flow noise is ~0.001A over 100 ms.

generally the one where the absorbance change is greatest on going from reactants to products. This may not always be the case, and one must sometimes be prepared to work at wavelengths away from the peak in the difference spectrum in order to work in regions of the spectrum where the apparatus is more sensitive.

3.3 The form of a simple progress curve: Making sure the apparatus is mixing and transferring reactants to the observation chamber rapidly

Whatever the mathematical form of the time course of the absorbance change that is reporting the reaction under study, it should commence with a short time period during which the absorbance remains constant. This represents the period of time between triggering the capture device and the time when the plunger of the stop syringe meets the stopping block. This time is usually about 2 ms when mixing is pneumatically driven at pressures ~10 bar. The 'flow period' should not be confused with the dead time of the instrument (see below) which is generally shorter.

The performance of the pneumatic system, mixing and transfer systems may be tested by using any reaction with a moderately short half time, say 10–20 ms, e.g. the reaction between 2,6-dichlorophenolindophenol, DCIP, and L-ascorbate. Here we illustrate using the reaction between myoglobin and carbon monoxide, but any reaction taking some 50 ms or less to complete may be substituted:

$$Mb + CO \underset{k_{-1}}{\overset{k_1}{\rightleftharpoons}} MbCO \qquad\qquad 2$$

Mb represents deoxy-Mb and MbCO the carboxy derivative.

213

Protocol 2

To determine the time between initiation of the signal recording and the stop of the flow of solution

Reagents and materials

- Prepare a solution of ferric (met) myoglobin by dissolving ~2 mg of the freeze-dried protein, Sigma Chemical Co., in 10 ml of buffer (say 0.1 M sodium phosphate pH 7.0). Add a few crystals (~2 mg) of solid dithionite ($Na_2S_2O_4$, BDH) to remove oxygen from solution and to reduce the myoglobin to form the ferrous deoxy species.

- Prepare the stock CO solution, 1 mM, by equilibrating water (or dilute buffer) at 20 °C with an atmosphere of pure CO at 1 atmosphere pressure (101 kP). Subject the buffer to several cycles of exposure to vacuum and equilibration with CO (at 101 kP pressure) accompanied by vigorous shaking. Gas equilibrates with a liquid phase faster the greater the interface area between them, thus many small bubbles are better than a few larger ones. Draw into a syringe a few ml of the CO solution and add a little sodium dithionite (~2 mg/10 ml). CO gas is toxic and these operations should be performed in a 'fume' cupboard.

Method

1 Load the Mb and the CO solution into the working syringes of the stopped-flow apparatus. Flush the apparatus through with some five shots to ensure the solutions have fully replaced water in the mixing chamber and optical cell.

2 Ensure there is no electronic noise filter and collect a further three to five traces, each of some 400 time points, and average these.

Figures 3 and 4 illustrate a typical result of such an experiment. The 'flow period' is clearly seen occupying the first 1.5 ms. Thereafter, once flow has stopped, the formation of the carbon monoxy complex is seen.

Should this profile be distorted it may indicate that there is a problem with the smooth and rapid flow of the reagents, possibly syringes sticking or the viscosity of the solutions not permitting the solutions to accelerate fully to the desired velocity. Alternatively the pressure is low. Check the pneumatic pressure, clean the system as recommended. Possibly the apparatus needs to be adjusted to deliver more of each solution by moving the stopping block.

3.4 Measurement of the dead time

The first question one might legitimately ask is 'why bother?'. Why should one check the dead time of the apparatus, what bearing has it on the interpretation of subsequent experimental results? The dead time is defined as the time required for the reactants, once mixed, to travel from the mixing chamber to the observation chamber. It is, therefore, a period of time in which the reaction is progressing, but not being recorded. This being so, it must follow that the

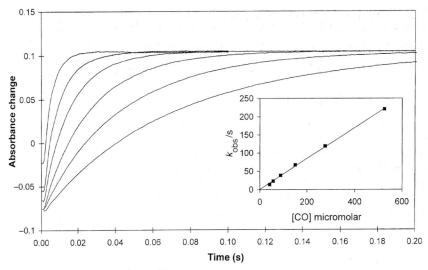

Figure 3 Reaction between deoxy myoglobin and carbon monoxide. The time courses show the binding of CO to deoxy myoglobin (10 μM) as a function of CO concentration (41, 56, 88, 150, 275 or 525 μM). The monitoring wavelength was 421 nm and the temperature 20°C. The reaction was carried out in sodium phosphate buffer, 0.1 M, pH 7.0. The initial flat portion of the progress curve (seen better in *Figure 4*) illustrates the flow period. The inset gives the dependence of the pseudo-first-order rate constant, k_{obs}, on CO concentration.

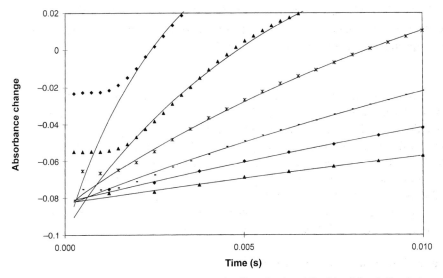

Figure 4 Analysis of time course to measure the 'flow time' and the 'dead time'. The first 10 ms of the traces in *Figure 3* are shown as individual data points. The solid lines are the exponential fits to the data extrapolated back to true time zero. The flow period is the flat initial portion of the curve and, in principle, may have any length and depends on the time between triggering data capture and the flow stopping. The dead time is the time between $t = 0$, given by the point of intersection of the curves, and when flow stops. In this experiment the dead time and flow period are approximately the same, but need not be so.

total amplitude of the absorbance change observed is smaller than expected from static spectroscopy, from which it is always possible to obtain the start and end points of a reaction, if not the kinetic progress curve connecting these. The fraction of the amplitude lost depends upon the rate constant for the reaction and the length of the dead time. For very rapid reactions, the majority and indeed the entire kinetic time course may be lost.

The principle of the method by which the dead time is determined is by measuring the fraction of the expected absorbance change of a well-characterized reaction that is lost in the time between mixing and observation. This is usually achieved through the use of a suitable second-order reaction, the rate of which can be controlled through the concentration of the reactants. This method is illustrated using the reaction given in *equation 2*, but any suitable reaction may be substituted. (e.g. the reaction between 2,6-dichlorophenolindophenol, DCIP, 1 mM and L-ascorbate in the concentration range 2–120 mM, concentrations before mixing (2)).

Protocol 3

Measurement of the 'dead time' of the instrument

Reagents

The myoglobin and CO solutions prepared as described in *Protocol 2* may be used.

Methods

1   Mix the deoxy Mb with 1 mM CO and record at least three traces (at least 200 points over 50 ms) and average.

2   Dilute the CO solution with buffer which has been degassed and equilibrated with N_2 and to which a small quantity of solid sodium dithionite has been added. To do this, without exposure to air, take a known volume of the CO solution into a glass graduated syringe (partially filling it) and connect this by its nozzle to a similar syringe containing the anaerobic buffer by using a short plastic tube. Push a known volume of the buffer into the CO solution. It is usually convenient to mix equal volumes thereby diluting the CO by a factor of two.

3   Record a further set of traces and average.

4   Repeat the CO dilution step and collect data over the appropriate time range a further three to four times. After each change of [CO] flush through some five shots to ensure the new solution has completely replaced the previous solution.

The results of such an experiment are given in *Figure 3*. As expected for a second-order reaction the rate depends on the concentration of the reactant in excess, here CO. The reaction being completed in a shorter time at high CO concentrations. Moreover it is seen that although the absorbance change must be the same at all CO concentrations because the myoglobin concentration is unchanged, the ΔA measured is smaller at high CO concentrations. This is because of the loss of absorbance due to the reaction in the dead time.

By extrapolating each curve back into the dead time there should be a time point where all the curves intersect, giving the true time zero of the reaction. *Figure 4* shows the extrapolation of a set of exponential curves fitted to the data presented in *Figure 3*. This is typical for such an experiment. The majority of the time courses intersect while one is to one side of this point. Nevertheless, the intersection point for the majority of the lines is a good estimate of time zero. The average time between this intersection point and the end of the flow period, i.e. where observation truly starts is approx. 1.3 ms and this is the dead time of the instrument. This may be made slightly shorter by cleaning the instrument, making sure the syringes are running smoothly or by slightly increasing the pressure of the pneumatic drive.

3.5 Amplitude of the signal

It is often essential to measure the amplitude of an absorbance change (ΔA_λ) associated with a reaction in order to calculate the change in concentration of a given species. For example, one may wish to determine the stoichiometry of the reaction, i.e. how much product is formed from known concentrations of re-actants. In principle, this is easy to do. The total amplitude change at a selected wavelength is given directly by all modern instruments, i.e. from the end of the flow period to the end point (ΔA_0). This may be corrected for loss of absorbance in the dead time (t_d) to give the true absorbance (ΔA_T). If the reaction follows an *exponential* time course this may easily be achieved from the reaction half-time ($t_{1/2}$) by using

$$\ln \left(\frac{\Delta A_T}{\Delta A_0} \right) = \ln 2 \left(\frac{t_d}{t_{1/2}} \right) \qquad\qquad 3$$

which may be transformed to

$$\log \left(\frac{\Delta A_T}{\Delta A_0} \right) \sim 0.3 \left(\frac{t_d}{t_{1/2}} \right) \qquad\qquad 4$$

Equation 3 yields

$$\ln (\Delta A_0) = \ln (\Delta A_T) - k t_d \qquad\qquad 5$$

that gives a straight line on plotting ΔA_0 versus k (the first-order rate constant, see Section 4.1), the slope of which is t_d, the dead time. (A similar, if more complex procedure, may be applied if the progress curve comprises more than one exponential.)

The absorbance change (ΔA_T) may now be converted to concentration change via the appropriate extinction coefficient, $\Delta \varepsilon_\lambda$ (see *equation 1*). However, it may be incorrect to use the value of this coefficient reported in the literature as this has been determined by conventional spectroscopy. It is often advisable to de-termine the *apparent* extinction coefficient in the stopped-flow apparatus using the instrumental settings employed in the experiment. This is because the monochromator slit width settings are often much wider than those used in a conventional spectrophotometer. This is especially the case when one is using highly absorbing solutions and one wishes to pass more light through the sample. The result of this is a much wider bandpass, i.e. the beam is no longer

Table 1 Dependence of the ΔA measured on the monochromator slit widths

Entrance slit (mm)	Exit slit (mm)	ΔA measured	$\Delta \varepsilon_\lambda$ *apparent* $\times$ **10^{-3}**
0.1	0.1	0.106	19.4
0.1	1	0.102	18.7
1	1	0.096	17.6
1	2	0.081	14.9
1	3	0.078	14.3

monochromatic but contains wavelengths on either side of that chosen to monitor the reaction. This hardly matters if the absorption band is broad, but if it is sharp, the absorption is averaged and thus attenuated. In such cases, it is well to measure $\Delta \varepsilon_\lambda$ *apparent* at the chosen wavelength. Here we illustrate using as an example cytochrome c but this calibration should be carried out for any reaction where ΔA_λ is required to be known accurately.

Cytochrome c possesses a sharp band at 550 nm when the central iron is in the ferrous state. The true $\Delta \varepsilon_\lambda$ at this wavelength on going from the ferric to ferrous form is 20 000 M^{-1} cm^{-1}. The table reports the ΔA values measured at 550 nm in a stopped-flow experiment in which 10.9 μM ferricytochrome c was mixed with 10 mM sodium ascorbate (0.1 M sodium phosphate pH 7.0). The concentration of cytochrome c after mixing was 5.45 μM and thus the expected ΔA was 0.109A. *Table 1* gives the values of ΔA measured at a number of settings of the monochromator slit width and reports the *apparent* $\Delta \varepsilon_\lambda$ for these settings. It is seen that as the slits open the absorbance measured falls because the monitoring beam becomes less monochromatic. The amplitude of the signal at the selected wavelength, once corrected for the absorbance lost in the dead time, should be compared with that expected from static experiments, i.e. the total absorbance change determined from the spectra of the starting materials and the spectrum of the reaction mixture at completion. *This is an important check because if there is a discrepancy this means that a fraction of the absorbance change has occurred very rapidly within the dead time in a process with kinetics different from those of the fraction that is observed.*

3.6 Assigning a signal

It is not always obvious how an absorbance change monitored at a single wavelength should be assigned. There is a danger of wrongly assigning such a signal in the light of preconceptions regarding the nature and mechanism of the reaction under study.

Suppose, for example, one wishes to determine the kinetics of the appearance of a species, C, formed from the reaction between A and B, i.e.

$$A + B \rightleftharpoons C \qquad\qquad 6$$

If the reaction is to be followed at a single wavelength, then this is chosen with reference to the spectra of the components in order to maximize the expected absorbance change. Suppose that absorbance increases on formation of C, and that on mixing A and B an increase in absorbance is indeed observed that follows an apparently sensible time course. It is tempting to assign this time course to the process under study, i.e. the production of C. However, one must be careful, and if the system has not been studied previously it is wise, if one is not to make embarrassing errors, to confirm this assignment. The way to do this is to monitor the reaction at a number of wavelengths and construct a 'kinetic difference spectrum'. If this spectrum matches in shape and amplitude that determined in a conventional spectrophotometer, then one is justified in assigning the time course and can thereafter proceed with determination of the kinetic parameters.

We have chosen to illustrate this problem by reference to the reaction between a ruthenium complex and nitric oxide (NO) because it poses a number of problems. This reaction is of interest because of the role that such complexes may play as NO scavengers in pathological conditions such as sepsis (3). The reaction is very rapid and the absorbance change associated with NO binding is small. This requires that the reaction be carried out at low temperature (here 7 °C) and at equal concentrations of reagents to yield the maximum absorbance change at the lowest rate. The absorption spectrum of the compound is in the UV region and approaches the lower end of the range accessible to instruments equipped with standard light guides. Furthermore, in this region there is the possibility of interference from side reactions, such as the production of nitrite. It is essential when studying such a reaction that assignments are carefully checked. Although this reaction is rather specific and unlikely to be widely studied, the problem it poses is typical of many reactions and the protocol below explains how to deal with these.

Protocol 4

Determination of a kinetic difference spectrum

Reagents

- Prepare in 50 mM potassium phosphate buffer, pH 7.4, an anaerobic solution of a ruthenate chelate, here potassium chloro[hydrogen(ethylenedinitrilo)ruthenate], and dilute this in the same anaerobic buffer to a final concentration of 50 μM of the complex (Solution A).

- Equilibrate NO gas at one atmosphere (101 kP) pressure and 20 °C with anaerobic water (see *Protocol 2* above for CO). This yields a solution ~2 mM in NO.

- Dilute, to a final concentration of 50 μM, the stock NO solution with the anaerobic buffer, using the procedure described for CO above (*Protocol 3*).

Method

1 Fill two optical cuvettes with the anaerobic solution of the ruthenium complex, and using one as a sample and the other as a reference take the baseline between 220 and 400 nm. Gently bubble a little NO gas through the sample solution and take the spectrum again. This is the difference spectrum between the Ru–NO complex and the Ru-chelate (see *Figure 5*).

Protocol 4 continued

2 Transfer the remainder of the solutions to the stopped-flow apparatus that has been previously flushed through with anaerobic buffer.

3 Set the monochromator to the wavelength of the maximum absorbance difference in the spectrum (here ~290 nm) and record a kinetic trace (*Figure 6*).

4 Record time courses at 5–10 nm intervals, being sure to pass through the isosbestic point indicated in *Figure 5*.

5 Plot the absorbance change (corrected at each wavelength) against wavelength, as in *Figure 5*. It is clear that the spectral distribution of the absorbance changes follows the absorption spectrum (there is a factor of two between the amplitude that is accounted for by the twofold dilution that occurs in the stopped-flow mixing).

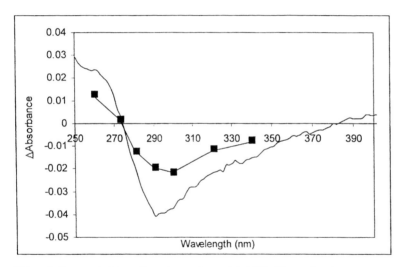

Figure 5 The spectral transition observed on mixing NO with a potassium chloro[hydrogen(ethylenedinitrilo)ruthenate]. The solid line is the difference spectrum between the complex in the absence and presence of excess NO. The square points represent the amplitude of the absorbance changes observed in the stopped-flow experiment mixing the Ru complex with NO and monitoring at the wavelengths depicted. The factor of two between the static and kinetic constants comes from dilution in the mixing experiment.

4 Determining rate constants

Time courses can follow a number of mathematical forms depending on mechanism. The software that runs stopped-flow spectrophotometers provides fitting programs to extract rate constants. The operator must however decide the likely form of the curve and the time range over which the curves should be fitted.

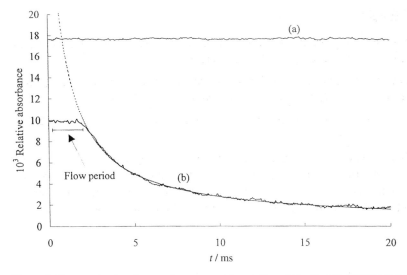

Figure 6 Changes in absorbance of an anaerobic solution of the Ru–NO complex in the presence and absence of NO. The Ru–NO complex (25 μM) in the absence (a) and presence (b) of NO (25 μM) in phosphate buffer (50 mM), pH 7.4 at 7.3°C. The fit is to the appropriate second-order rate equation 14. The trace is the average of six individual time courses.

4.1 First-order processes

Reactions of the form

$$A \underset{k_{-1}}{\overset{k_1}{\rightleftharpoons}} B \qquad\qquad 7$$

give rise to exponential time courses of the following form.

$$[A] = [A]_0 e^{-(k_{obs})t} \quad [B] = [A]_0(1 - e^{-(k_{obs})t}) \qquad\qquad 8$$

where $k_{obs} = k_1 + k_{-1}$; k_1 and k_{-1} are the first-order rate constants and have units s^{-1} and the equilibrium constant, K, is given by $K = k_1/k_{-1}$. As the concentrations and absorbances are linearly related the absorbance changes used to monitor the reaction are described by the same equations.

Where the equilibrium lies far to the right, the value of k_{-1} may be ignored and $k_{obs} \sim k_1$. An example of such a reaction is provided by pH-dependent transitions, common in proteins, and which may be studied by 'pH jump' experiments. In such experiments, a pH-dependent equilibrium is perturbed by mixing the system equilibrated at one pH in a weak buffer with a strong buffer at a different pH. An example is afforded by the spectral change induced in cytochrome c by increasing the pH. Here we illustrate with a chemical derivative of cytochrome c in which the methionine that is normally coordinated to the iron is carboxymethylated so that it can no longer play this role (4). In this modification the ferrous haem c exhibits a pH dependent spin state transition accompanied by a large spectral change.

Protocol 5

An experiment to determine the effect of a 'pH jump' on the absorption of carboxymethylated horse heart cytochrome *c*

Reagents

- Prepare a solution of ~5 ml of the protein (~10 μM) in a weak buffer (say 10 mM sodium acetate pH 5.5).

- Prepare a stronger buffer at the desired pH value, e.g. 0.1 M sodium phosphate pH 7.

Method

1. Fill one drive syringe with the protein solution and fill the other drive syringe with the strong buffer solution.

2. Set the wavelength to give the maximum absorbance difference for the pH-dependent change, determined separately as described above.

3. Discharge several 'shots' to flush out old reagents/buffer and then record and average the data from several reactions.

The time course approximates closely to an exponential increase in absorption at 417 nm and has been fitted as such to yield a pseudo-first-order rate constant of 35 s^{-1}. In *Figure 7* the difference (multiplied by ten and moved upwards by 0.1*A*) between the experimental time course and the fit is also shown. The residuals are small, indicating a good fit, but also show a non-random distribution which amounts to a small ripple ~10^{-3}*A* in amplitude which is instrumental in origin, see *Figure 2*. The initial few ms are distorted by the 'flow' period.

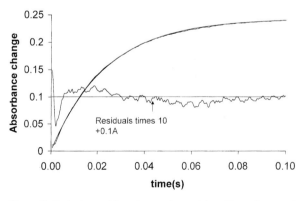

Figure 7 Optical transition observed on mixing 10 μM ferrous Cm-cyt *c* at pH 5.5 (0.01 M sodium acetate) with sodium phosphate (0.1 M) buffer pH 7, temperature 20°C. The fit is to a single exponential and the residuals are shown multiplied by a factor of ten and moved by 0.1*A* for clarity.

4.1.1 Multi-phasic time courses composed of the sum of exponentials

It is very often the case that time courses are not simple but comprise more than one exponential phase. This may be because there are parallel first-order reactions proceeding simultaneously. In principle it is very easy to extract the separate rate constants and the amplitudes of the absorbance change associated with each reaction. All instruments are provided with software that will fit experimental data to the sum of exponentials. However, this simplicity of the fitting procedure belies the underlying difficulty inherent in this analysis. This may be illustrated by reference to the reaction of Cm-cyt c described above but at pH 9. Now the time course is the sum of two first-order processes, one fast with $t_{1/2} \sim 2$ ms and the other, slower, with $t_{1/2} \sim 10$ ms, see *Figure 8*. The trace is easily fitted, but in order to do so the user must select the time range over which the fitting should take place. This selection should be made to ensure that approximately equal numbers of data points representing the kinetic process are collected (best done by using a split time base). However, this may not always be easy to judge and it is best to be aware that rather different values for rate constants may be obtained depending on the 'fit-range' one chooses as shown in *Figure 8*.

4.2 Second-order processes

Many reactions of biochemical interest are of the mechanistic form

$$A + B \underset{k_{-1}}{\overset{k_1}{\rightleftharpoons}} C \quad \text{or} \quad A + B \underset{k_{-1}}{\overset{k_1}{\rightleftharpoons}} C + D \qquad\qquad 9$$

i.e. second-order processes.

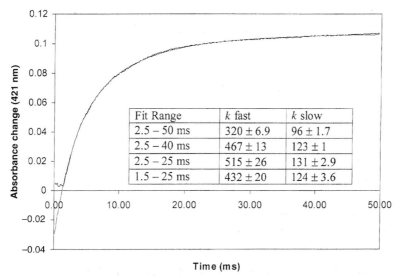

Fit Range	k fast	k slow
2.5 – 50 ms	320 ± 6.9	96 ± 1.7
2.5 – 40 ms	467 ± 13	123 ± 1
2.5 – 25 ms	515 ± 26	131 ± 2.9
1.5 – 25 ms	432 ± 20	124 ± 3.6

Figure 8 The same experiment as shown in *Figure 7* except the final pH was 9. The inset table illustrates that with biphasic traces, here two exponentials, the value of rate constants (k s^{-1}) determined by the fit depend critically on the fit range chosen.

Integration of the rate equations for the above yield complex and unwieldy expressions and if true second-order reversible reactions are to be analysed it is often best to use global analysis to retrieve the rate constants, see below. However, in many instances the back reaction may be neglected and the reaction considered as quasi-irreversible. Under these circumstances the rate equation may be written:

$$\frac{dx}{dt} = k_1 ([A_0] - x)([B_0] - x) \qquad 10$$

where x is the concentration of the product(s) and $[A_0]$ and $[B_0]$ are the initial concentrations of A and B mixed in the stopped-flow spectrophotometer. The second-order rate constant (k_1; units $M^{-1}\,s^{-1}$) has the form

$$k_1 t = \frac{1}{([A_0] - [B_0])} \, l_n \left(\frac{[B_0]([A_0] - x)}{[A_0]([B_0] - x)} \right) \qquad 11$$

There are a number of ways to use this equation in the analysis of data but the simplest way to proceed is to use this equation directly with the fitting program provided with the apparatus. There is one problem to bear in mind, however, if such a fitting procedure is used, namely, the routine will be executed *using the measured absorbance values* and not the *concentration of the reactants*. This means that the second-order rate constants will have the units of $\Delta A^{-1}\,s^{-1}$ and thus must be converted by use of the appropriate $\Delta\varepsilon_\lambda$ to $M^{-1}\,s^{-1}$.

This equation may be considerably simplified if one conducts the experiment under conditions in which one of the reactants, say A, is in large excess (>10-fold over the other). Under these conditions the concentration of the component in excess can be considered to remain constant during the reaction, i.e. $[A_0] >> x$. Under these circumstances x collapses to

$$x = [B_0](1 - \exp(-k_{obs}t)) \qquad 12$$

that has the same form as that describing a first-order process. Here k_{obs}, *the pseudo-first-order rate constant* has the form $k_{obs} = k_1[A_0]$, i.e. the second-order rate constant multiplied by the concentration of the component in excess.

Under pseudo-first-order conditions ($[A_0] >> [B_0]$) reversible second-order reactions also yield the same exponential forms as above but now

$$k_{obs} = k_1[A_0] + k_{-1} \qquad 13$$

Thus under these conditions both k_1 and k_{-1} may be found from experiments in which the pseudo-first-order rate constant (k_{obs}) is determined at a number of concentrations of the reactant in excess. The data presented in *Figure 3*, which was used to measure the dead time, are the results of such an experiment. In the inset to *Figure 3* the pseudo-first-order rate constants (k_{obs}) derived from exponential fits to the data are plotted against the CO concentration. The slope gives the second-order rate constant k_1, ~$4 \times 10^5\ M^{-1}\,s^{-1}$, and the intercept k_{-1}, $0.6\ s^{-1}$. The latter value is rather high compared with the literature value of $0.04\ s^{-1}$ and illustrates the limitation of the method when k_{-1} is very small (5).

The rate constant of any first-order process is related to the half-time of the reaction (i.e. the time taken to complete half the reaction) by the expression $k_{obs} = l_n2/t_{1/2} \sim 0.693/t_{1/2}$.

Essentially irreversible second-order processes may also be analysed easily if experiments are performed under condition in which the starting reactants are at equal concentration, i.e. $[A_0] = [B_0]$. Under these circumstances the solution of the rate equation gives

$$\frac{x}{([A_0] - x)} = k_1[A_0]t \qquad 14$$

This simple analytical equation may be used in fitting procedures (see *Figure 6*). The rate constant is related to the half-time of such a reaction by the expression $k_1 = 1/[A_0]t_{1/2}$.

4.2.1 More complex processes

First and second-order processes are often coupled together through common chemical intermediates to form more complex mechanisms. There are numerous examples, e.g. the coupling between second-order complex formation, or electron-transfer reactions, and first-order protein conformational changes. The analysis of some of these is given elsewhere (see Chapter 9). Here we consider a mechanism common to many proteases in which the enzyme and its substrate come together in a second-order process which leads to acylation of the enzyme and release of a product, P_1. The acyl enzyme is now hydrolysed to yield a second product, P_2, and the original enzyme that enters a second cycle. This may be written as follows:

$$
\begin{array}{ccccc}
& k_1[S] & & k_2 & \\
& \downarrow & & \downarrow & \\
E + S & \rightleftharpoons & ES' & \rightleftharpoons & E \qquad 15 \\
& \searrow & & \searrow & \\
& P_1 & & P_2 &
\end{array}
$$

It can readily be appreciated that if $k_1[S]$ is much larger than k_2 then on mixing E and S there is an initial rapid production of P_1, the 'burst', the concentration of P_1 being stoichiometric with the initial concentration of the enzyme $[E_0]$. Indeed as it is often easier to measure $[P_1]$ than $[E_0]$ this provides an excellent way to measure the latter and is the basis of active site titrations. Once the enzyme has been acylated it decays to form P_2, and in the presence of excess substrate the cycle is repeated, but in the second and subsequent cycles the enzyme cannot react with S faster than the enzyme deacylates. Provided $[S] >> [E_0]$ this leads to a linear time course for P_1 formation, with rate dependent on the value of k_2.

Where $k_1[S]$ and k_2 are comparable the situation is more complex and analysis of the mechanism leads to the conclusions depicted in *Figure 9*. This diagram gives the basis of active site titrations in which the enzyme is mixed with

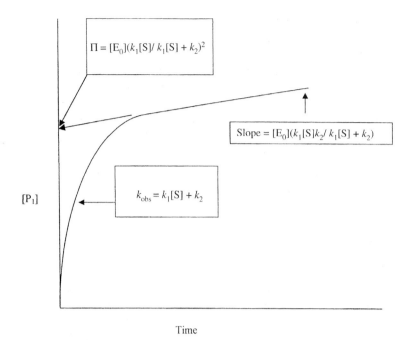

Figure 9 Schematic diagram illustrating the 'burst' seen in the concentration of P_1 on mixing a protease with its substrate. The amplitude of the 'burst' and the rates are determined by the kinetic parameters given in the text as are the concentrations of the enzyme and substrate.

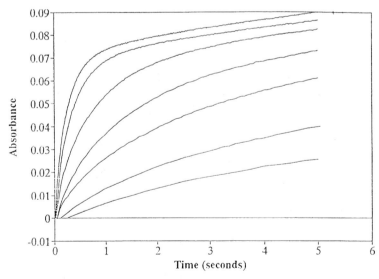

Figure 10 The reaction between rat tissue Kallikrein (5 μM) and p-nitrophenylguanidino benzoate. Kallikrein (5 μM) was reacted with 6.25, 12.5, 25, 50, 100, 250 and 500 μM p-nitrophenylguanidino benzoate. The monitoring wavelength was 402 nm where p-nitrophenol, P_1 absorbs. The temperature was 25°C.

increasing concentrations of S and the amplitude of the 'burst', Π, is measured. As [S] becomes very large Π tends to $[E_0]$, k_{obs} to $k_1[S]$ and the slope of the linear portion tends to $k_2[E_0]$ (6).

An example of 'burst' kinetics is given in *Figure 10* in which rat tissue kallikrein is reacted with a substrate, *p*-nitrophenylguanidino benzoate. In this system P_1 is *p*-nitrophenol that absorbs strongly at 402 nm allowing the reaction to be monitored easily. The figure shows the time course at increasing concentrations of substrate. At high values of [S], the 'burst' phase followed by a linear time course is seen. The maximum amplitude, together with the extinction coefficient for *p*-nitrophenol, $14\,000$ M^{-1} cm^{-1}, gives the enzyme concentration as 5 μM.

5 Multi-wavelength detection: diode array 'rapid scan methods'

We have concentrated above on aspects of monitoring the kinetics of a reaction by capturing and analysing time courses at a single wavelength using a photomultiplier as detector. In contrast to this a diode array detector can register simultaneously absorbance changes at a large number of wavelengths. This feature is especially valuable when a complex reaction, involving a number of spectral species, is to be analysed. The method relies on passing white light through the optical cell where the reaction is taking place. The emergent light beam is now dispersed, using a diffraction grating onto a linear array of light sensitive diodes (usually $\sim$100), such that a small range of the spectrum ($\sim$3 nm wide) impinges on each diode. As the reaction proceeds, the changing composition of the reaction mixture absorbs light at those wavelengths specific for the products, while progressively allowing light absorbed by the reactants to pass through. These changes are recorded by the diode array, and by comparing the voltage output of each diode with that when only water (or buffer) is in the cell, an absorbance spectrum may be constructed. This spectrum may be updated at known time intervals, thus providing the experimenter with spectra recorded throughout the time course of the reaction. Although the precision in the measurement of absorbance amplitudes is better in the single wavelength mode monitored by a photomultiplier, it may be appreciated that the ability to monitor kinetics simultaneously at some 100 wavelengths and monitor the spectral changes throughout a reaction, all from one 'shot', holds enormous advantages.

The data set generated by a single rapid scan experiment is very large. Typically, for an experiment that involves the collection of 200 time points, e.g. approximately 20 s of a reaction in which data are collected with logarithmic periodicity, in a spectral range 370–720 nm (with one data point every 3.3 nm), a file of around 220 kbytes will be created. The content of these raw data files can be represented by a matrix **A** in which each column represents a spectrum obtained at a given time point and each row represents a time course at a given wavelength. Capital bold letters denotes matrices. Thus $\mathbf{A} = a_{ij}$

	t_1	t_2	t_3	t_m
λ_1	a_{11}	a_{12}	a_{13}	$\ldots$
λ_2	a_{21}	a_{22}	a_{23}	$\ldots$
λ_3	a_{31}	a_{32}	a_{33}	$\ldots$
λ_n	$\ldots$	$\ldots$	$..$	a_{nm}

16

It is often the case that the reaction analysed is not simple and may possibly contain more than one substrate (or product) and one or more intermediates on the reaction pathway. This increased number of unknowns in the system can be successfully determined with the additional information provided by a multi-wavelength experiment. Obviously, the accuracy of this determination will depend on the amount of information available, and it will be more precise the more data points are taken in the spectral region or time period of interest. Thus, a whole process is composed of partial reactions, each with its own rate constant and spectral components. This information is contained in matrix $\mathbf{A}$, together with random noise inherent in any such experiment, and can be extracted using a deconvolution procedure, based on matrix algebra methods, called *singular value decomposition* (SVD), followed *by global fitting*. What is required is that we analyse the matrix in such a way that we can determine the minimum number of processes contributing to the overall spectral change, i.e. the number of independent time courses and spectral components which together can combine to form the whole data set. Then, using this information one fits the data set to a mechanistic model of the reaction thereby yielding the rate constants and spectra of all species.

In what follows we give a theoretical outline of the methods and then take the reader through the steps using a practical example. For further information and examples the reader may consult references (7–9).

5.1 SVD (singular value decomposition)

'Although SVD performs no magic, it does efficiently extract the information contained in data sets (i.e. matrix $\mathbf{A}$) with a minimum number of input assumptions' (Henry and Hofrichter; 7).

The goal of SVD is to provide an objective assessment of the *minimum number* of spectra and time courses (formally vectors) that through linear combination can generate the whole data set. What these vectors mean in this context can be understood by reference to the simple reaction: A → B. At any time during the course of the reaction the spectrum collected is a certain linear combination of the spectra corresponding to species A and B. Thus, the concentration of A at time t is given by: $[A] = [A_o]e^{-kt}$ (see above). Correspondingly, the concentration of B at any time t is: $[A_o] - [A]$, if $[B_o] = 0$. Therefore:

$$[B] = [A_o] - [A_o]\,e^{-kt} = [A_o]\,(1 - e^{-kt})$$

17

As the absorbance is defined as the concentration times the molar extinction, we multiply each of the concentrations by the respective molar extinction coefficient ε_A and ε_B, the sum of the absorbancies corresponding to A and B at any time t is $\varepsilon_A[A_o]e^{-kt} + \varepsilon_B[A_o](1 - e^{-kt})$. As $[A_o]$ is a constant, we can write $\varepsilon_A e^{-kt} + \varepsilon_B(1 - e^{-kt})$. This shows that the total absorbance can be expressed as a linear combination of ε_A and ε_B, and the minimum number of linearly independent vectors that can be used to obtain matrix **A** is two. Alternatively, if the last expression is regrouped, $\varepsilon_A e^{-kt} + \varepsilon_B - \varepsilon_B e^{-kt}$, we obtain $e^{-kt}(\varepsilon_A - \varepsilon_B) + \varepsilon_B$, which again gives us two vectors $(\varepsilon_A - \varepsilon_B)$ and ε_B. This time, although the number of vectors is the same, one of them is the difference between the two.

A similar conclusion can be reached from the system:

$$A \overset{k_1}{\to} B \overset{k_2}{\to} C \qquad\qquad 18$$

As before,

$$[A] = [A_o]\, e^{-k_1 t} \qquad\qquad 19$$

In this case, however, the solution for [B] (or [C]) is not immediately evident:

$$[B] = [A_o]\left[\left(\frac{k_1}{k_2 - k_1}\right) \times (e^{-k_1 t} - e^{-k_2 t})\right] \qquad\qquad 20$$

and, as $[A] + [B] + [C] = [A_o]$:

$$[C] = [A_o]\left[1 + \left(\frac{k_1 e^{-k_2 t} - k_2 e^{-k_1 t}}{k_2 - k_1}\right)\right] \qquad\qquad 21$$

The absorbance at any time point is expressed as the sum $\varepsilon_A[A] + \varepsilon_B[B] + \varepsilon_C[C]$, and the linearly independent vectors will be ε_A, ε_B and ε_C. As before, certain combinations of these vectors, e.g. ε_A, ε_B and $(\varepsilon_C - \varepsilon_B)$ or $(\varepsilon_A - \varepsilon_B)$ and ε_B and ε_C, are equally valid. In any case, the minimum number of linearly independent vectors in this system is three.

The number of linearly independent vectors is, however, not necessarily the same as the number of spectrally distinct species. This is made clearer by consideration of the next system, which can be defined by three linearly independent vectors, even though it contains four spectrally different species:

$$\begin{array}{c} A \to B \\ C \to D \end{array} \qquad\qquad 22$$

As in the first example,

$$\begin{array}{c} [A] = [A_o]\, e^{-k_1 t} \\ [B] = [A_o]\, (1 - e^{-k_1 t}) \end{array} \qquad\qquad 23$$

and for the second reaction,

$$\begin{array}{c} [C] = [C_o]\, e^{-k_2 t} \\ [D] = [C_o]\, (1 - e^{-k_2 t}) \end{array} \qquad\qquad 24$$

Hence, the absorbance at any time point is represented by $\varepsilon_A[A] + \varepsilon_B[B] + \varepsilon_C[C] + \varepsilon_D[D]$, and substituting for the expressions above:

$$\varepsilon_A[A_o]\, e^{-k_1 t} + \varepsilon_B[A_o]\, (1 - e^{-k_1 t}) + \varepsilon_C[C_o]\, e^{-k_2 t} + \varepsilon_D[C_o]\, (1 - e^{-k_2 t}) \qquad\qquad 25$$

As the initial concentrations are constants, this can be expressed as:

$$\varepsilon_A \, e^{-k_1 t} + \varepsilon_B \, (1 - e^{-k_1 t}) + \varepsilon_C \, e^{-k_2 t} + \varepsilon_D \, (1 - e^{-k_2 t}) \qquad 26$$

and regrouped as:

$$(\varepsilon_A - \varepsilon_B) \, e^{-k_1 t} + (\varepsilon_C - \varepsilon_D) \, e^{-k_2 t} + (\varepsilon_B + \varepsilon_D) \qquad 27$$

Hence, this system, which has four spectrally different species, can be represented by only three linearly independent vectors: $(\varepsilon_A - \varepsilon_B)$, $(\varepsilon_C - \varepsilon_D)$ and $(\varepsilon_B + \varepsilon_D)$.

A similar example is:

$$A \rightarrow B \, ; \, C \rightarrow D \, ; \, E \rightarrow F \qquad 28$$

The absorbance at any time point is:

$$\varepsilon_A[A] + \varepsilon_B[B] + \varepsilon_C[C] + \varepsilon_D[D] + \varepsilon_E[E] + \varepsilon_F[F] \qquad 29$$

Substituting the concentrations by suitable kinetic expressions and regrouping terms and ignoring the initial concentrations (constants) gives,

$$(\varepsilon_A - \varepsilon_B) \, e^{-k_1 t} + (\varepsilon_C - \varepsilon_D) \, e^{-k_2 t} + (\varepsilon_E - \varepsilon_F) \, e^{-k_3 t} + (\varepsilon_B + \varepsilon_D + \varepsilon_F) \qquad 30$$

Therefore, this system can be represented by only four linearly independent vectors, even though it has six spectrally distinct species:

$$(\varepsilon_A - \varepsilon_B), \, (\varepsilon_C - \varepsilon_D), (\varepsilon_E - \varepsilon_F) \text{ and } (\varepsilon_B + \varepsilon_D + \varepsilon_F) \qquad 31$$

In other words, to define the spectrum at any time, one needs three difference spectra, the final spectrum of the reaction mixture and their time courses.

The value of the SVD procedure is that it allows us to obtain, in an objective manner, *the number of linearly independent vectors* required to describe the system, thus providing valuable information about mechanism. Once the number of linearly independent vectors is determined, models can be formulated, as above, and these used to fit the time courses. Any particular mechanism is evaluated on the basis of its ability to reproduce the original data. From this fitting procedure, both the spectral components and the rate constants corresponding to the partial reactions can be obtained.

Models may be eliminated by recourse to the statistics of the fitting procedure, but also by appealing to biochemical common sense, e.g. no absolute spectrum can have negative absorbance at any wavelength.

5.1.1 The decomposition of the data set to determine pseudo (basis) spectra, singular values and time courses (SVD)

The matrix **A**, corresponding to the raw data, can be expressed as a product of two matrices. One, **E**, represents the linearly independent vectors referred to above, containing spectra or difference spectra of the contributing species. Another, **C**^{T}, corresponds to the time courses of these species, the superscript indicates that the matrix has been transposed, i.e. the columns becomes rows and the rows columns. Thus

$$\mathbf{A} = \mathbf{E} \mathbf{C}^T \qquad 32$$

Our ultimate objective is to determine the contents of these two separate matrices (i.e. to gain knowledge of all the spectra and their time courses), but this is impossible to do directly. A useful first step is to employ SVD analysis. This is based on a theorem that states that a matrix of numbers $\mathbf{A}$ (i.e. the raw data matrix) can be expressed as a unique product of three other matrices (*Figure 11*), usually referred to as $\mathbf{U}$, $\mathbf{S}$ and $\mathbf{V}^T$, where $\mathbf{U}$ and $\mathbf{V}$ are orthogonal matrices, i.e. $\mathbf{U}\mathbf{U}^T = \mathbf{V}\mathbf{V}^T = \mathbf{I}$, the identity matrix. Thus:

$$\mathbf{A} = \mathbf{U}\mathbf{S}\mathbf{V}^T \qquad\qquad 33$$

Luckily, the matrices $\mathbf{E}$ and $\mathbf{C}^T$ are ultimately related to $\mathbf{U}$ and $\mathbf{V}$, from whence they can be obtained, as explained below. The matrix $\mathbf{U}$ has the same dimensions as $\mathbf{A}$ ($m \times n$), whereas $\mathbf{V}^T$ and $\mathbf{S}$ are square ($n \times n$), see *Figure 11*. When plotted, the columns of $\mathbf{U}$ have features that are reminiscent of spectra and are termed pseudo spectra or basis spectra, a linear combination of these can give any spectrum captured during the course of the experiment. The columns of $\mathbf{V}$ are the time courses. In fact, the pairs of matrices ($\mathbf{U}$, $\mathbf{E}$) and ($\mathbf{V}^T$, $\mathbf{C}^T$) are related, because the entire matrices $\mathbf{E}$ and $\mathbf{C}^T$ can be generated by linear combination of the columns in $\mathbf{U}$ and $\mathbf{V}^T$, respectively.

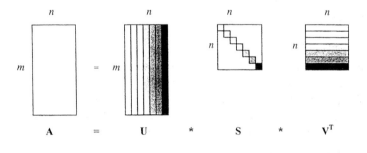

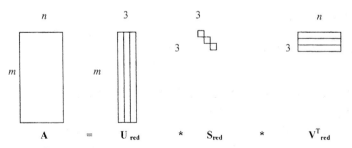

Figure 11 Illustration of SVD. The m by n data matrix $\mathbf{A}$ may be written uniquely as the multiple of three other matrices $\mathbf{U}$ ($m \times n$), $\mathbf{S}$ ($n \times n$) and the transpose of $\mathbf{V}$ ($n \times n$). The chemical information is confined to the first three columns, the remainder containing noise (increasing shading). Once the rank has been decided as three by examination of the s values and autocorrelation coefficients the data matrix may be reconstituted without noise by multiplication of the reduced matrices. The significant information may be stored economically as the reduced matrices.

The minimum number of linearly independent vectors that can generate matrix **A** is equal to the rank of the matrix **A**, which in turn is also the rank of the matrices **U** and **V** and the number of columns in the matrices **E** (which are represented by the linearly independent vectors referred to above) and **C**^{T}. This is to say that this rank is also the number of linearly independent vectors in the matrices **U** and **V**^{T}. In consequence, linear combinations of these are enough to represent **E** and **C**^{T}, respectively.

Here is where the matrix **S** comes into play. In the matrix **S**, all the elements are zero, except for the diagonal elements, which are called *singular values*. The singular values are a measure of the contribution the associated columns in the **U** and **V** matrices make to the experimental data set in matrix **A**, that is, they can be considered as a kind of weighting. These values are relatively large in the first columns but tend rapidly to zero in the later columns. In an ideal situation, the rank of the matrix **A** would be just the number of non-zero values in the diagonal of **S**. In practice, however, this is not the case because of the noise in the system and it can be difficult to locate the cut-off between values associated with signal and those with noise.

One procedure to aid this decision is to plot the logarithm of the different values s_i as a function of i (see example later). The plot shows an abrupt change in slope that can be used as a cut-off. Another method is the use of auto-correlation coefficients for the columns of **U** and **V**. This coefficient is obtained by summing over the whole column the products of x_i and x_{i+1}, where x_i is one of the elements of the column of **U** (or **V**), formally $c(U) = \Sigma U_{ji} U_{j+1,i}$ from $j = 1$ to $n - 1$. Since the matrices **U** and **V** are orthogonal, the values of the elements contained in their columns lie between 1 and -1. According to an empirical rule, columns with autocorrelation coefficients with values ≥ 0.5 are more likely to contain real signal rather than noise.

The rank can also be deduced from the columns in the matrix **V** or **U**. When the columns of these matrices are represented graphically, the amount of noise (generally) is seen to increase with the column number. The cut-off, the number of columns before noise overcomes the signal, can also be used to estimate the rank. Thus, the rank will be the number of rows (or columns) in which the signal is still perceived by visual inspection (see example later). As in any noise removal procedure, some information is lost when these columns are eliminated. The effect of noise, which comes from many sources, is the introduction of uncertainty about the true rank of the matrix.

5.1.2 Removal of noise and data compression

Once the rank of the matrix has been decided, i.e. the number of columns in the **U**, **V** and **S** matrices that contain useful information, the remaining columns may be discarded. This leads to two useful consequences that are illustrated in *Figure 11*. Suppose one has collected, in matrix **A**, 200 spectra each consisting of 100 wavelength points, i.e. we have 2×10^4 absorbance values in a matrix comprising 100 rows (time courses) and 200 columns (spectra). Let us further suppose that following SVD we decide that only three columns of the **U** and $\mathbf{V}$ matrices contain significant information, the remaining containing noise. Now all the information is contained in 3×100 values of $\mathbf{U}$ and 3×200 of $\mathbf{V}$ and 3 of $\mathbf{S}$ (see *Figure 11*). This means that we need to store only 903 data points instead of the original 2×10^4 data points. This gives a saving of over 20-fold in disk storage space. In addition, if we now reconstruct matrix $\mathbf{A}$ from the three columns of $\mathbf{U}$, $\mathbf{V}$ and $\mathbf{S}$, then we have the experimental data *filtered of noise*.

The software that is provided with the commercial stopped-flow spectro-photometers contains the SVD algorithm and the analysis takes a few seconds. Generally about eight columns of the matrices are stored to ensure that no useful data is discarded with the noise.

5.2 Fitting to a mechanism

The rank of the matrix, as we have seen, is the number of linearly independent vectors that we need to reconstruct the original matrix, and is a function of both the number of chemical components and the mechanism of the reaction. This also means that the knowledge of the rank of the matrix will enable us to restrict the possible mechanistic models to a handful.

The following step is the evaluation of the models represented by our chosen rank. The procedure involves first the truncation of the matrices $\mathbf{U}$, $\mathbf{S}$ and $\mathbf{V}$, so that they contain only the first r columns (*Figure 11*) and is based on the following algebraic manipulation of the matrices.

As stated above:

$$\mathbf{A} = \mathbf{USV}^\mathrm{T} = \mathbf{EC}^\mathrm{T}$$
$$\mathbf{USV}^\mathrm{T}\mathbf{C} = \mathbf{EC}^\mathrm{T}\mathbf{C}$$
$$\mathbf{USV}^\mathrm{T}\mathbf{C}(\mathbf{C}^\mathrm{T}\mathbf{C})^{-1} = \mathbf{EC}^\mathrm{T}\mathbf{C}(\mathbf{C}^\mathrm{T}\mathbf{C})^{-1} \qquad 34$$
$$\mathbf{USV}^\mathrm{T}\mathbf{C}(\mathbf{C}^\mathrm{T}\mathbf{C})^{-1} = \mathbf{E}$$

Let

$$\mathbf{H} = \mathbf{V}^\mathrm{T}\mathbf{C}(\mathbf{C}^\mathrm{T}\mathbf{C})^{-1} \qquad 35$$

hence

$$\mathbf{E} = \mathbf{USH} \qquad 36$$

substituting:

$$\mathbf{USV}^\mathrm{T} = \mathbf{USHC}^\mathrm{T}$$
$$\mathbf{U}^\mathrm{T}\mathbf{USV}^\mathrm{T} = \mathbf{U}^\mathrm{T}\mathbf{USHC}^\mathrm{T}$$
$$\mathbf{SV}^\mathrm{T} = \mathbf{SHC}^\mathrm{T} \qquad 37$$
$$\mathbf{S}^{-1}\mathbf{SV}^\mathrm{T} = \mathbf{S}^{-1}\mathbf{SHC}^\mathrm{T}$$
$$\mathbf{V}^\mathrm{T} = \mathbf{HC}^\mathrm{T}$$

Hence:

$$\mathbf{V} = \mathbf{CH}^\mathrm{T} \qquad 38$$

The aim of the fitting process is to obtain a concentration matrix $\mathbf{C}^*$ by varying the values of the rate constants, so that the matrix $\mathbf{V}^*$ obtained from the relationship is as close as possible to $\mathbf{V}$ obtained from the data, i.e. $\mathbf{V}^* - \mathbf{V} \to 0$. For each set of estimated rate constants, a matrix $\mathbf{V}^*$ is produced: $\mathbf{V}^* =$

$C^*(C^{*T}C^*)^{-1}C^{*T}V$, where C^* is the concentration matrix obtained from the solution of the differential equations derived from the mass action law. When the right set of rate constants are obtained, the relation is satisfied and the matrix H can be computed from the relationship:

$$H = V^T C^+ \qquad\qquad 39$$

where C^+ is $C^*(C^{*T}C^*)^{-1}$ and it is called the pseudo-inverse. H can then be substituted in $E = USH$ to obtain E (spectral reconstruction). Instead of using the relationship $V = CH^T$, we could use $A = EC^T$ so that $(A^* - A) \to 0$. However, with this substitution the procedure is generally slower, as noise is included and larger matrices would be involved.

A helpful hint that can help during the selection of the best mechanism is that the plot of the V columns is characteristic of a particular mechanism. To this end, one can construct synthetic spectra and simulate the whole data set A for every particular mechanism. Each synthetic matrix A can then be decomposed into the matrices U, V and S. The plot of these synthetic V columns can

Protocol 6

Reduction of the metmyoglobin–cyanide complex by sodium dithionite

Reagents

- Horse skeletal muscle myoglobin, supplied as the ferric form MbFe(III) (Sigma)
- Phosphate buffer, 0.2 M, pH 6.4
- Potassium cyanide (Sigma)
- Sodium dithionite (BDH)

Method

1 Dissolve MbFe(III) in buffer. To measure the Mb concentration, add sodium dithionite to an aliquot of the solution of MbFe(III) to obtain deoxy Fe(II) ($\varepsilon_{560} = 13\,800$ M^{-1} cm^{-1}).

2 Prepare a fresh solution of KCN (1 mM) in phosphate buffer at pH 6.4.

3 Add KCN to the MbFe(III) solution to obtain the MetMb–CN complex. Final concentrations of 200 μM KCN and 100 μM Mb, approximately. After allowing 1 min to allow complex formation gently deoxygenate the solution by passage of N_2.

4 Prepare fresh sodium dithionite solutions adding the solid to degassed phosphate buffer at concentrations 100, 25 and 6.2 mM.

5 Mix the MetMb–CN complex with the sodium dithionite solution in a stopped-flow apparatus.

Syringe A: 100 μM MetMb–CN complex in phosphate buffer at pH 6.4.

Syringe B: Sodium dithionite at concentrations 100, 25 and 6.2 mM in phosphate buffer at pH 6.4.

6 Collect the spectra in the range 400–600 nm using a diode array.

then be compared to the experimentally obtained **V** column. This can be used to identify the correct mechanism.

In practice, the data is analysed using global analysis, in which all the data (the truncated matrix **A**) are simultaneously fitted to a selected reaction mechanism. The inputs for this analysis are the partial reactions describing the system, in which each species is represented by one letter. The initial concentration for each species is then introduced together with the guessed rate constants for each one of the partial reactions. This software package is provided by manufacturers of the stopped-flow instruments.

The output of global analysis is visualized in two ways. In one, the spectra of the various species that compose the system are plotted. Even in the case that the fit is good, the identification of realistic features in the spectra corresponding to the different species can be used to discriminate between a good and a bad model. The goodness of the fit can also be assessed by comparing the fitted time courses at different wavelengths to those obtained experimentally. When using such powerful and esoteric mathematical procedures one may feel totally dependent on the computer. This can be dangerous and it is as well for the experimentalist to keep a tight grip on reality. It is entirely possible to obtain very good fits to a biochemically nonsensical mechanism—the computer just doesn't understand this!

The theoretical steps that have been discussed above may be illustrated using the practical example of the application of SVD and global fitting to a simple reaction containing a spectral intermediate given in Protocols 6 and 7.

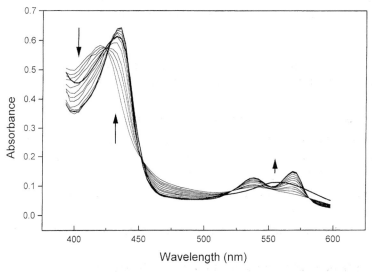

Figure 12 The reaction between the cyanide complex of ferric myoglobin (horse heart) and sodium dithionite. A subset of the total number of spectra collected is shown. An initial increase in absorbance is seen at 435 nm and in the visible region, spectra displayed every 20 ms, followed by a slower decrease, only the final spectrum (thicker line) is shown for clarity.

The options for data collection are:

(1) simple, with constant time interval between spectra;

(2) split, with two different time scales; or

(3) logarithmic.

A logarithmic scale allows one to obtain data from processes that are complete in less than a second and also from slower reactions that extend over seconds or even hundreds of seconds, in the same experiment.

Figure 12 shows only spectra corresponding to the first observed transitions. For the sake of clarity, spectra obtained during a second transition have been omitted, although the final spectrum, at the end of the experiment, is also shown. The figure shows the formation of an intermediate, with bands at 515 nm and 585 nm in the visible region. The changes involved in the two transitions can also be observed by inspection of single-wavelength traces, which show a fast process followed by a slower process that leads to the formation of the final product MbFe(II).

The whole data set can be represented by a matrix **A** with as many columns as time points and as many rows as wavelengths. This matrix is analysed according to the following protocol:

Protocol 7

The SVD procedure

Method

1 Decompose the matrix **A** into the matrices **U**, **V** and **S** using the option provided in the stopped-flow software, Matlab or other suitable package.

2 Determine the rank of the matrix using any of the methods described above. Depending on the rank of the matrix a mechanism is chosen to fit the data.

3 Plot the first six columns of the **V** matrix versus time and of the **U** column versus wavelength.

Plots of the **U** columns (*Figure 13*) and the **V** columns (*Figure 14*) show that the rank is either 3 or 4. The same conclusion is reached by looking at the logs plot (*Figure 15*) or at the *s* values and autocorrelation coefficients for the columns in **V** and **U** (*Table 2*). For the remaining analysis the rank was taken as three, the fourth *s* value being much smaller than the third. A very small fourth component is in fact present and was identified as being due to a small fraction of denatured myoglobin reacting differently from the bulk protein.

As the rank is three, mechanisms compatible with this were chosen for global fitting to the truncated (reconstructed from the first three columns) matrix **A**. Only the sequential mechanism A → B → C gave a sensible solution, i.e. all kinetic and spectral profiles fitted well, and non-negative absolute spectra for all

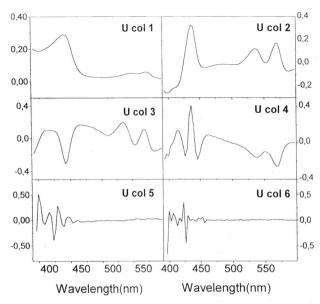

Figure 13 The first six columns, pseudo-spectra, of the **U** matrix following SVD of the total data set; matrix **A**. Columns 5 and 6 appear as noise.

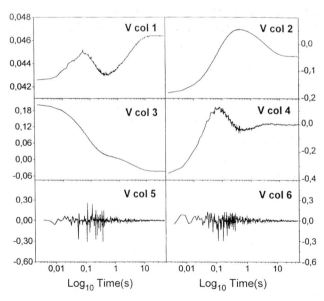

Figure 14 The first six columns, time courses, of the **V** matrix following SVD of the total data set; matrix **A**.

components. The results of the fitting process are shown together with the dithionite concentration dependencies of the two processes in *Figure 16* and *17*. Thus the results of this analysis may be summarized as follows:

$$\text{Mb Fe(III)–CN + dithionite} \rightarrow \text{Mb Fe(II)–CN} \rightarrow \text{Mb Fe(II) + CN} \qquad 40$$

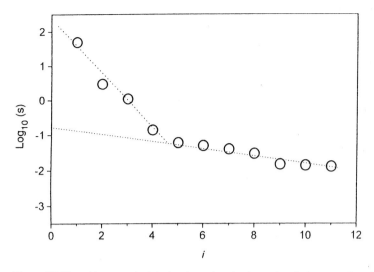

Figure 15 Plot of $\log s_i$ against i, showing a break where signal gives way to noise.

Table 2 s values and autocorrelation coefficients (AC) of the first six columns in the **U** and **V** matrices

s	AC V	AC U
47.9	0.973	0.998
2.74	0.913	0.982
1.14	0.898	0.978
0.14	0.888	0.605
0.06	0.193	−0.02
0.05	−0.596	−0.02

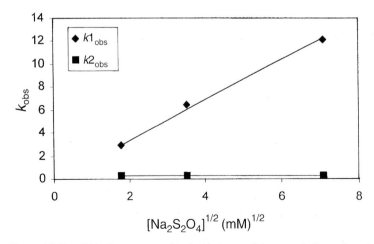

Figure 16 The dithionite concentration dependence of the pseudo-first-order rate constants of the fast and slow processes seen in the reduction of Mb–CN.

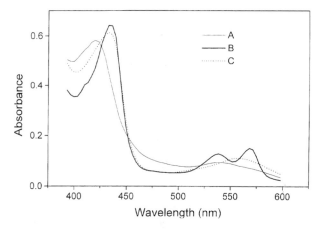

Figure 17 Global analysis of the reduced (rank = 3) matrix **A** to a simple sequential model generates the spectra of the chemical species A, B and C, see text.

Dithionite first reduces the MbFe(III)–CN complex (species A) in a rapid second-order process. The pseudo-first-order rate constant for this is directly proportional to the concentration of the true reductant, SO_2^-, which is in rapid equilibrium with dithionite ($S_2O_4^{2-} \leftrightarrow 2SO_2^-$), thus resulting in the linear dependence on $[S_2O_4^{2-}]^{1/2}$, *Figure 16*. As cyanide has a low affinity for ferrous iron the complex now dissociates in a dithionite independent process to yield, finally, deoxyMb (species C), *Figure 16*. The spectrum of the short lived intermediate is provided by the global fitting and shows the intermediate to be a low-spin ferrous cyanide complex (species B), as expected, which on CN dissociation gives deoxy Mb, recognized by its characteristic spectrum, *Figure 17*. It may be noted that the spectra in *Figure 17* are somewhat less noisy than those in *Figure 12* as the noise has partially been removed by discarding the majority of the columns in the **U** and **V** matrices.

Acknowledgements

We thank K. Tzouvara for help with the experiments and SVD analysis of the MbCN system.

References

1. Eccleston, J. F. (1987) In *Spectrophotometry and spectrofluorimetry: a practical approach* (ed. D. A. Harris and C. L. Bashford), p.137. IRL Press, Oxford and Washington DC.
2. Tonomura, T. (1978). *Anal. Biochem.* **84**, 370.
3. Davies, N. A., Wilson, M. T., Slade, E., Fricker, S. P., Murrer, B. A., Powell, N. A. and Henderson, G. R. (1997) *Chem. Commun.* 47.
4. Brunori, M., Wilson, M. T. and Antonini, E. (1973) *J. Biol. Chem* **247,** 6076.
5. Antonini, E. and Brunori, M. (1971) *Haemoglobin and myoglobin in their reactions with ligands.* North Holland, Amsterdam and London.
6. Fersht, A. (1977) In *Enzyme structure and mechanism*, p.123. W. H. Freeman, Reading.
7. Henry, E. R. and Hofrichter, J. (1992) *Methods in enzymology* **210**, 129.
8. Shrager, R. I. and Hendler, R. W. (1982) *Anal. Chem.* **54**, 1147.
9. Ownby, D. W. and Gill, S. J. (1990) *Biophys. Chem.* **37**, 395.

Chapter 9

Stopped-flow fluorescence spectroscopy

Michael G. Gore* and Stephen P. Bottomley[†]
* Institute of Biomolecular Sciences, Division of Biochemistry,
School of Biological Sciences, University of Southampton, Bassett Crescent East,
Southampton, SO16 7PX
[†] Department of Biochemistry and Molecular Biology, Monash University, Clayton,
Victoria 3168, Australia

1 Introduction

Biochemical reactions, such as substrate or coenzyme binding to enzymes are usually completed in no more than 50–100 ms and thus require rapid reaction techniques such as stopped-flow instrumentation for their study. Fortunately, many such reactions can be followed by changes in the absorption properties of the substrate, product or coenzyme, and examples of these have been described in Chapters 1, 7 and 8. An alternative possibility is that during the reaction there is a change in the fluorescence properties of the substrate, coenzyme or the protein itself. Some reactions, particularly those involving the oxidation/ reduction of coenzymes, involve both changes in absorption and changes in fluorescence emission intensity. In many cases, the fluorescence properties of the ligand or protein itself may change when a complex is formed, even in the absence of a full catalytic reaction occurring, e.g. the protein fluorescence emission of most pyridine or flavin nucleotide-dependent dehydrogenases is quenched when NAD(P)H or FADH (respectively) binds to them, due to resonance energy transfer from the aromatic amino acids of the protein to the coenzyme. Conversely, the fluorescence emission from the reduced-coenzymes is usually enhanced on formation of the complex with these enzymes (1–3).

The principles behind both fluorescence and stopped-flow techniques have been described in preceding chapters (2 and 8, respectively) and therefore readers should familiarize themselves with these chapters for some of the background information. In this chapter, we discuss the use of stopped-flow fluorescence spectroscopy and its application to a number of biochemical problems.

2 Instrumentation

A typical stopped-flow system is assembled from modular components of a conventional spectrophotometer/fluorimeter, a device permitting rapid mixing of the components of a reaction and a data recording system with a fast response. Commercially available instruments offer facilities for the observation of changes in absorption and/or fluorescence emission after rapid mixing of the

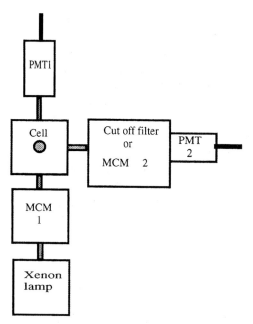

Figure 1 A diagram of the optical arrangement of a stopped-flow system capable of simultaneous observation of changes in absorbance and fluorescence. The light from the xenon lamp is diffracted by monochromator 1 (MCM1) to select the excitation wavelength. Usually quartz optical fibres conduct the light to the observation cell and absorption is detected at 180° and fluorescence emission (wavelength selected by a cut-off filter or MCM2) is detected at 90° relative to the incident light.

reagents. These measurements can often be made simultaneously due to the different optical requirements of the two spectroscopic techniques. *Figure 1* gives a generalized diagram of the geometry of a stopped-flow system able to simultaneously measure changes in absorption and fluorescence intensity of a reaction. Note that the flow-cell is illuminated by light selected by monochromator 1 (MCM1) and that the change in light absorption at this wavelength is detected at 180° relative to the light beam. Changes in fluorescence intensity are detected at 90° relative to the excitation light beam via a second monochromator (MCM2), a narrow band pass filter or a cut-off filter to select the emission wavelength.

2.1 Data collection

Since the reactions under study are very rapid, the short period of time available for data sampling and recording places extra demands upon the system. However, the high sampling rate of current on-line computerized data logging devices , typically 100 kHz, does permit a sample of the signal to be collected approximately every 10 μs allowing several samples to be averaged per ms and going some way to improving the quality of the detected signal. The number of signal samples taken and averaged (n) per stored data point is described by $n = ft/d$, in which t is the observation period in seconds, d is the number of data points used to describe the reaction profile and f is the sampling rate of the A/D converter. For example, if 1000 data points are used to describe an observation period of 100 ms and the effective sampling rate permitted by the A/D converter is 100 kHz, then 10 samples of the signal will be averaged for each datum point. The signal:noise ratio will improve by the square root of the number of samples in the average, and thus it is clear that a reaction progress curve will have less noise the slower the reaction and, therefore, the greater the number of signal sampling events for the same number of data points (1000 in this example). For very fast reactions the result must usually be improved by averaging the data from several reactions. However, the amount of noise observed on a reaction profile will also depend upon the amplitude of the signal change upon reaction and how careful the experimenter has been in selecting the conditions for the experiment, the setting up of the instrument and the preparation of the reagents. *All of the recommendations made in previous chapters regarding the preparation of solutions are particularly important in stopped-flow fluorescence studies.*

2.2 Instrument calibration, stability and dead time

It is important to test the reliability of the stopped-flow instrument using control experiments that test a range of parameters such as the dead time, mixing efficiency and signal output. In general, these tests will be the same for the instrument in the configuration for fluorescence studies as that for absorbance studies, and have been discussed in Chapter 8.

2.3 Measuring mixing efficiency

It is recommended that the mixing efficiency of the instrument be checked periodically to ensure that the reagent lines, typically 2 mm (i.d.), have not become obstructed by precipitated reagents such as proteins. This can be readily assessed by mixing together two buffer solutions of different pH (one containing a pH indicator) with a suitable pK_a, so that the final pH is somewhere between those of the starting solutions. The change in ionization of the indicator will occur over a very short time period relative to the mixing time of the instrument and can be detected by absorbance or fluorescence changes. An example of the latter is 4-methylumbelliferone (pK_a approximately 7.6; see *Figure 2*) which emits fluorescence at 450 nm when excited by light at 390 nm.

Protocol 1

To check the mixing efficiency of the stopped-flow instrument

Equipment and reagents

- 0.1 M sodium phosphate buffer pH 6.5 (buffer A)
- 0.1 M sodium phosphate buffer pH 8.5 (buffer B)
- Pipettes and measuring cylinders
- 1 ml of a stock solution of 10 mM 4-methylumbelliferone in water
- Disposable 5-ml syringes and a few glass beakers

Method

1 Mix 50 ml of buffers A and B to give 100 ml of buffer, approximately pH 7.1 to give C.

2 Add 0.1 ml of the stock solution of 4-methylumbelliferone to 19.9 ml of the mixed buffer C to give D.

3 Add 0.2 ml of the stock solution of 4-methylumbelliferone to 19.8 ml of buffer A to give E.

4 Equilibrate all solutions at the operating temperature of the stopped-flow observation cell.

5 Use solution C to wash out the contents of the two drive syringes of the stopped-flow apparatus. Refill the two syringes with C and discharge five or more reactions to wash out and fill the observation cell.

6 Set the excitation wavelength to 388 nm and use a cut-off filter to select emitted light at 410 nm and above.

7 Flush out the system with solution D and adjust the sensitivity of the instrument to give a working fluorescence signal. The working signal should be the same as that obtained in the following procedure.

8 Replace the contents of one drive syringe with solution E and that of the other drive syringe with solution B.

9 Discharge five or more 'shots' to replace the solutions in the observation cell.

10 Then discharge several more and record any change in signal intensity with time.

Results

On mixing the two solutions E and B, the pH of the reaction mixture will change to that of solutions C and D (pH 7.1) within the dead time of the instrument, and the degree of ionization of the fluorophore (originally at pH 6.5) will alter. Observation of the fluorescence intensity of the solution at 410 nm and above should yield an 'instantaneous' increase in fluorescence intensity (within the dead time of the instrument) followed by a stable signal intensity. Any change in fluorescence intensity over the first 10–50 ms after mixing suggests that poor mixing is occurring.

2.4 Sample preparation

Thorough sample preparation is critical to the success of any spectrophotometric experiment whether using absorbance or fluorescence detection systems. Even the rapid mixing of two samples of buffer in a sensitive instrument can result in what appears to be a reaction progress curve. This may be caused by simple effects such as the rapid compression and decompression of an air bubble in the flow path, the mixing of two solutions at different temperatures or the effect of the stopping process upon small dust particles present in the solutions. Therefore it is essential that solutions be prepared thoroughly before

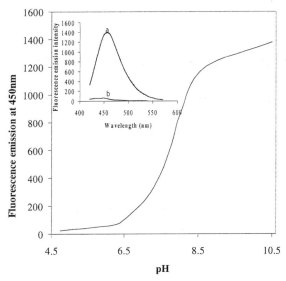

Figure 2 The effect of pH on the fluorescence emission intensity of 4- methylumbelliferone. 10 μM 4-methylumbelliferone was dissolved in 50 mM potassium phosphate buffer at the indicated pH and its fluorescence was excited by light at 388 nm and detected at 450 nm. The inset shows the fluorescence emission spectrum of 4-methylumbelliferone at pH 10 (a) and at pH 4.5 (b).

use. Degas buffers to remove dissolved air that may otherwise come out of solution following rapid decompression of the reaction solutions and filter all solutions using an appropriate filter, e.g. Millipore 0.45 μm. Sometimes sample loss may occur during filtration in which case centrifugation of the sample may help eliminate suspended particles without compromising the sample.

2.5 Artefacts

It is important to verify that the signal change being monitored is due to the reaction under study and not due to artefacts. Comparison of signal changes noted under equilibrium conditions should be compared with the kinetic difference spectrum obtained in the stopped-flow (see Chapter 8). If deviations are noted then the observed change in fluorescence may well be due to an artefact and not reflect the reaction under study.

2.6 Temperature effects

Good temperature control is essential to the success of any experiment, particularly those using fluorescence techniques since the quantum yield of a fluorophore is very sensitive to temperature (see Chapters 2 and 12). The quantum yield of fluorescence decreases with increasing temperature due to the increased molecular vibration leading to loss of excited state energy as heat rather than light. Furthermore, the pH values of several buffers (especially Tris) are dependent upon temperature, and it is always advisable to adjust the pH of the buffer to be used at the temperature required in the experiments. *Table 1* in Chapter 13 gives a list of different buffers and the effects of temperature on their pK_a value. The temperature of the reaction solutions in the stopped-flow apparatus should be maintained at all times using a water bath system that circulates the water rapidly, so that there is fast heat exchange between the water bath and the observation cell.

Finally, the mixing of some solutions may lead to a rapid change in temperature taking place, e.g. if a solution of protein in 6 M Gdn-HCl is mixed with aqueous buffers to initiate refolding of a protein, a decrease in temperature occurs due to the dilution of the denaturant. This decrease may cause a small increase in fluorescence emission intensity from the protein that will reduce as the mixed solution returns to the equilibrated temperature of the reaction vessel. This may lead to an observable short-lived artefact in the reaction progress curve, which may be minimized by opting to mix the solution of protein in denaturant 1:10 v/v with aqueous buffer.

2.7 Density differences between the two solutions

Mixing two solutions of different density can produce turbulence effects. Again this is a common problem in protein unfolding studies in which a viscous solution of 6 M Gdn-HCl is mixed with a protein solution. Varying the mixing ratio can again limit the problem. Alternatives are to make sure that the two solutions are matched, for example, using an inert salt. The problem may be identified as described in *Protocol 1* by including the salts of interest with the pH indicator, e.g. 6 M Gdn-HCl in one buffer. It is important that the 'flow path' be thoroughly rinsed with distilled water after experiments including high concentrations of salt. Any minute leaks around valves or syringe pistons will lead to the formation of crystals that may scratch PTFE seals and valves or block the narrow flow tubes.

3 Factors affecting the sensitivity of the optical system

3.1 Slit width

The quantum yield of fluorescence from a fluorophore is the ratio of the number of photons of light emitted to the number of photons of light absorbed by the molecule. Thus it follows (at the level of light intensity available in laboratory fluorimeters) that the intensity of the fluorescence emission will be directly related to the intensity of light available for the excitation process. Since it is not usually possible to routinely alter the nature of the light source, the light intensity falling upon the reaction flow-cell can only be varied by adjustment of the slit width of the excitation monochromator. However, this has repercussions on the range of wavelengths allowed to fall upon the reaction mixture (see *Figures 11–12*, Chapter 1). In addition, xenon lamps (the most common form of lamp for use in stopped-flow fluorescence instruments) do not emit equal light intensities at all wavelengths. Thus a shift in the wavelength selected by the excitation monochromator may well be accompanied by a change in light intensity with a resultant change in sensitivity (see *Figure 6*, Chapter 2).

3.2 Selection of wavelength of emitted light

For fluorescence detection it is necessary to select both the excitation and the emission wavelength. To achieve this it is usual to use a monochromator to select the excitation wavelength and a narrow bandpass or a cut-off filter to select the emission wavelength. The latter can, of course, be selected by a second monochromator but this option is not often chosen since these transmit light less efficiently than a filter leading to a significant loss of sensitivity. However, success relies upon the availability of a suitable filter for the desired wavelength required, and on the difference in wavelength between the excitation and the emission wavelengths of the fluorophore (Stoke's shift). If this is very small, then the task is more difficult since overlapping excitation and emission band-widths will lead to increased transmission of scattered light. A cut-off filter is usually used that absorbs light at the excitation wavelengths (typically these will have an absorbance of about 4 at the non-permitted wavelengths) but allows emitted light at higher wavelengths to pass through. The limitation of such a filter is that any wavelength above the cut-off value is allowed to pass. This may lead to a high background light intensity from stray light, upon which the fluorescence emission is superimposed. An alternative is to use a filter with a very narrow bandpass at an appropriate wavelength as far removed from the excitation wavelength as possible. Such filters usually have a bandpass between 1 and 20 nm. However, suitable bandpass filters for non-routine experiments are not always readily available in the laboratory and are usually more expensive to buy than cut-off filters. The manufacturers of stopped-flow devices supply suitable filters for their instruments, or filters can be obtained from specialist companies, e.g. Omega Optical Inc. (http://www. omegafilters.com).

3.3 Voltage applied to PMT

The sensitivity of a PMT depends upon the voltage applied to its dynodes. *Figure 3* shows the amplified signal coming from a PMT when it is illuminated by fluorescence emission at 360 nm arising from 15 μM warfarin in 100 mM phosphate buffer, pH 7.4 (excited at 330 nm). The observed signal is plotted against the voltage applied to the PMT. It can be noted that the output signal is very dependent on the voltage applied to the PMT and on the fluorescence intensity allowed to fall upon the PMT (controlled indirectly in this case by changing the slit width of the excitation monochromator).

It is sometimes found that a non-linear response to light intensity may occur if the light intensity is very high and only a low dynode voltage is required to obtain a reasonable working signal. Under these circumstances it is better to reduce the concentration of the fluorophore or, if this is not possible, to excite the fluorescence at a wavelength that is lower than the excitation maximum

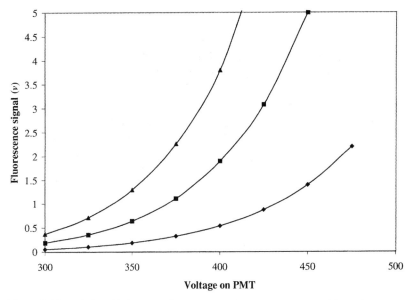

Figure 3 The relationship between fluorescence signal (volts) and the high voltage applied to the PMT. The observation cell contained a solution of 15 μM warfarin in 100 mM potassium phosphate buffer at pH 7.4, 20°C. The fluorescence signal in volts (V) at 360 nm is plotted against the voltage applied to the PMT of an Applied Photophysics SX.17MV stopped-flow fluorimeter. The fluorescence was excited by light at 330 nm through 0.5 mm (♦), 1.0 mm (■) and 1.5 mm (▲) slit widths of MCM1 (see *Figure 1*).

(use of a higher wavelength may lead to increased light scattering problems). Alternatively, the same result may be achieved by observation of the reaction at a higher wavelength than that of the emission maximum. If high light intensity falls upon a PMT then 'saturation' may occur, i.e. there is no further increase in output signal from the PMT when the light intensity increases. At high levels of light, permanent damage to the PMT may be caused. *In general, PMTs should never be exposed to daylight particularly while a voltage is applied to its dynodes.* The linearity of response of the PMT/system to light intensity may be tested by flowing very dilute solutions of a fluorophore through the observation cell. The fluorescence signal obtained should be proportional to the concentration of the fluorophore. Note that corrections for the inner filter effect may have to be made if the concentration of the fluorophore (hence absorption of the excitation and emitted light) is high (see Chapter 2). This correction will depend upon the geometry of the observation cell and the optical pathlength for the excitation and emission light.

Figure 4 shows the noise levels of fluorescence signals of similar magnitude arising from warfarin in phosphate buffer (the same solutions described in *Figure 3* above). The different traces were obtained by using two different intensities of excitation light (adjusted by changing the slit widths of the excitation monochromator) and different voltages applied to the PMT. The lower trace has the least noise and was generated by using an 8-nm bandwidth and a relatively

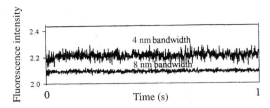

Figure 4 Signal intensity and noise level; the effect of slit width and voltage to PMT. A plot showing the recorded traces from an Applied Photophysics SX.17MV stopped-flow fluorimeter when 20 µM warfarin (final concentration) in 10 mM potassium phosphate buffer at pH 7.4, 20°C, was mixed with buffer. The fluorescence intensity at 360 nm was measured using an excitation wavelength of 330 nm. The sensitivity of the instrument was adjusted to give a similar signal intensity by varying the slit width and the high voltage to the PMT (see *Figure 3*).

low dynode voltage (350 V). This is generally the case and hence the amount of light available and the optical alignment of the instrument are very important considerations for achieving high sensitivity.

4 Selection of reporter group

The choice of a suitable fluorescent reporter group is essential to the success of any experiment. The ideal reporter group should have a high molar extinction coefficient and a high quantum yield (so that it is easily detected by the system) and should allow faithful monitoring of the reaction. It must also report a change in its environment/polarity or pH and must respond to these immediately and proportionally. Often constraints are placed on the choice of reporter due to the nature of the sample being studied. However, when a choice is available, the time taken to chose the best reporter will be well spent and maximize the chances of successful experimentation. The choice of reporter group falls into two main categories, intrinsic reporter groups and extrinsic reporter groups.

4.1 Intrinsic reporter groups

Intrinsic reporter groups are already present in an essential component of the reaction under study. They may be located within the protein, a prosthetic group or cofactor, such as flavins (3), NADH (2), pyridoxal-5'-phosphate (4), or a substrate or other ligand (5).

The indole group of tryptophan (Trp) has a higher molar extinction coefficient than the phenolic and phenyl side chains of tyrosine (Tyr) and phenylalanine (Phe), respectively (see *Table 1*, Chapter 12). Thus although its quantum yield is similar to that of Tyr, its fluorescence emission is much more intense. Furthermore, its excitation spectrum overlaps the emission spectrum of Tyr and therefore fluorescence resonance energy transfer (FRET, see Chapters 2 and 3) from Tyr to Trp occurs readily when both residues are in close proximity (i.e. located in the same protein molecule) and favourably orientated. The intrinsic fluorescence of a protein is therefore dominated by the contribution from Trp residues, although proteins lacking Trp residues or with many Tyr will show fluorescence emission from Tyr residues.

The quantum yield and emission maximum of the Trp residue is highly dependent upon the microenvironment of the residue. Trp residues in the buried, apolar interior of a protein can have emission maxima as low as 308 nm and those exposed to aqueous solvent have an emission maximum close to 350 nm (see Section 4.3 and references 37, 38, and 40 in Chapter 12). It is this sensitivity to environmental changes that makes the Trp residue such an ideal reporter group. Furthermore, the judicious use of site-directed mutagenesis, particularly with the aid of structural knowledge of the protein, allows the removal of unwanted Trp residues or the addition of unique Trp residues to a protein or to a domain of a protein (6–8). Recent developments in molecular biology now allow modified Trp residues (e.g. 5-hydroxy-L-tryptophan (9)) with red-shifted excitation and emission spectra to be located at key positions in proteins. These residues can be selectively excited, permitting spectral changes to be identified from specific parts of the protein without having to risk some destabilization of the protein by the removal of other Trp residues by mutagenesis. Thus Trp residues can be used as extremely sensitive conformational reporter probes.

4.2 Examples of the use of protein fluorescence to follow binding reactions

4.2.1 Protein folding/unfolding

Proteins usually unfold upon the addition of denaturant; the time taken ranges from ms to hours depending upon the concentration of the denaturant used and the stability of the protein. During the unfolding process, endogenous Trp residues become exposed to solvent, usually resulting in red shifts in their emission wavelength maxima. Typically, this shift is accompanied by a simultaneous increase or decrease (10–12) in fluorescence intensity depending upon the environment of the Trp residues before and after unfolding of the protein. For example, proximity to a disulphide bridge quenches the fluorescence emission of Trp residues and therefore unfolding, resulting in movement away from the disulphide bridge, may result in an increase in quantum yield (10).

Refolding reaction progress curves can take place over a few μs to a few minutes. The observed order of the reaction will depend upon the stage at which the change in fluorescence occurs and whether, for example, association of subunits is a rate-limiting step preceding a change in fluorescence.

A plot of the logarithm of the rates of unfolding and folding against the concentration of denaturant (see *Figure 5*) can give valuable information in protein folding studies (10). Deviation from linearity in either the refolding or unfolding part of the curve suggests intermediates may be present in these processes. Comparison of the two curves in *Figure 5* suggests that an intermediate may exist in the unfolding of the protein consisting of three consecutive short consensus repeat domains of human complement receptor 1 (SCR1–3), but not for the unfolding of a single domain protein (SCR3) derived from the third

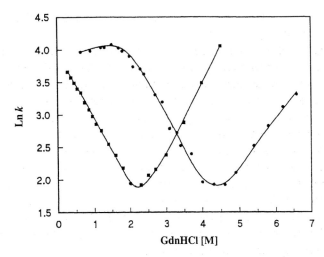

Figure 5 The dependence of the rates of folding and unfolding of non-reduced short consensus repeat (SCR) domains from human complement receptor 1 on the concentration of Gdn-HCl. Unfolding reactions were achieved by mixing 1:10 v/v 50 μM of a protein consisting of the first three domains, SCR1–3 (●), or a protein consisting of the single third domain, SCR3 (■) of complement receptor 1 with various concentrations of Gdn-HCl in 20 mM potassium phosphate buffer, pH 7.4 at 25°C. Refolding reactions were achieved by mixing the proteins in 3–7 M Gdn-HCl 1:10 v/v with buffer containing various concentrations of the denaturant to give the final concentrations of Gdn-HCl shown above. This figure has been reproduced from Clark, N. S., Dodd, I., Mossakowska, D. E., Smith, R. A. G. and Gore, M. G. *Protein Engineering* (1996) **9**, 10, 877 by permission of Oxford University Press.

domain of complement receptor 1. The minima of the curves occur at the concentration of denaturant where the rate of refolding is the same as the rate of unfolding. That is the concentration of denaturant that gives the midpoint of an unfolding/folding transition where half of the protein is in each of the unfolded and folded forms (see *Figure 4*, Chapter 12). These rates should be compared, whenever possible, to rates of unfolding determined by stopped-flow CD measurements (far UV) which reflect changes in the backbone structure of the polypeptide chain (see Chapters 4, 10 and 12).

4.2.2 Protein conformational changes and protein–protein or protein–ligand interactions

The use of quenching of protein fluorescence to monitor ligand binding is a very sensitive technique. This high sensitivity arises from the fact that proteins in general are highly fluorescent, not only because they may contain Trp residues, but also because Tyr residues absorb light and fluoresce or pass on the energy to Trp residues by FRET. Furthermore, aromatic side chains are usually buried in the interior of a protein and are shielded from deactivating collisions with other ions or polar species resulting in high quantum yields (see Section 4.3, Chapter 12). If a ligand, with an appropriate absorption spectrum, binds to the protein then significant decreases in protein fluorescence may occur due to

FRET. However, the fractional quenching of protein fluorescence by FRET to a bound ligand is usually a complex relationship. The saturation of the binding sites for the ligand on the protein is not linearly related to the decrease in protein fluorescence if the protein has more than one ligand binding site per molecule. Holbrook (1, also see 2) has shown that the protein fluorescence of an oligomer with multiple equivalent binding sites (n) for a ligand is proportional to $(1 - \alpha(1 - x))^n$ where α is the fractional saturation of the ligand binding sites and $x = F_p^{1/n}$ where F_p is the protein fluorescence when $\alpha = 1$. Exceptions to a non-linear relationship occur if no FRET occurs between individual subunits of an oligomer and they therefore behave spectroscopically as monomers; if binding of a ligand induces the protein to dissociate into subunits; or if the ligand binding exhibits strong positive or negative cooperativity, i.e. the binding of one ligand molecule results in all of the other sites for the ligand on that protein molecule becoming saturated or the binding of one ligand molecule prevents the binding of others to the same protein molecule, respectively.

Therefore, reaction curves obtained by monitoring the changes in protein fluorescence may not directly correlate with the saturation of the binding sites by ligand. Thus, it follows that experiments in which ligand binding is detected by excitation of protein fluorescence at 280 nm and observation of the fluorescence emission from the bound ligand (excited by FRET) will also be subject to the same limitations.

Nevertheless, many different examples (8, 13–15) exist of the successful use of stopped-flow techniques based upon changes in protein fluorescence to monitor conformational changes or binding reactions. For example, site directed mutagenesis was used to place Trp reporter groups into the L-lactate dehydrogenase from *Bacillus stearothermophilus* in order to examine its reaction mechanism (8). The enzyme catalyses the interconversion of L-lactate and pyruvate coupled to the reduction/oxidation of NAD^+/NADH, respectively. The rate-limiting step in the mechanism is a conformational change induced by coenzyme and substrate binding, involving the closure of a peptide loop over the active site. A Trp reporter group (Trp106) was placed into the peptide loop region and the rate of the conformational change was determined from the changes in the fluorescence emission properties of the Trp as it moved with the peptide loop from a polar to a less polar environment.

Studies using changes in protein fluorescence have also revealed detailed information about protein–protein binding mechanisms (13, 14). *Figure 6* shows the change in fluorescence intensity from a reaction between a 78-residue single immunoglobulin (Ig)-binding domain of protein L (SpL) from *Peptostreptococcus magnus* and human Ig-kappa light chains. The two curves shown were obtained using two mutant forms of SpL with engineered unique Trp residues at positions 39 (upper trace) or 64 (lower trace). On forming complexes with kappa chain both mutants of SpL show rapid increases in fluorescence intensity at 335 nm that reflect the formation of initial 'encounter complexes'. The observed rates of these reactions depend upon the concentration of the ligand (in these experiments the concentration of SpL was varied and that of kappa

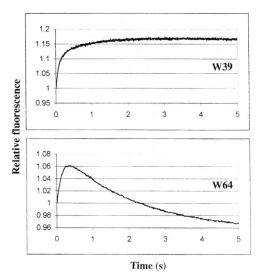

Figure 6 The binding curves obtained on mixing κ chains with a mutated immunoglobulin-binding domain of protein L (SpL) from *Peptostreptococcus magnus*. The curves show the changes in protein fluorescence intensity from unique tryptophan residues (W39 or W64) as each domain binds to κ chains at pH 8.0 in potassium phosphate buffer, 20°C. The concentration of κ chain was 3 μM and the concentration of each mutant of SpL was 30 μM. These data were kindly provided by Dr J. Beckingham and N. Housden.

chain kept constant). In each case, the rapid phase of change in fluorescence is followed by a slower, single exponential process with a rate that is *independent* of the total concentration of SpL used. This is usually indicative of a structural change occurring in the complex or in either reactant.

There are several possible models that may, in principle, be able to explain these results. In order to analyse what is happening in this or any reaction we must first identify the step or steps in the reaction that lead to signal changes; in this case the stages at which the rapid and slow changes in fluorescence occur.

One possible model (a) is given below;

(a)
$$\text{SpL} \underset{k_r}{\overset{k_f}{\rightleftharpoons}} \text{SpL}^* + \kappa \underset{k_{-1}}{\overset{k_1}{\rightleftharpoons}} \text{SpL}^*.\kappa \qquad\qquad 1$$

In this model, an equilibrium exists between two conformations of (SpL and SpL*); only SpL* is able to bind to κ chain; there is a change in fluorescence intensity as SpL isomerizes to SpL* (an increase for W39 or decrease for W64 mutants); an increase of fluorescence occurs (for both mutant proteins) when SpL* binds to κ chain. If we vary the concentration of κ but keep it in large excess over the total concentration of SpL (+ SpL*) present and if $k_f << k_1[\kappa]$, then the observed *initial* reaction will follow an exponential curve of rate = $k_1[\kappa] + k_{-1}$ providing some SpL* exists at the time of mixing. As SpL* is removed from the equilibrium with SpL, by formation of the complex with κ chains, some SpL will isomerize to SpL*. This gives rise to a further change in fluorescence intensity, an increase for the reaction involving W39 mutant and a decrease for the reaction involving the W64 mutant. If k_f is not much smaller than k_1 [κ] then more complex analysis is needed. *Figure 7* gives an example of reaction conditions that permit observation of the second phase of the reaction in virtual isolation from the initial phase. In this experiment a high concentration of SpL (a non-mutated domain containing no Trp residues) was reacted with κ chain and the change in fluorescence from Tyr residues was monitored at 305 nm (λ_{ex} = 280 nm). The high rate of the initial phase of the reaction (shown by a decrease in fluorescence intensity noted when this domain binds to κ chain) is virtually completed at the point indicated by the arrow (approximately 200 ms). These conditions effectively separate out the two parts of the reaction, and allow the rate of the slower second phase to be determined without much contribution from the fast phase of fluorescence change, thus facilitating curve fitting algorithms (see Chapter 8, Section 4.1.1 and *Figure 8*).

The exact form of binding curves for multi-phase reactions such as this will be complex. Amongst other considerations, the relative amplitudes of the two phases of signal change will depend upon the position of the equilibrium between SpL and SpL*. If SpL predominates then the amplitude of the slow change in fluorescence will be larger than if SpL* predominates. Furthermore, the

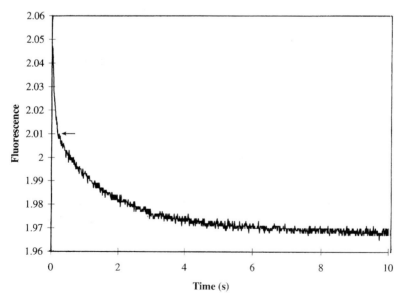

Figure 7 The reaction progress curve obtained on mixing κ chains with a non-mutated SpL domain lacking a Trp residue. The curve show the changes in protein fluorescence intensity at 305 nm (excited at 280 nm) from tyrosine residues as the domain binds to κ chains at pH 8.0 in potassium phosphate buffer, 20°C. The concentration of κ chain was 4 μM and the concentration of SpL was 70 μM. These data were kindly provided by N. Housden.

change in fluorescence signal as SpL isomerizes to SpL* will almost certainly be of different amplitude than the signal change that occurs when SpL* interacts with κ chain.

Alternatively, let us consider another possible model; that the rapid increase in fluorescence occurs on the formation of the binary complex SpL.κ and that the slow change in fluorescence occurs as SpL.κ isomerizes to [SpL.κ]*. This model is described by scheme (b) below and has been proposed for a number of protein–ligand systems (e.g. see 3, 14). Note that it is not known which component in the complex changes conformation or whether both proteins are involved.

(b) $$ \text{SpL} + \kappa \underset{k_{-1}}{\overset{k_1}{\rightleftharpoons}} \text{SpL.}\kappa \underset{k_r}{\overset{k_f}{\rightleftharpoons}} \text{SpL.}\kappa^* \qquad\qquad 2 $$

Here, k_1 ($\text{M}^{-1}\,\text{s}^{-1}$) is the second order rate constant for the formation of the binary complex and k_{-1} (s^{-1}) is the first order rate of dissociation of the complex. The rate of the fast change of fluorescence intensity is proportional to the concentration of κ chain and represents the rate of formation of SpL.κ. Providing that the rate of formation of the encounter complex SpL.κ is rapid compared to the unimolecular isomerization step (i.e. the two parts of the process can be considered as two separate reactions), k_1 and k_{-1} may be derived from the slope and intercept, respectively, of a plot of k_{obs} against the final concentration of κ chain. The K_d for the encounter complex is approximated by k_{-1}/k_1.

The observed forward rate for the conformational change is the sum of the forward and reverse rates of the conformational change between SpL.κ and SpL*κ. This is given by $fk_f + k_r$ in which f is the fraction of the SpL present as SpL.κ. This is readily calculated by considering the initial part of the reaction i.e.

$$ \text{SpL} + \kappa \underset{k_{-1}}{\overset{k_1}{\rightleftharpoons}} \text{SpL.}\kappa $$

$$ K_d = \frac{[\text{SpL}][\kappa]}{[\text{SpL.}\kappa]} = \frac{k_{-1}}{k_1} \qquad\qquad 3 $$

In which [SpL], [κ] and [SpL.κ] are the concentrations of these components at equilibrium. Hence [SpL.κ] = [SpL][κ]/K_d and the fraction of SpL present in the complex

$$ = \frac{[\text{SpL.}\kappa]}{[\text{SpL}] + [\text{SpL.}\kappa]} \qquad\qquad 4 $$

Substituting [SpL.κ] in the above equation by [SpL][κ]/K_d and simplifying we get the fraction of SpL in the form SpL.κ

$$ = \frac{[\kappa]}{[\kappa] + K_d} \qquad\qquad 5 $$

255

Thus the observed forward rate of the conformational change is given by

$$k_{obs} = k_r + \frac{k_f[\kappa]}{[\kappa] + K_d}$$

6

Examination of this equation indicates

(1) If K_d is very low, then $k_{obs} = k_r + k_f$.

(2) If the concentration of κ chain = 0, then $k_{obs} = k_r$.

Thus, from (2) above we can see that if a preformed complex of SpL.κ* is rapidly diluted by buffer so that the free concentration of κ chain is negligible, then the observed reaction will depend upon the relative values of rates k_{-1} (the dissociation of the encounter complex SpL.κ) and k_r (the reverse conformational change from SpL.κ* to SpL.κ). In this example, k_r is rate limiting and is therefore observed.

   In general, if the observed rate of dissociation of a complex is shown to be independent of the final free concentration of ligand (in this case κ chain) then it is indicative that the dissociation process is rate limited by a conformational change and supports model (b) above for the SpL − κ chain reaction (16).

   In the reaction between SpL and κ chain, k_r can be measured directly by 'displacement' techniques. In these, preformed complexes of W39 SpL and κ chain are rapidly mixed with a large excess of wild-type SpL (which lacks Trp residues and therefore gives no change in fluorescence intensity at 335 nm on binding to κ chain). Thus the decrease in fluorescence at 335nm noted from the reaction of wild-type SpL with the W39 mutant complex is due to the dissociation of the W39 mutant from its binary complex with κ chain. Providing that the concentration of the SpL.κ species is very low then the observed rate for this reaction equals k_r, since in these experiments, the reaction in the forward direction (k_f) involving mutant proteins is competitively inhibited by a large excess of the wild-type SpL that lacks Trp residues. By subtracting the value of k_r from the value of k_{obs} for forward the isomerization reaction (equation 6) it is possible to calculate the value of k_f.

   For this system, the K_d for the encounter complex is k_{-1}/k_1 and at equilibrium, the macroscopic K_d for the reaction (K_d^{eq}) is dependent upon all four rate constants and is described by *equation 7* (14).

$$K_d^{eq} = \frac{k_{-1}}{k_1} \times \frac{1}{1 + k_f/k_r}$$

7

Which is another form of the equation (3)

$$K_d^{eq} = \frac{k_{-1}}{k_1} \cdot \frac{k_r}{k_f} \times \frac{1}{1 + k_r/k_f}$$

8

Note that if k_f is very large compared to k_r then these equations simplify to

$$K_d^{eq} = \frac{k_{-1}}{k_1} \cdot \frac{k_r}{k_f}$$

9

Resolving whether scheme (a) or (b) is the correct model for such a reaction is often difficult because one can rarely be certain where the changes in fluorescence actually occur. Other supporting experimental data are often required. It should be emphasized that as the rates of a biphasic process become closer then the method of analysis is more complex and readers should refer to other texts (see Section 4.3 and references therein).

Diffusion controlled rates of bimolecular association reactions are usually around 10^7–10^8 M^{-1} s^{-1} in agreement with theoretical modelling studies. These high rates are often not obtained experimentally which implies that something else, such as an isomerization in either the reactant or the complex, is limiting the rate. However, there are other possible reasons for association rates lower than expected and these have been reviewed in (17, 18).

Finally, a cautionary note. Reaction progress curves exhibiting more than one rate may not be due to complex kinetic behaviour in the reaction under investigation, but may be caused by a combination of real spectral changes due to the chemical reaction under study plus spectral changes due to another unknown process. For example, it is very common for a protein to become photodegraded when illuminated by intense UV radiation such as that given out by a xenon lamp. This is particularly evident in fluorescence experiments where high intensity excitation light is used to increase the sensitivity of the system; thus, slow decreases in fluorescence intensity may be noted over long observation periods. Alternatively, of course, the ligand may be similarly unstable. Control experiments in which each reagent is mixed with buffer in the stopped-flow should always be carried out to check for such degradation processes and to determine the expected fluorescence intensities of the control at the start and end point of the reaction.

4.3 Use of ligand fluorescence to monitor binding reactions

Reduced pyridine-nucleotide coenzymes (NADH and NADPH) are fluorescent molecules excited by light at 340 nm (absorption by the reduced nicotinamide ring) and fluorescing at approximately 460 nm in aqueous buffers. However, when bound to proteins, the fluorescence emission is usually shifted to other wavelengths and becomes more intense (2). The bound NADH can act as a reporter for the binding of another ligand to form a ternary complex (19–21). There are many other examples of fluorescent ligands.

Figure 8 shows data gained from stopped-flow fluorescence experiments in which warfarin's (W) fluorescence emission (excited at 330 nm and detected at 360 nm and above) is used to study its binding to human serum albumin (A). This reaction proceeds by two, consecutive reversible steps. A fast, diffusion limited bimolecular reaction by which the initial complex is formed (A.W), followed by a unimolecular conformational change in the protein following complex formation (A*W).

$$A + W \underset{k_{-1}}{\overset{k_1}{\rightleftharpoons}} A.W \underset{k_r}{\overset{k_f}{\rightleftharpoons}} A^*.W \qquad 10$$

The critical differences between this reaction and model (b) for the reaction between SpL and κ chain (described above) is that the change in fluorescence is limited to the unimolecular step (5) and the forward rate of the conformational change (k_f) is relatively fast. The inset to *Figure 8* shows that as the concentration of warfarin is increased, the observed rate of change of fluorescence also increases. Over this range of concentration of ligand, the bimolecular reaction is slower than the unimolecular conformational change and is therefore rate limiting. However, at high concentrations of warfarin, the rate of the bimolecular step exceeds that of the conformational change and the observed rate of reaction becomes limited by the latter step.

In a two step reaction such as this with only one source of signal change at the isomerization step the observed rate of the reaction is as described by *equation*

$$k_{obs} = \frac{k_f}{1 + \dfrac{K_d}{[W]}} + k_r \qquad 11$$

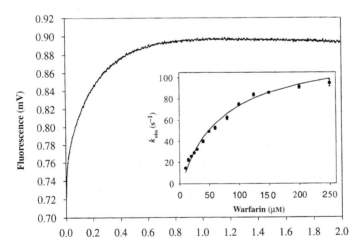

Figure 8 A reaction progress curve obtained when 20 μM warfarin was mixed with 1 μM HSA. In this reaction, 1 μM albumin was reacted with 10 μM warfarin (final concentrations) in 100 mM potassium phosphate buffer at pH 7.4 also containing 0.8% NaCl, at 25 °C. The fluorescence of the warfarin was excited at 330 nm, and the emission intensity at 360 nm and above was selected by a cut-off filter. The inset shows a plot of the pseudo-first-order rates, k_{obs}, for the reactions between various concentrations of warfarin and 1 μM HSA against the final concentration of warfarin used. We are indebted to Miss Sue Twine for the provision and use of these data.

in which $K_d = k_{-1}/k_1$ and [W] is the concentration of warfarin. From *equation 11* it can be seen that a plot of k_{obs} against [W] gives a hyperbolic curve as shown in *Figure 8*. The maximum value of k_{obs} at high concentrations of warfarin can be obtained from the plot shown in the inset to *Figure 8* and equals $k_f + k_r$. Rearranging *equation 11* shows that a plot of $1/(k_{obs} - k_r)$ against $1/[W]$ gives a straight line of slope K_d/k_f and an intercept on the ordinate of $1/k_f$. The value of k_r can be determined by a displacement reaction using a non-fluorescent compound that binds at the same site on albumin as warfarin, e.g. phenylbutazone. For a detailed kinetic analysis of this reaction see (5).

Protocol 2

Determination of k_{obs} for the binding of warfarin to human serum albumin

Equipment and reagents

- A stock solution of 100 μM human serum albumin (HSA) in 100 mM sodium phosphate buffer pH 7.4 containing 0.8% NaCl
- A stock solution of 3 mM warfarin in the same buffer

- A 360 nm cut-off filter to select the emission wavelength
- Some disposable 5-ml syringes and some small beakers/tubes

Method

1 Dilute some of the warfarin solution to 50 μM using the phosphate buffer.

2 Dilute the HSA to 2 μM in phosphate buffer.

3 Set the excitation and emission wavelengths to 330 nm and 360 nm (and above), respectively.

4 Fill both of the stopped-flow drive syringes with buffer and discharge 5–10 'shots' of these to wash the observation cell.

5 Then replace the buffer in one drive syringe with the 50 μM warfarin. After discharging five 'shots' to flush out the buffer in that side of the system, discharge five more. After these adjust the slit width of the monochromator selecting the excitation wavelength and the high voltage on the PMT to obtain a reasonable working signal. Record the signal voltage (unbound HSA). Note that this will be lower than that obtained when warfarin is in the presence of HSA.

6 Refill the syringes with phosphate buffer and discharge 10 'shots' and record the signal voltage (buffer baseline).

7 Now fill one drive syringe with warfarin and the other with 2 μM HSA, discharge five to six 'shots' to flush out the system. During the last one or two reactions you should notice that the signal voltage increases as the HSA and warfarin mixture reaches the observation cell.

8 Discharge five reactions and record the reaction progress curves. These should be single exponential curves with k_{obs} about 35 s^{-1} at 25 °C.

Protocol 2 continued

9 The warfarin solution can then be replaced with another at a higher concentration and the process repeated to obtain a curve such as that shown in the inset to *Figure 8*. Remember that if the system has a 1:1 v:v mixing ratio then the final concentrations of the reagents will be half that used to fill the syringes.

A cautionary note: It is very important to realize that in reversible reactions involving more than one step, the observed rates of reaction may contain several contributing rate constants. Furthermore, there are several possible variables to be identified, e.g. the relative values of forward and reverse rate constants, and at which stage or stages of a reaction the signal change occurs that is being used to follow the reaction progress. These all give rise to different analytical requirements beyond the scope of this chapter and readers should refer to other specialist texts for more information (22–24).

4.4 Extrinsic probes

Extrinsic probes must be used when the system under study has no useful intrinsic reporter groups or a reaction produces no fluorescence change from intrinsic reporter groups. In such cases one may use an appropriate fluorescent group which is added to the system to report the reaction. Extrinsic reporter groups come in many forms such as non-covalently and covalently bound fluorescent labels. Many of the latter can be covalently attached to the protein of interest at a specific site.

4.4.1 Non-covalent modification of proteins

The most widely used compounds for non-covalent attachment to proteins are 1-anilinonaphthalene-8-sulphonate (1–8 ANS) and bis-ANS. These compounds are virtually non-fluorescent in aqueous solution (Φ_F ANS = 0.004), but fluoresce intensely in apolar solvents (Φ_F ANS = 0.4 in ethanol) or when partitioned into hydrophobic clefts or the core of a protein (Φ_F ANS = 0.7 (25)). The emission wavelength maxima (around 480 nm and 500 nm in methanol for ANS and bis-ANS, respectively) are also highly dependent upon the polarity of the environment. They have found use in protein folding/unfolding studies (26, 27), especially in the detection of molten globule-like intermediates, which are able to bind more equivalents of the dye because their expanded structure gives increased access to the hydrophobic core. ANS and bis-ANS have also been used to study protein conformational changes (28, 29) and ligand binding (30).

4.4.2 Covalent modification of proteins and ligands

Chemical modification reagents for labelling reactive amino acid side chains (predominantly cysteine or lysine) have been available for many years and used to probe for residues close to or in the active site of enzymes. Although fluorescent derivatives of these reagents have been used less frequently, they occasionally reward persistent experimentation by offering more detailed information than simple chemical modification and the identification of reactive residues.

Both proteins and ligands can be labelled by fluorescent compounds to report binding interactions either directly, or in the case of fluorescent ligands, by inclusion in competition experiments. For example, the binding properties of the liver fatty acid binding-protein and the activity of phospholipase A_2 have been investigated by the use of 11-(dansylamino)undecanoic acid (DAUDA), a dansylated analogue of a fatty acid (31). This binds to fatty acid binding-protein and fluoresces intensely. Other, non-labelled fatty acids added, or liberated from phospholipids by the action of phospholipase A_2, will compete for the site and this can be monitored by a decrease in fluorescence as the DAUDA leaves the site. Dansyl sarcosine and dansylglycine have been used to study similar binding reactions with albumin (32–34).

Reactive groups on proteins, such as the thiol side chains of cysteine residues and the ϵ-amino groups of lysine residues, are ideal for reaction with fluorescent reagents, although guanidino, imidazole, alcohol (serine and tyrosine) and carboxyl side chains may also be labelled by various reagents. Suitable chemical reagents for the modification of cysteine side chains include maleimide derivatives of, for example, pyrene, fluorescein and eosin, and iodoacetamide derivatives of, for example, fluorescein or NBD (7-chloro-4-nitro-2,1,3-benzo-oxadiazole). The fluorescence of these groups is strongly quenched by water and is therefore a very sensitive probe for conformational changes in which increased accessibility to solvent occurs. Isothiocyanate and succinimidyl esters of most fluorophores are good reagents for labelling the ϵ-amino groups of lysine. o-Phthaldialdehyde is a very reactive compound that cross-links proteins by reacting with one equivalent each of a thiol and a suitably placed amino group, and provides a fluorescent isoindole group. Alternatively, it can be used to label an amino group in a protein and the reaction completed by the addition of a soluble thiol compound such as β-mercaptoethanol. The spectral properties of these fluorophores are given in *Table 1*, Chapter 12. An immense range of well-characterized fluorescent reagents are commercially available and interested readers should consult specialist companies such as Molecular Probes (http://www.probes.com).

Such labelling techniques can be very useful as long as the label allows the binding reaction to be monitored and does not significantly affect the structure or activity of the protein. It must be remembered that few chemical reagents are absolutely specific for one type of amino acid side chain and that side reactions will most probably occur. Thus it is recommended that some attempts are made to characterize the stoichiometry and site(s) of modification on a protein. Furthermore, some caution must be used in the interpretation of reaction progress curves obtained using proteins labelled with fluorescent reagents. Care must be taken to ensure that these reagents, usually with hydrophobic fluorescent groups, have not partitioned into hydrophobic regions of a protein. Such binding will lead to inaccurate calculations of the stoichiometry of labelling and, unless prevented (or the offending reagent removed after the modification reaction), may lead to spurious signal changes during experiments. It has been found in this laboratory that passage of a labelled protein down a short gel-filtration column to remove excess reagent is preferable to dialysis, as many of the fluorescent reagents will adsorb to the column matrix and leave the protein.

The presence of fluorescent groups such as pyrene, fluorescein, eosin etc. may also give characteristics to the protein or ligand that were previously not present. For example, such groups on ligands generally decrease water solubility and sometimes cause the compound to adsorb to the walls of plastic containers and cuvettes.

4.5 Examples of reactions monitored by changes in fluorescence of covalently attached fluorophores

NBD-Cl reacts at pH 7 with a cysteine residue of the Ca^{2+}–Mg^{2+}-ATPase from rabbit muscle sarcoplasmic reticulum. It was found that the fluorescence emission from the NBD label increases 17% on addition of saturating Ca^{2+} ions, and this feature has been exploited in experiments to examine the binding mechanism of Ca^{2+} ions to its complex with the protein (35). Several studies giving further details of the interaction of this protein with various ligands have been achieved using various coumarin reagents (36).

The pyrene group of pyrene-maleimide becomes fluorescent (λ_{ex} = 330 nm, λ_{em} = 378 nm and 400 nm) on reaction with cysteine residues. When 3 μM reagent is incubated with 3 μM recombinant bovine brain inositol mono-phosphatase, one equivalent of Cys218 becomes modified over 10–15 min at pH 8.0 at 20°C without loss of enzyme activity (37). Pre-equilibrium and equilibrium fluorescence studies showed that in this position the pyrene group is able to report the binding of essential Mg^{2+} ions at the active site of the enzyme (37, 38). *Figure 9* shows that a plot of the observed rates, k_{obs}, against the final concentration of the Mg^{2+} ions produces a straight line of slope k_1 M^{-1} s^{-1} and intercept on the ordinate of k_{-1} s^{-1}. The reaction is characterized by a high value for k_{-1} which makes the approach to equilibrium (k_{obs}) very fast (since k_{obs} = $k_1[Mg^{2+}]$ + k_{-1}).

It can be noted from *Figure 9* that the slopes of the lines (reflecting k_1) are very similar, whereas the intercepts on the ordinate (giving values of k_{-1}) increase with decreasing pH. The K_d for the equilibrium is therefore very sensitive to pH (38), and this is almost entirely due to changes in the value of k_{-1}. The data suggest that the binding of the metal ion to its coordinating ligands on the protein is not in direct competition with protons and that the rate of binding is limited by something else, possibly a conformational change. However, the reverse process, described by k_{-1}, is affected by the proton concentration, and a plot of k_{-1} against pH yields an approximation of the pK_a of the amino acid side chain(s) to which the metal binds. The determination of the rate of dissociation of the metal ions from the complex at various pH values can be achieved by rapidly diluting a preformed complex, of pyrene-labelled enzyme and Mg^{2+} ions at pH 8.0, with buffer (1:10 v/v) at the experimental pH containing 10 mM EDTA.

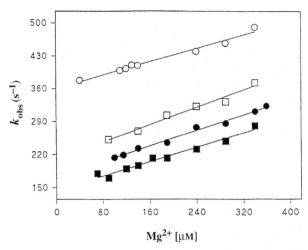

Figure 9 The effect of pH on k_1 and k_{-1} for the binding of Mg^{2+} to pyrene-labelled recombinant inositol monophosphatase. Enzyme (3 μM final concentration) was reacted with various concentrations of Mg^{2+} ions in 50 mM Tris–HCl buffer at pH 9.0 (■), pH 8.0 (●), pH 7.3 (□) and pH 6.5 (○) at 20 °C. The k_{obs} for each reaction is plotted against the final concentration of Mg^{2+} ions used. Reproduced with permission from Thorne, M. R., Greasley, P. J. and Gore, M. G. (1996) *Biochemical Journal*, **315**, 989–994. © Biochemical Society.

The advantage of this method is that the protein is kept at pH 8.0 and only exposed to low or high pH solution for a few ms.

It is interesting to note that in this case it is the high rate of dissociation of the metal ion (k_{-1}) that limits the ability of the stopped-flow technique to determine k_1 at high concentrations of Mg^{2+} ions at pH 7.0 or below. The data in *Figure 9* suggest that k_{-1} is approximately 360 s^{-1} and k_1 is approximately 3.7 × 10^5 M^{-1} s^{-1} at pH 6.5. Therefore in the presence of 300 μM Mg^{2+} ions the forward rate $(k_1[Mg^{2+}])$ will be 111 s^{-1} and k_{obs} will be approximately 461 s^{-1}. Thus the half-time of the reaction will be only 1.5 ms, and much of the reaction will take place within the dead time of the instrument.

References

1. Holbrook, J. J. (1972) *Biochem. J.* **128**, 921.
2. Holbrook, J. J. Yates, D. W., Reynolds, S. J., Evans, R. W., Greenwood, C. and Gore, M. G. (1972) *Biochem. J.* **128**, 933.
3. Chaiyen, P., Brissette, P., Ballou, D. P., and Massay, V. (1997) *Biochem.* **36**, 2612.
4. Ikushiro, H., Hayashi, H., Kawata, Y., and Kagamiyama, H. (1998) *Biochem.* **37**, 3043.
5. Wilting, J., Kremer, J. M. H., Ijzerman, A. D. P. and Schulman, S. (1982) *Biochim. Biophys. Acta* **706**, 96.
6. Gore, M. G., Greasley, P. J., Knowles, M. R., Gee, N., McAllister, G. and Ragan, C. I. (1993) *Biochem. J*, **236**, 3, 811.
7. Waldman, A. D. B, Clarke, A. R., Wigley, D. B., Hart, K. W., Chia, W. N., Barstow, D., Atkinson, T., Munro, I. and Holbrook, J. J. (1987) *Biochim. Biophys. Acta* **913**, 66.
8. Waldman, A. D. B, Hart, K. W., Clarke, A. R., Wigley, D. B., Barstow, D., Atkinson, T., Chia, W. N. and Holbrook, J. J. (1988) *Biochem. Biophys. Res. Comm.* **150**, 752.

9. Ross, J. B. A., Senear, D. F., Waxman, E., Kombo, B. B., Rusinova, E., Huang, Y. T., Laws, W. R. and Hasselbacher, C. A. (1992) *P.N.A.S. USA* **89**, 24, 12023.

10. Clark, N. S., Dodd, I., Mossakowska, D. E., Smith, R. A. G., and Gore, M. G. (1996) *Prot. Eng.* **9**, 10, 877

11. Bottomley, S. P., Popplewell, A. G., Scawen, M., Wan, T., Sutton, B. J. and Gore, M. G. (1994) *Prot. Eng.* **7**,12, 1463.

12. James, E., Whisstock, J., Gore, M. G. and Bottomley, S. P. (1999) *J. Biol. Chem.* **274**, 14, 9482.

13. Beckingham, J., Bottomley, S. P., Hinton, R., Sutton, B. J. and Gore, M. G. (1999) *Biochem. J.* **340**, 143.

14. Wallis, R., Moore, G. R., James, R. and Kleanthous, C. (1995) *Biochem.* **34**, 13743.

15. Lee, A. G., Baker, K., Khan, Y. M. and East, J. M. (1995) *Biochem. J.* **305**, 1, 225.

16. Benson, S. W. (1960) *The foundation of chemical kinetics*. McGraw-Hill, New York.

17. Pontius, B. W. (1993) *Trends Biochem. Sci.* **18**, 181.

18. Schreiber, G. and Fersht, A. R. (1993) *Biochemistry* **32**, 5145.

19. Piersma, S. R., Visser, A. J. W. G., de Vries, S. and Duine, J. A. (1998) *Biochem.* **37**, 3068.

20. Spencer, P. D., Slade, A., Atkinson, A. and Gore M. G. (1990) *Biochem. Biophys. Acta.* **1040**, 130.

21. Clarke, A. R., Wigley, D. B., Chia, W. N., Barstow, D., Atkinson, T. and Holbrook, J. J. (1986) *Nature* **324**, No. 6098, 699.

22. Strickland, S., Palmer, G. and Massey, V. (1975) *J. Biol. Chem.* **250**, 11, 4408.

23. Gutfreund, H. (1975) *Enzymes: physical principles*. Wiley, London.

24. Fersht, A. (1984) *Enzyme structure and mechanism* (2nd Edn). W. H. Freeman, New York.

25. Slavik, J. (1982) *Biochim. Biophys. Acta.* **694**, 1, 1.

26. Jones, B. E., Beecham, J. M. and Mathews, C. R. (1995) *Biochem.* **34**, 1867.

27. Teschke, C. and King, J. (1995) *Biochem.* **34**, 6915.

28. Walmsley, A. R., Martin, G. E. M. and Henderson, P. J. F. (1994) *J. Biol. Chem.* **269**, 17009.

29. Gibbons, D. L. and Horowitz, P. M. (1995) *J. Biol. Chem.* **270**, 3, 7335.

30. Lin, Z., Scharz, F. P. and Eisenstein, E. (1995) *J. Biol. Chem.* **270**, 3, 1011.

31. Wilton, D. C. (1990) *Biochem. J.* **266**, 435.

32. Muller, N., Lapique, F., Drelon, E. and Netter, P. (1994) *J. Pharm. Pharmacol.* **46**, 300.

33. Sudlow, E., Birkett, D. J. and Wade, D. N. (1975) *Mol. Pharm.* **11**, 824.

34. Epps, D. E., Raub, T. J. and Kezdy, F. J. (1995) *Anal. Biochem.* **227**, 342.

35. Henderson, I. M. J., Starling, A. P., Wictome, M., East, J. M. E. and Lee, A. G. (1994) *Biochem. J.* **297**, 3, 625.

36. Dalton, K. A., East, J. M., Mall, S., Oliver, S., Starling, A. P. and Lee, A. G. (1998) *Biochem. J.* **329**, 3, 637.

37. Greasley, P. J., Hunt, L. G. and Gore, M. G. (1994) *Eur. J. Biochem.* **222**, 453.

38. Thorne, M. R.., Greasley, P. J., and Gore, M. G. (1996) *Biochem. J.* **315**, 989.

Chapter 10

Stopped-flow circular dichroism

Alison Rodger* and Michael J. Carey[†]

* Department of Chemistry, University of Warwick, Coventry CV4 7AL
[†] Applied Photophysics Ltd., 203/205 Kingston Road, Leatherhead, Surrey KT22 7PB

1 Introduction

As is apparent from previous chapters (Chapters 6, 8 and 9), understanding the kinetics of chemical and biological processes is extremely important. Questions we often consider, explicitly or implicitly, include: Has something happened 'instantaneously' or will it take 20 years? Does changing the conditions or available reagents affect either the end product or the rate of a process? What intermediates are produced during a reaction? Can we characterize any intermediates? Do we need to remove them to prevent side reactions?

If some or all of the reactants or products are chiral, then circular dichroism (CD) detection may be the ideal tool for following the kinetics of a reaction, and if the half-life of the reaction is of the order of milliseconds to seconds or even minutes then stopped-flow mixing of the reagents will almost certainly be the appropriate choice of sample handling method. For reactions with half-lives of a few minutes to tens of minutes the reagents can be mixed by hand in a normal cuvette and the signal monitored at an appropriate wavelength. CD is not well suited to kinetics on timescales of hours due to the baseline drift that does occur (see Chapter 4, Section 2.5). Some CD spectropolarimeters have the useful facility of being able to perform a wavelength scan at pre-set intervals as well as monitoring continuously (except during the wavelength scan) at a chosen wavelength, thus facilitating the characterization of any intermediates.

In this chapter we shall highlight some of the considerations of the stopped-flow technique that are particularly relevant to CD experiments. Particular problems may be encountered when performing CD (as opposed to other detection methods) stopped-flow experiments. The measured signals are very small (typical CD intensities are 0.1% or less of the absorbance signal), and the noise level observed is particularly sensitive to any inhomogeneities or turbulence in the samples. Also, as one of the main applications of stopped-flow CD is in the study of protein folding and unfolding, samples are often very viscous and/or corrosive, have significant absorbances due to buffers etc., and the experiments often require wide and variable mixing ratios.

2 Instrumentation considerations

In establishing a stopped-flow CD system for a particular laboratory a number of different factors need to be taken into consideration. The most important of these are:

(1) available budget;

(2) sensitivity required—the ease of averaging over multiple scans is an important factor here;

(3) minimum dead time required;

(4) corrosiveness or adsorbance of the samples to be studied;

(5) type of flow system: stepper motor or pneumatic driven;

(6) size and design of optical cell;

(7) number of syringes and mixing stages required;

(8) mixing ratios required and whether these need to be varied from experiment to experiment;

(9) sample viscosity;

(10) wavelength range for detection, wavelength scanning and bandwidth; and

(11) software.

In addition to the above, there is also the important question of whether the instrument will be a stand-alone stopped-flow CD system, or a wavelength scanning CD spectropolarimeter with stopped-flow attachment, or a stopped-flow instrument where one of the detectors is a CD detector. Recent instrument advances have complicated rather than simplified the decision about instrumentation. The importance of the above factors will be summarized before a brief discussion of particular manufacturers and their current instrumentation is given below.

2.1 Available budget

Budget is often the starting point in instrument choice. The cheaper options involve a stopped-flow accessory attachment to existing wavelength scanning instrumentation. (One has almost certainly been performing steady state CD measurements if one is considering stopped-flow experiments.) The main disadvantages of this approach is that access to multi-purpose instruments is almost always more restricted than to dedicated instruments and changing the instrument from one mode to another may be non-trivial. The question of whether sensitivity and accuracy are lost by coupling two instruments is not a simple one, since, at least until recently, the optics of wavelength scanning CD instruments were significantly better than those of dedicated stopped-flow CD instruments. In addition, wavelength scans on samples used for kinetic determinations were significantly better from a combined instrument. It is no longer clear that this is the case.

2.2 Sensitivity required

Usually, the required signal-to-noise ratio can be achieved by averaging together a number of kinetic records (though it is important to remember that the noise only decreases with the inverse *square* of the number of data accumulations). However, ultimately, the optics of the instrument and the delivery of a homogeneous sample into the light beam will be limiting factors. The mixing issue is more important for stopped-flow CD than for other stopped-flow spectroscopic techniques, as turbulent flow increases the noise significantly and may even be birefringent thus giving a false CD signal. Chiral reagents are also typically more expensive than achiral ones so longer term budget considerations may favour a more expensive instrument to conserve required sample volume by minimizing the number of data accumulations. A typical system requires 200 μl of sample per shot plus of the order of 2 ml to prime a reagent line; sample concentrations are as for normal CD, which usually means an absorbance of ~1 in the stopped-flow cuvette, see Chapter 4, Sections 2.2 and 2.3.

2.3 Minimum dead time required

The minimum dead time of a stopped-flow system determines the starting point of a kinetics experiment. The dead time depends on certain physical aspects of the system, such as how fast the drive mechanism is able to deliver the reagents to the stopped-flow mixer (i.e. final velocity). In addition, the physical distance between the point of mixing and the point of observation is of paramount importance together with the type of flow that occurs immediately after mixing (i.e. how much turbulence is created). Submillisecond dead times may be claimed by manufacturers for their instruments but invariably such claims refer to ideal or special conditions. Dead times for CD kinetic measurements may be considerably longer, especially if a high viscosity reagent or asymmetric mixing is involved.

2.4 Corrosiveness and adsorbance of the samples to be studied

Acids, bases and organic solvents are the obvious samples where the inertness of the component parts of the stopped-flow system needs to be considered. Whether reagents might adsorb to tubing and other component parts is also important, not only for concentration accuracy in a particular experiment, but also to avoid contamination in subsequent experiments if new reagents prove to be more effective than the cleaning solvents at removing adsorbed species. These problems have been fully addressed in the context of high performance liquid chromatography, so almost any HPLC sales representative will be able to tell you what components are required for any particular application. For example, PEEK (polyether ether ketone) has excellent chemical resistance, is not particularly expensive, is flexible and is biocompatible. It is, therefore, suitable for flow tubing and other components that will come into contact with reagents. However, PEEK is attacked by some chemicals including concentrated acids, bromine, chlorine and fluorine. Titanium is a possibility for these substances. The flowlines in some stopped-flow systems are made of stainless steel. Although medical grade stainless steel is relatively inert, nonetheless, one should avoid using corrosive or reactive reagents. Some biochemical measurements can be compromised by the presence of ferrous ions. Whilst considering the interaction between reagents and stopped-flow components, one should also think about the material used in the loading and drive syringes. Plastic may adsorb many analytes and there is also the possibility of plasticizers and other species leaching out, especially in acidic media. Glass has similar problems and is particularly prone to leaching sodium.

2.5 Type of flow system: stepper motor or compressed air driven

A key component of any stopped-flow system is the drive unit for injecting reagents into the reaction/observation cuvette. The consensus of opinion seems to be that pneumatic drive systems provide the more rapid acceleration and stopping of reagents with greater reproducibility. However, the disadvantage of such systems is that only a limited range of mixing ratios is currently possible without changing syringes and their volume. Manufacturers of stepper motor systems claim the same dead times for their systems as pneumatic driven ones, and that the ratios of reagents may be changed extensively and with direct control from the computer. This allows significantly more flexible experiment design but is unnecessary for most applications. It should be noted that in order to produce a final velocity commensurate with a low dead time, there can be very significant pressure forces exerted on the stopped-flow cell during the stopped-flow drive and in particular when flow is stopped. These forces can result in small distortions of the cell window thus introducing birefringence which will affect the efficiency with which the CD signal can be measured. If birefringence effects are present, they will undoubtedly become more severe in the far UV (i.e. below 230 nm). Certainly, it is always advisable to acquire suitable reagent blanks which can then be used to compensate for baseline shifts due to pressure effects.

2.6 Size and design of the optical cell

The optical cell is where the chemical change whose kinetics are to be followed by CD takes place. All stopped-flow CD cells are flow-through cells, usually with a vertical orientation. A small optical flow cell (width, length and depth) reduces the system dead time and also reduces the amount of sample required to ensure the cell is filled. However, it reduces the distance available for observation and also requires a small light beam, both of which reduce sensitivity. In addition, mixing and flow constraints preclude the use of a cell where the bore is too narrow whereas a wide cell will have more back-mixing problems. The consensus seems to be that ~2 mm pathlength is optimal. The length of a cell needs to be such as to ensure that there is no back or forward mixing during the kinetics run.

The issue of forward mixing during a kinetic data acquisition is particularly important when high concentrations of one reagent are in use (i.e. experiments with 8 M urea or 6 M guanidine hydrochloride). If measurements are to be made below 200 nm then, just as for steady-state CD measurements, the cell compartment must be nitrogen purged so as to optimize light throughput.

Although ~5 mm length cell is often assumed to be adequate to avoid forward or back mixing, 10 mm is certainly safer. Such a cell requires (for a 2 mm square cross section cell) ~40 μl of sample in the cell. It also requires the light beam to be focused to about 1 mm by 8 mm in the centre of the cell to avoid light scattering off the sides of the cell. If the beam is not properly focused onto the centre of the cell and the light is scattered, then the signal will be distorted as discussed in Chapter 4. In practice, this means that a wavelength scan performed in the stopped-flow cuvette may show spectral features shifted in wavelength. The orientation of the cell and flowline input to the cell should also be such that gravity effects (i.e. high density reagent being able to run downhill to produce a concentration change in the cell after a few seconds) are avoided. Such an event can have a marked effect on the shape of the kinetic record.

2.7 The number of syringes and mixing stages required

Most kinetics experiments can be designed with two reagents, one or both of which may itself be a multi-component mixture. However, at least an additional syringe to contain washing solvent is required. More flexible and interesting experiments will require three or more reagent syringes. Also consider the volumes of the syringes and how often they will need to be reloaded. If each run requires ~200 μl reagent, then 1 ml only allows five accumulations with no provision for washing lines.

2.8 Mixing ratios required and whether these need to be variable from experiment to experiment

For single mix experiments, ratios of up to 10:1 are typically employed. These are routinely available on commercial instruments regardless of drive type. For certain multi-sequential experiments (e.g. 10:1, 1:1, 1:10) the flexibility of stepper motor systems may be the only option. The programmability of stepper motor systems can also help during experimental design. Some experiments require the concentrations of one reagent to be varied over a continuous wide range (e.g. urea concentrations in certain protein folding experiments). A stepper driven system may allow such experiments to be performed more quickly, as the concentration ramp is generated by altering the relative volumes delivered from the drive syringes, rather than the time consuming preparation of a whole range of reagent concentrations and their subsequent loading into the instrument.

2.9 Sample viscosity

If the samples are viscous, mixing efficiency can be affected. Incomplete mixing will result in signal changes unrelated to chemical events in the early part of the kinetic record. Mixer options or even premixers can be employed to counter such difficulties but are likely to compromise the instrumental dead time.

2.10 Wavelength range for detection, wavelength scanning and bandwidth

Several key aspects of instrument operation relate to spectral performance. For example, quality magnesium fluoride optics and nitrogen purging is required if the wavelength range for detection is required to extend below 200 nm. CD spectrometers require the use of a high intensity light source which has a significant far-UV content. For CD scanning instruments, the xenon light source, with its intense continuum, works adequately. For kinetic measurements at certain specific wavelengths, a mercury–xenon lamp may offer a worthwhile improvement in signal-to-noise ratio. However, such a lamp is not ideal when recording steady-state CD spectra nor does it provide any improvement for kinetic measurements below about 215 nm.

Wavelength scanning capability serves two purposes for the CD kineticist: first, it enables the CD spectrum of the solution at the end of the reaction to be compared to that of the starting material, and, second, it is the easiest way to confirm correct cell alignment with the optical beam and to calibrate the instrument. Incorrect cell alignment can produce effects due to light scattering, such as distortion of spectral features in a scanned CD spectrum. Proteins containing tryptophan residues (e.g. lysozyme) have sufficiently resolved peaks in the aromatic region (~280 nm) to be used for this purpose. It should be noted that scattering problems may be masked if the bandwidth of the instrument is very wide.

For kinetic measurements, a wider spectral bandwidth than normally employed for steady-state measurements is usually required to improve the signal-to-noise ratio of the kinetic records. Without the increased bandwidth, a very large amount of signal averaging may be needed to produce clean enough kinetic data from which reliable kinetic information can be derived. This is a very real problem when precious materials are being used. Conversely, a wide bandwidth may obscure kinetics particularly where different components of the sample have relatively small spectral separations.

2.11 Software

The software question relates to ease of use, degree of control over the various instrument functions, whether the required extent of data analysis is possible, ease of access to data, and ease of transfer of data in a format that can be readily used on other computer platforms and software packages. For kinetics in particular, automation of repetitive procedures such as multi-shot averaging is desirable. Flexibility in the control of data acquisition allows kinetic records to be optimized according to the event being studied. For example, a full range of linear and logarithmic time bases covers the requirements of simple kinetic changes to more complicated multi-step reactions. In addition, by making full use of the ADC's (analogue to digital converter) conversion rate (oversampling), data signal:noise quality can be significantly improved, especially as the acquisition period increases.

3 Currently available instrumentation

The following five manufacturers are active in stopped-flow CD.

1. Applied Photophysics Ltd, Leatherhead, Surrey UK (http://www.apltd.co.uk/).
2. Aviv Instruments Inc., Lakewood, New Jersey, USA (http://www.avivinst. com/).
3. Biologic–Science Instruments SA, Claix, France (http://www.bio-logic.fr/).
4. Hi-Tech Scientific Ltd, Salisbury, UK (http://www.hi-techsci.co.uk/).
5. OLIS, Bogart, Georgia, USA (http://www.olisweb.com/).

The manufacturers should be consulted to find details of current instrumentation. What follows is a brief discussion of some of the systems available in the context of the above 'considerations' list. It should not be considered definitive as advances are constantly being made but, hopefully, it may provide a starting point for making comparisons. The biggest decision (especially as regards to price) is whether to have a stand-alone stopped-flow system (either new or an upgrade of an existing one) or to add a stopped-flow sample attachment to existing wavelength scanning instrumentation. A summary of instrument costs is given in *Table 1*.

3.1 Stopped-flow attachment for CD spectropolarimeters

Before going ahead with the purchase of an attachment for a wavelength scanning instrument, it is advisable to check that the electronic response of your instrument is fast enough for millisecond time constant acquisition of kinetic data—the Jasco J-720, for example, needs a minor modification to remove a

Table 1 Approximate costs of instrumentation discussed in the text. Systems with an entry in the 'attachment required' column must be fitted to another instrument

System	Supplier	Attachment required	Price range
RX 2000	Applied Photophysics	Wavelength CD	$
SHU-61CD	Hi-tech	Jasco J-700 series	$$
SFM-3	Bio-Logic	Wavelength CD	$$–$$$
SF 305	Aviv	Aviv wavelength CD	
Olis module	Olis	Olis wavelength CD	
CD.2C	Applied Photophysics	SX.18MV	$$$
MOS-400 + SFM-3	Bio-Logic		$$$$
π*-180	Applied Photophysics		$$$$$

$ = less than 5k USD; $$ = 5k to 25k USD; $$$ = 25k to 50k USD; $$$$ = 50k to 100k USD; $$$$$ = 100k to 125k USD.

stabilizing feedback feature (which in fact seems to have no effect on normal instrument sensitivity).

The most cost effective (see *Table 1*) stopped-flow accessory for use with a standard spectropolarimeter is the RX.2000 from Applied Photophysics. This accessory is available with a manual or a pneumatic drive. The observation cell with integral mixer is at the end of a flexible and thermostatted umbilical. It is designed to fit sample holders that accommodate 12×12 mm standard cuvettes. Asymmetric mixing is possible by choosing drive syringes of an appropriate size. The dead time of this unit is in the range of 5 to 10 ms depending on type of drive and the drive syringe ratio in use.

An alternative stopped-flow attachment, designed for use with Jasco J-700 series spectropolarimeters, is the HI-TECH Scientific SHU-61CD with pneumatic drive. Its temperature range is 5–80 °C, its pathlength is 2 mm, it has a 3.5 mm window size and a cell volume of 19 μl. The mixer is integral to the cell. 200 μl is required per shot with 2 ml to prime the flowlines. Drive syringes may be altered to allow variable mixing. Hi-Tech do not offer a dedicated stopped-flow circular dichroism system.

The Bio-Logic SFM-3 system is a three-syringe unit with each syringe being driven by an independent stepper motor thus enabling independent programming of each syringe and easy variation of mixing ratios. For samples of differing viscosity/density, the specially designed mixer unit (HDS) is claimed to perform extremely well. It has a dead time of a few milliseconds (dependent on the flow rate chosen for the experiment). The system has been designed for Jasco, Aviv and Jobin-Yvon CD spectropolarimeters. A range of cuvettes may be used, though the 2 mm pathlength, 54 μl volume one is recommended for CD applications. The instrument may be manufactured from inert materials.

The approach adopted by Aviv for CD kinetic measurements was to design a dedicated module (SF 305 stopped flow system) to work with their existing CD spectropolarimeter. The SF 305 uses a servo motor drive with a range of motor options, which allow up to triple mixing with delay lines as required. It is designed to operate with samples of differing viscosities. Absorbance and fluorescence stopped-flow facilities may be integrated into the unit. A dead time of 1 ms is claimed with the T mixer. A range of mixing ratios are possible using the programmable stepper motor drive. It has a temperature range of 10 °C to 60 °C. The standard pathlength is 1 mm with a cell volume of 10 μl.

OLIS produces a CD version of its spectrometer which uses a novel RSM monochromator design to deliver rapidly changing monochromatic light to the sample. In addition, they claim increased sensitivity because of their ability to use both beams produced by the polarizing beam splitter, whereas other manufacturers use only one of the beams. Furthermore, OLIS state that this phase-coherent design removes the need to periodically check instrument calibration since absolute differential absorption of left and right circularly polarized light is correctly measured at any point in time. Whilst it is capable of operating in rapid-scanning, slow-scanning and fixed wavelength modes, the instrument's real strength is in its rapid scanning capability where, using real-time signal averaging, CD spectra are seen to 'come out' of the background noise. For CD stopped-flow, OLIS attaches its stopped-flow module to the RSM instrument. Although this combination will perform adequately where CD changes are of good intensity, one should be cautious if the need is to measure fast reactions accompanied by small changes in CD.

In addition to the RX.2000 stopped-flow accessory referred to above, Applied Photophysics provides a CD detection option, CD.2C, as an upgrade to its SX.18MV research stopped-flow spectrometer. In this instrument, the standard optical light guide coupling between the monochromator and the sample handling unit (used in the absorbance and fluorescence configurations) is replaced by a purpose built optical module comprising lens coupling, calcite polarizer, photo-elastic modulator and sample housing. A second monochromator is added to the spectrometer in order to optimize performance in the 225 nm region. The stopped-flow CD cuvette is at the end of a short umbilical which is attached to the SX.18MV stopped-flow sample handling unit. The umbilical mounted cell slides in or out of the sample holder thus allowing ready interchange between the stopped-flow cell and rectangular spectrometer cuvettes. A mercury–xenon lamp can be substituted for the standard xenon lamp used in the SX.18MV spectrometer, so as to improve CD kinetic performance at selected wavelengths. The stopped-flow sample handling unit uses a pneumatic drive. Ratio mixing thus requires selection of appropriately sized drive syringes. The viewport on the CD stopped-flow cell has the dimensions $10 \times 3 \times 2$ mm (i.e. 60 µl). The dead time for the instrument varies according to the drive ratio (about 4 ms for a 1:1 mix and 8 ms for a 1:10 mix). For a research group already using an SX.18MV spectrometer, the CD.2C provides a convenient upgrade path for the measurement of CD kinetics.

3.2 Integrated stopped-flow CD systems

The Bio-Logic system MOS-400 plus the SFM-3 system is the simplest stand-alone stopped-flow CD system. It is capable of being configured as a multi-functional spectrometer allowing absorption, fluorescence and CD measurements to be made. Although the instrument is optimized for single wavelength kinetics, the addition of motorized monochromators permits steady-state wavelength scanning. When configured for stopped-flow operation, the stopped-flow observation chamber can be removed and replaced by a simple 1×1 cm spectrometer cuvette holder. Simultaneous absorption and CD measurements can be made and, for samples where it is permissible to use the same wavelength for fluorescence excitation as for the CD measurement, simultaneous fluorescence and CD data acquisition is allowed.

The Applied Photophysics π^*-180 kinetic spectropolarimeter, is a fully integrated kinetic and steady state CD instrument. The π^*-180 instrument is designed in a modular format being built around a high-performance direct coupled optical bench with built in control and acquisition electronics. Alternate sample handling units, both kinetic and steady state, can be coupled to its optical interface. The principal difference between the above mentioned CD.2C and the π^*-180 instrument is that the π^*-180 is purpose designed both for high-performance steady-state CD scanning and quality CD kinetics as well. The optical design allows CD measurements to be made down to 180 nm. It uses the same stopped-flow cell as the SX.18MV instrument ($10 \times 2 \times 1$ mm; 20 µl volume) which is mounted in a thermostatted, rotatable cell block so that either 2 mm or 10 mm optical paths are readily accessible (changing the optical path in this way conveniently allows the same protein concentration to be used for kinetic measurements in the near UV as for the far UV). Fluorescence may be directly observed from the stopped-flow cell via a suitable cut-off filter and both fluorescence and CD detectors can be mounted at the same time. Research groups interested in protein studies will find switching between CD and fluorescence, by software alone, particularly convenient. The dead time on this instrument is approximately 1 ms for a 1:1 mix and 3 ms for a 10:1 mix. The base instrument also provides high quality absorbance detection and a range of interesting upgrade paths, including fluorescence emission scanning and anisotropy. The software suite provides complete instrument control as well as many novel digital optimizations to improve data quality and acquisition efficiency. The software is multi-tasking so that activities such as data manipulation and analysis can be pursued whilst the next set of data is being collected.

4 Additional experimental considerations

4.1 Parameters

The choice of parameters for kinetic experiments and particularly for stopped-flow experiments are based on the same considerations as for wavelength scans (see Chapter 4). Signal to noise ratio increases with: $\sqrt{n}$, where n is the number of times the data (in this case the kinetics plot) is accumulated and the data averaged; $\sqrt{\tau}$, where τ is the time over which the machine averages each data point; and $\sqrt{I}$, where I is the intensity of the light beam (which is in turn influenced by the bandwidth). For kinetics, in contrast to wavelength scans, a new sample is required for each new scan, thus increasing n may be expensive as well as time consuming. Ideally each kinetics run should be reviewed before being included in an average, as experiments may be prone to 'glitches' due to poor mixing or air in the cell. τ should be chosen as large as possible consistent with the kinetics of your run—thus if millisecond resolution is required, choose τ to be 1 ms (0.001 s).

Almost all stopped-flow CD experiments are performed at a single wavelength. The wavelength chosen should be one where a significant change in CD is observed between reactants and products. Usually one chooses a (positive or negative) maximum or a minimum of one species. It is thus difficult to design a stopped-flow CD experiment without access to a wavelength scanning instrument. Unless the bands are very narrow, a large bandwidth, b, (say 2 nm or larger) can be chosen to optimize the intensity of light incident on the sample.

The data interval, d, should be chosen to be approximately the same as τ the time constant. If this means you are collecting too many data points for the length of kinetics run you wish to perform, then either reconsider your choice of τ or use a variable (such as logarithmic) time axis. If d is significantly longer than τ, then only every few data points measured is recorded. If it is significantly shorter, then more-or-less identical data points are being recorded.

4.2 Zero-time and dead-time calibration

The time $t = 0$ value chosen for a stopped flow kinetic experiment may be determined by a number of different methods, none of which is perfect. The simplest is to assume that the first measurement is the appropriate value. However, one has to be sure that the mixing and flow characteristics of the system are not giving a spike at the beginning of the data record. If the kinetics are first order, then it does not matter if $t = 0$ is incorrectly defined. However, if the kinetics are not known to be first order, then an alternative approach is required. Unfortunately, measurement of the CD signal of the reactants and then estimating the expected $t = 0$ value does not seem to work. If necessary, the best option seems to be to mimic the kinetics experiment but remove a key reagent each time and replace it with a non-reacting non-CD active species (ideally of the same viscosity etc.) then add the component CD intensities. On a multi-function instrument, a good estimate of the true $t = 0$ point can be determined by using a reaction monitored by absorption or fluorescence (see Chapter 8).

4.3 Baseline

The acquisition of blank (baseline) runs for comparison with the kinetic data set is to be recommended. Data obtained from suitable blank runs should be subtracted from the main kinetic data in order to correctly define the CD change of interest (i.e. to correct for any CD offset together with any pressure induced artefact present immediately after mixing).

4.4 System tests

Before performing stopped-flow kinetic measurements, the instrument calibration should be checked using the spectral bandwidth that is to be used for the set of experiments. An aqueous solution of (+)-10-camphorsulphonic acid (CSA) is used for calibrating in the near-UV at 290.5 nm and D-(−)-pantoyllactone can be used to calibrate the far-UV at 219 nm. A simple and convenient reaction for performing initial checks on a stopped-flow CD instrument is the hydrolysis of glucuronolactone (1 mg/ml) by sodium hydroxide (1 M). The change in CD signal is measured at 225 nm over 0.5 s (*Figure 1*).

In relation to protein folding, lysozyme (*Figure 2*) and cytochrome c (*Figure 3*) provide convenient test systems. Lysozyme, for example, is prepared in 6 M

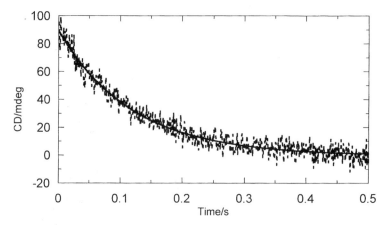

Figure 1 Kinetics of the hydrolysis of glucuronolactone (Sigma G-8875, 1 mg/ml in water) by potassium hydroxide (1 M) in 1:1 mixing ratio. Data shown is from a single run measured on a π*-180. The pathlength is 2 mm, bandwidth 4 nm, the light source is a Xe lamp and the detection wavelength is 225 nm. A blank has been subtracted. Data can be fitted to a single exponential: $(89.7 \pm 0.6)\exp[(-8.41 \pm 0.13)x] - (0.37 \pm 0.34)$.

Gdn-HCl (*Figure 2*) with 10 mM phosphate buffer (about pH 7). Dilution by buffer (1 in 11, i.e. 1 volume of sample plus 10 volumes of buffer) will produce a biphasic change in the CD signal at 225 nm as folding takes place. The signal amplitude at 225 nm is approximately 34 mdeg when the final lysozyme concentration is 200 µg/ml (final Gdn-HCl concentration will be 0.55 M). For protein folding measurements (Figures 2 and 3), it is usual to average at least 16 records together before performing kinetic analysis. With some instruments, it may be necessary to average well in excess of 16 records in order to produce a kinetic record of suitable signal-to-noise to permit accurate data analysis. Recent instrumentation, however, can produce an analysable kinetic record using a single stopped-flow drive as illustrated in *Figure 2*.

Because CD performance drops off quite rapidly in the far-UV, it is advisable to check the absorption spectra of the starting material together with any other reagents known or suspected to be present during a run. Although CD signals will increase as the sample concentration increases, if the total absorption (including *all* chiral and achiral species) increases above 0.8 AU, data quality will start to degrade rapidly as far less light will then be transmitted through the sample. Other species that absorb at the wavelength of interest should be as dilute as possible. This causes particular problems with some buffers and denaturing agents in protein folding/unfolding experiments, as one is often trying to probe at wavelengths below 230 nm and many reagents required at quite high concentrations absorb light in this region (see *Figure 4*).

All solutions for stopped-flow measurements should ideally be degassed, though if cavitation (as shown by spikes in the signal due to 'bubbles' caught in the flow cell) does not occur it may not be necessary. A simple degassing system is to sonicate samples under slight vacuum as illustrated in *Figure 5*.

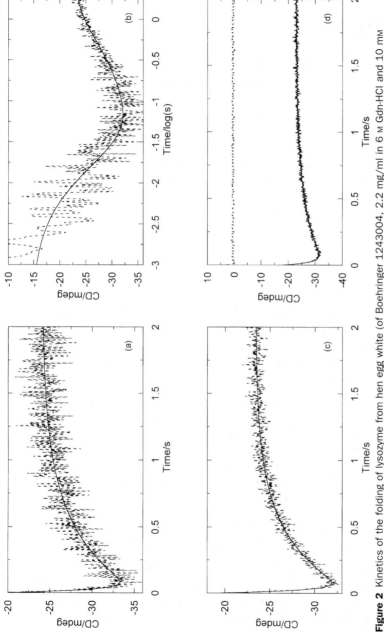

Figure 2 Kinetics of the folding of lysozyme from hen egg white (of Boehringer 1243004, 2.2 mg/ml in 6 M Gdn-HCl and 10 mM phosphate buffer pH 6.85) diluted 1:10 by phosphate buffer (10 mM phosphate buffer pH 6.85). Pathlength is 2 mm, bandwidth 4 nm, the light source is an HgXe lamp and the detection wavelength is 225 nm. Data shown are from a π*-180. (a) Single shot, (b) single shot with log time base, (c) average of eight shots, (d) eight shots also showing blank run. Data from the eight-shot run can be fitted to a double exponential: $(16.0 \pm 0.3)exp[(-42.2 \pm 1.5)x] - (11.03 \pm 0.12)exp[(-2.54 \pm 0.05)x] - (23.39 \pm 0.033)$.

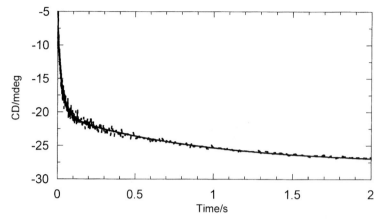

Figure 3 Kinetics of the folding of cytochrome *c* from horse heart (Sigma C-2506, 2.2 mg/ml in 6 M Gdn-HCl and 10 mM phosphate buffer pH 6.85) diluted 1:10 by phosphate buffer (10 mM phosphate buffer pH 6.85). Pathlength is 2 mm, bandwidth 4 nm, the light source is an HgXe lamp and the detection wavelength is 225 nm. Data shown are from a π*-180 averaged over 16 repeat iterations. Data can be fitted to a double exponential: $(14.48 \pm 0.19)\exp[(-41.4 \pm 0.9)x] + (7.08 \pm 0.05)\exp[(-1.233 \pm 0.03)x] - (28.80 \pm 0.05)$.

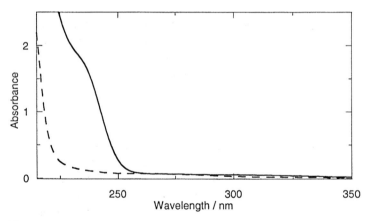

Figure 4 Absorption spectra of Gdn-HCl (6 M in water) of the same stated purity from different batches and/or different suppliers. The sample with the significant long wavelength absorbance is unsuitable for spectroscopic experiments. See, for example, (1).

5 Examples

A literature search with key words such as <stopped-flow> or <kinetics> and <CD> or <circular dichroism> and <millisecond> together with words for particular types of systems, such a protein, will reveal many examples of stopped-flow experiments. One area where stopped-flow CD has been a key technique is that of protein folding and unfolding. In this section we shall briefly review the role of stopped-flow CD in elucidating folding pathways for

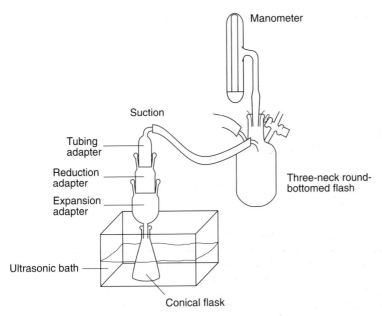

Figure 5 A simple system for degassing solutions.

lysozyme. The chapter concludes with a final example where the kinetics are due to the simultaneous interaction of three species one of which is a catalyst. Varying the concentration of the catalyst in the experiments takes them from the time domain where manual mixing is feasible to millisecond and faster timescales.

5.1 Stopped-flow CD and lysozyme folding

The folding of lysozyme has been extensively studied by a wide range of techniques and the progress that has been made in understanding this subject has been in large part due to complementary information from different techniques. Stopped-flow CD has played a pivotal role in the whole story, although the final answer has yet to be achieved. Our first lysozyme experiments (2) gave traces similar (but with worse signal:noise ratios) to that of *Figure 2*. The sharp increase in the negative signal at short times seemed to be an artefact. This impression was reinforced by the fact that the kinetics probed in the far-UV region seemed to be different from that indicated by the near-UV region. However, the result has since proved to be real and illustrates the importance of determining the wavelength dependence of the kinetics in complex systems (3–5 and references therein). Data from complementary techniques is also important for determining kinetic mechanisms. In this instance, the apparent discrepancy between the near- and far-UV kinetics data arises from the fact that the far-UV CD (see Chapter 4, Sections 5.3 and 5.4) is dominated by the formation of secondary structural units (such as the helices), whereas the near-UV region probes mainly individual tryptophans whose CD signal arise from coupling with the folded protein in their neighbourhood. Therefore, the near-UV CD is a more direct probe of tertiary structure.

5.2 DNA as a catalytic template

One of the earliest kinetic reactions to be followed by CD was the racemization of $[Fe(1,10\text{-phenanthroline})_3]^{2+}$. This takes place with a half-life of approximately 20 min at room temperature. $[Co(1,10\text{-phenanthroline})_3]^{3+}$ when synthesized under fairly rigorous oxidation conditions (we found chlorine gas was effective[1] (6)) is essentially stable at room temperature. Literature claims of rapid racemization, e.g. (7), are in fact the result of the presence of catalytic amounts of the labile Co(II) complex; electron transfer occurs readily between the different oxidation states of the tris-phenanthroline complexes.

When enantiomerically pure $[Co(1,10\text{-phenanthroline})_3]^{3+}$ and racemic $[Co(1,10\text{-phenanthroline})_3]^{2+}$ are mixed in the presence of DNA, the DNA causes

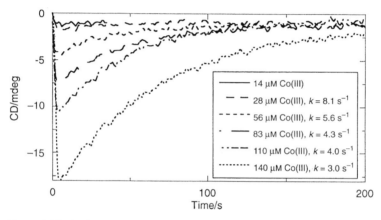

Figure 6 Single-shot stopped-flow kinetics traces from the mixing of $[Co(1,10\text{-phenanthroline})_3]^{3+}$ with a solution of $[Co(1,10\text{-phenanthroline})_3]^{2+}$ and calf thymus DNA (450 μM and 158 μM, respectively, in the reacting mixture). The +3 charged Co(III) complex binds much more strongly to DNA than does the Co(II) complex, hence its smaller concentrations. The third syringe was filled with water. Data were collected using a Jasco J-720 spectropolarimeter with a Biologique SFM-3 attached between the sample compartment and the photomultiplier tube. The light beam was focused with a single lens. Data were collected with a band width of 2 nm, response time of 1 s (so comparatively slow for stopped-flow) at 307 nm just outside the DNA absorption region.

[1] The racemization of $[Fe(1,10\text{-phenanthroline})_3]^{2+}$ is temperature dependent and the enantiomers can be resolved with antimonyl tartrate at 0°C. Antimonyl d-tartrate is commercially available and most resolutions of metal complexes using it involve collecting the precipitate as one enantiomer, and precipitating what remains in solution as the perchlorate to give the other enantiomer. The enantiomeric purity of the more soluble isomer is usually much lower than that of the less soluble salt. Potassium antimonyl l-tartrate is not commercially available but it may be synthesized by heating an aqueous slurry of Sb_2O_3 + KOH + l-tartaric acid (in molar ratio 1:1:2) under reflux for 2–3 h. Slow evaporation in a fumehood gives the required compound in crystalline form. $[Co(1,10\text{-phenanthroline})_3]^{3+}$ may also be resolved using antimonyl tartrate isomers.

a significant increase in the catalytic activity of Co(II) (8). The CD change as a function of time in an individual run follows a simple exponential decay (*Figure 6*); however, the dependence on the concentration of each reagent is not straightforward. To probe the concentration dependence of the racemization with only three syringes requires one syringe to be pre-loaded with two of the reagents. Some illustrative data are shown in *Figure 6*. As the concentration of the chiral reacting species, Co(III), increases, the initial CD signal increases but the rate decreases. The Co(II) and DNA concentrations also affect the reaction in a manner that does not correlate simply with their concentration. These observations can be understood when it is realized that the DNA is forming a template to facilitate the electron transfer reaction between two cobalt tris-phenanthroline complexes of different oxidation states. The DNA may also be playing a more active role in the electron transfer. Increased concentrations of Co(II) make such encounters on the DNA more likely, until it is in such excess that it is inhibiting Co(III) binding. Conversely, decreasing the DNA concentration increases the likelihood of an encounter between the reacting species molecules until the same saturation of the kinetics occurs.

References

1. Mo, J, Holtzer, M. E. and Holtzer, A (1991) *Biopolymers* **31**, 1417–1427.
2. Radford, S. and Rodger, A. (1988) *Unpublished data.*
3. Kulkarni, S. K., Ashcroft, A. E., Carey, M., Masselos, D., Robinson, C. V. and Radford, S. E. (1999) *Protein Science*, **8**, 35.
4. Weissman, J. S. and Kim, P. S. (1991) *Science*, **253**, 1386.
5. Evans, P. A. and Radford, E. S. (1994) *Current Opinion in Structural Biology* **4**, 100.
6. Lee C. S. (1966) *Inorg. Chem.* **5**, 1397.
7. Rehmann, J. P. and Barton, J. K. (1990) *Biochemistry* **29**, 1701.
8. Bates, P. J. and Rodger, A. (1992) *Unpublished results.*

Chapter 11

Spectrophotometry and fluorimetry of cellular compartments and intracellular processes

C. LINDSAY BASHFORD

Department of Biochemistry and Immunology, Cellular and Molecular Sciences Group, St George's Hospital Medical School, Cranmer Terrace, London SW17 0RE

1 Introduction

Optical spectroscopy, spectrophotometry and fluorimetry can be used to monitor processes occurring in living cells provided that suitable chromophores are present which 'report' on the events in which they participate. The advantages of optical techniques are manifold. Firstly they can be fast—with appropriate apparatus events in the pico- and nano-second domains can be studied by fluorescence spectroscopy. Secondly they are continuous—instant feedback from the experimental system can guide the most complex of experimental protocols, and allow the experimenter to adjust system parameters as necessary. Thirdly they are convenient, and most laboratories have access to equipment that can provide quantitative analysis of optical signals; examples include conventional spectrophotometers/fluorimeters, dedicated instruments (e.g. for fluorescence lifetime and polarization measurements), cameras, microscopes and plate readers. Significantly detectors from one apparatus can often be used on others to open up new experimental protocols. Fortunately the principles underlying the use of such a diverse array of optical devices are straightforward and universal—they apply just as much to laboratory 'work-horse' instruments as they do to the most specialized, laser-illuminated fluorescence microscope. The availability of fast laboratory computers with large storage capacities means that most modern spectrometers are microprocessor controlled and digitization of signals opens up the full range of possibilities of data accumulation, storage, analysis and interpretation.

The main problem with optical measurements is not the acquisition but rather the interpretation of the data obtained. Straightforward analysis of the results depends on the clarity of the experimental design and the appropriate choice of chromophore. This chapter describes some of the problems that can be addressed by spectroscopic techniques and attempts to give guidance on good experimental design.

2 **Experimental design**

2.1 **Apparatus**

Optical spectroscopy requires either spectrophotometers, to measure absorbance, fluorimeters, to measure fluorescence, or microscopes, which can measure fluorescence or absorbance of single cells or small groups of cells. Fluorimeters and spectrophotometers usually require solutions or suspensions of material in conventional cuvettes; microscopes provide two-dimensional images from smears, slices or surfaces. Other devices that record signals resolved in two-dimensions include gel scanners and microplate readers. Essentially these devices sample the 'object' in an organized manner (detectors can be set up to record absorbance or fluorescence) and information is stored in an electronic array that maps precisely the physical layout of the original object.

Fluorimeters and spectrophotometers may have attachments that permit recording of signals from surfaces, or can be adapted to record such signals using fibre optic 'light guides'. In this case, the guide (a bundle of optical fibres) conducts the appropriate illumination to the surface and both reflected and fluorescent light are collected by other fibres and conducted to a detector, usually a photomultiplier (1). The reflected light has two components: specularly reflected light, this is the mirror-like reflection, and diffusely reflected light. Only the latter contains information (analogous to absorbance) concerning the chromophores present at the surface.

Light output from the image plane of an optical microscope can be assessed with the light metering system of a camera attached to the microscope. The time taken correctly to expose a film of given speed (ASA/DIN) is inversely related to the brightness of the image. The exposure time, that is the interval between the opening and the closing of the shutter, is measured with a stopwatch. For accurate measurements the exposure time should lie in the 10–100 s range; shorter exposures are difficult to time with appropriate precision and longer exposures are compromised by stray light entering the objective (see Section 2.1.2 for discussion of stray light) and by photobleaching. The correlation between intensity and reciprocal exposure time is linear over a very wide range (1), so this procedure provides a handy 'rule-of-thumb' comparison of the brightness of images. The main disadvantage of this simple approach is that no allowance is made for the differing brightness of objects in the field of view, indeed most of the background may be black. Most conventional camera meters weight light in the centre of the field of view more heavily than that in the periphery.

A digital approach can resolve the difficulty of signal variation across the image and CCD (or similar 'digital') cameras are now routinely found in the camera port of optical microscopes. They provide digital images that can be analysed pixel by pixel. A wide range of image analysing software, not all of it specific to particular commercial packages, is available. In such processing, the key decisions that have to be taken are the zero level, sometimes called the 'black level', and the gain employed. If zero and gain are not chosen correctly a very distorted impression of relative levels can be achieved. At high magnifications photobleaching, that is a time-dependent loss of intensity, usually occurs and it is necessary to arrange for short exposure times and minimal exposure of the field to the excitation beam.

2.1.1 Artefacts

Optical investigations of biological systems are beset with hazards and it cannot be too strongly emphasized that very rigorous criteria must be applied before interpreting the experimental data. Artefacts are more common in fluorescence experiments and a few of the most easily avoided ones are considered here.

2.1.2 Stray light and turbid solutions

Stray light is a general term for any light that reaches the detector which is irrelevant to the experiment in progress, e.g. either not of fluorescence origin or, in an absorbance experiment, of a wavelength different from that intended. It may include ambient light and 'leaks' through the monochromators and filters from the source. It is avoided by correct apparatus construction, the correct choice of blocking filters (to exclude higher order diffraction and specular reflection from the grating in the monochromator) and the appropriate use of dark rooms and/or black curtains. It is important to understand that most detectors provide an output (usually a voltage) even when no light is present. Operators should aim to make this 'dark current' equivalent to 'zero light' (0% transmission) or the background ('black') level. With some devices this selection of true zero may be affected by automatic regulation of gain and/or contrast; in such cases, the most precise measurements may only be possible in the manual operation mode where the observer can choose for themselves the correct setting.

Suspensions of biological materials are turbid. It is especially important in measurements of fluorescence to resolve fluorescence from scattered light. Since scattered light usually exceeds fluorescence by a substantial margin, careful choice of filters and slit widths is essential to record just fluorescence (2). It is important to be aware that substantial changes in turbidity, and hence in the absolute value of fluorescence or absorbance, occur when membrane vesicles are exposed to sudden changes in osmolarity or are exposed to permeant salts (3). In order to maximize the chances of 'capturing' the emitted/transmitted photons, the cuvette is usually placed as close as possible to the detector.

Turbid solutions can be useful when the chromophore has low extinction or can only be used at very low concentration. This arises because the multiple internal reflections amongst the particles greatly increases the effective path-length of the absorbance experiment and hence increases the amount of light absorbed (a consequence of Lambert's Law). In such circumstances, the *dual wavelength* optical arrangement is strongly recommended: absorbance at one wavelength is measured *in the same cuvette* with respect to light of a slightly different wavelength; changes in turbidity affect both beams equally and are thus cancelled out (1, 4). The ideal situation is for the reporter group to show a large response at one wavelength and a small (or oppositely directed) one at a neighbouring wavelength; this is most likely to occur in compounds which have intense, narrow ('sharp') absorbance bands.

2.1.3 Inadequate excitation

The excitation light may be absorbed by non-fluorescent material, such as occurs when monitoring the surface fluorescence of tissues or organs. Blood completely absorbs the light used to excite fluorescence due to pyridine nucleotides and flavoproteins, hence an increase in the amount of blood in the field of view, due to vasodilation, will decrease fluorescence and lead to an erroneous conclusion that the oxidation–reduction state of the tissue mitochondria has changed. This problem is obviated by recording the ratio of the pyridine nucleotide (PN) and flavoprotein (FP) fluorescence: as mitochondria become reduced PN fluorescence increases and FP fluorescence decreases, as the mitochondria become oxidized the changes of fluorescence are reversed. Hence the FP/PN fluorescence ratio is a sensitive indicator of the mitochondrial oxidation-reduction state (5). It is not severely affected by masking problems as both signals are affected equally so that their ratio remains unaltered. An alternative procedure for correcting for the interference of blood is to obtain the 'corrected' fluorescence, namely the difference between the measured fluorescence and the measured reflectance (of the excitation light). In this mode, loss of excitation, due to absorbance by the blood pigments, leads to a loss of reflectance which compensates for the loss of fluorescence. The adequacy of this procedure has been demonstrated in perfused systems where the blood can be replaced by a transparent perfusate (6), but its applicability *in vivo* is less certain. In many applications the ratiometric technique is the method of choice to overcome the problem of heterogeneous distribution of the indicator molecules. In very high resolution optical microscopy, the slight difference in refraction of excitation beams of differing wavelength may mean that an image perfectly focused for one wavelength is marginally out of focus for the other. Fortunately this discrepancy is usually sufficiently small to lie within the range of experimental error of most quantitative measurements.

The excitation light may be absorbed so strongly by the fluorophore that fluorescence is reduced along the excitation path, one aspect of the inner filter effect (1, 2). Such an effect should be suspected if dilution of a sample leads, initially, to a fluorescence increase. This could arise, for example, if the detector 'observes' a part of the sample reached by the excitation beam only in dilute solutions. The expected linear relationship between fluorescence and concentration occurs only with dilute ($<10^{-3}$ M) preparations (1). Remember that it is not only the fluorophore that may absorb the excitation and that it is the *overall* absorbance that must be kept low to avoid any inner filter effects.

Inner filter effects can be reduced by altering the geometry of the fluorescence experiment: fluorescence from the surface of highly absorbing material is much less sensitive to the effect, because neither the excitation nor the emission has to traverse an opaque medium in order for a fluorescence signal to be recorded. Indeed, if a microscope with an epifluorescence attachment is used to excite the material, the effective pathlength is of the order of a few microns and little filtering occurs. This can lead to confusion if results obtained with a microscope are compared with those obtained in a conventional fluorimeter. Cyanine dyes are avidly accumulated by mitochondria *in vivo* (7) with a resultant quenching of their fluorescence when assessed in cell suspension (8); however, in the fluorescence microscope the mitochondria appear as brightly fluorescent, filamentous structures (7). Similarly acridine orange and 9-aminoacridine are accumulated by acidic endosomes with a quenching of their blue fluorescence; in the fluorescence microscope the vesicles appear as yellow or orange vesicles, this being the characteristic emission of very concentrated solutions of these dyes (1).

2.2 Light sources

The selection of light source is determined by the nature of the experiment. For most determinations of fluorescence the most important factor is providing sufficient excitation to generate enough fluorescence to detect. The most intense sources are lasers, and they have the added advantages that they have very well-characterized wavelengths containing essentially no 'stray' light of other wavelengths and their output level is very steady. The latter is particularly important when monitoring small changes of fluorescence with time—ultimately the sensitivity of such experiments is limited by the time-dependent variation in excitation level of the source. The disadvantage of lasers is that they only produce a limited range of excitation wavelengths which are not suitable for all potential fluorophores. The widespread availability of 488 nm, 543 nm and 610 nm lasers means that many potentially useful fluorophores have been deliberately synthesized to have spectral properties that match those wavelengths. 488 nm and 543 nm are the wavelengths used for the 'fluorescein' ('green' channel) and 'rhodamine' ('red' channel) of fluorescence microscopes.

After lasers the next most intense sources are arc-lamps, usually mercury or xenon. Mercury arcs provide a very intense output at 366 nm with reasonable levels at other wavelengths; xenon arcs provide a more balanced output across the ultraviolet and visible regions of the spectrum, although users should be aware that there are a number of significant peaks even in the xenon arc output. Arc lamps run at high temperatures and generate ozone so they must be well-ventilated and cooled. In addition the ultraviolet output is so strong that they should not be observed with the naked eye—indeed no high intensity light source should be so observed. For some fluorescence applications pulsed arc lamps are particularly useful if the detector is synchronized to the source. This arises because the power output during the pulse is very high while overall output is much lower. A 30 watt pulsed xenon source can be as effective as a 150 watt continuous source for this reason.

In absorbance experiments where transmission is monitored in the presence and absence of sample a very steady light output is required, often not of particularly high intensity. Filament lamps are usually perfectly satisfactory in this regard.

2.3 Wavelength selection

Choice of the correct wavelength is essential for all optical measurements. Usually the procedure is straightforward: for absorbance choose the wavelength available from your source that most closely matches the absorbance maximum of the material you are monitoring; use the same criterion when selecting an excitation wavelength in a fluorescence experiment. If the molecule you wish to study has spectral properties that overlap with other material in your sample, it may be necessary to choose an alternative wavelength, as a rule of thumb it is wise to select an absorbance/excitation where your probe contributes at least 90% of the total absorbance at that wavelength. Remember that when single wavelengths are employed a change in signal may arise either from a change in the absolute level of absorbance/fluorescence or from a *shift* in the absorbance/ fluorescence spectrum. Indeed it is always good practice, where possible, to record the spectrum of the response to help in the interpretation of the results.

In fluorescence measurements, the emission wavelength must be sufficiently different from the excitation wavelength that scattered and 'stray' light do not contribute to the output. Again inspection of excitation and emission spectra will guide the choice of wavelength. Fluorophores with very high quantum yields, such as fluorescein, have strongly overlapping excitation and emission spectra. To provide enough excitation and emission without the problems of scattered/stray light it is usually necessary to choose an excitation wavelength to the blue of the excitation maximum and an emission wavelength to the red of the emission maximum.

2.4 Chromophore selection

Useful chromophores may be endogenous, or 'intrinsic', that is, already present in the system of interest, or exogenous, 'extrinsic', pigments provided by the observer. In the former category, pigments which absorb light in the visible region of the spectrum (400–700 nm) are the most useful because they are restricted both in abundance and in the range of processes in which they participate. Porphyrins, flavins and carotenoids exhibit characteristic spectra with rather narrow, intense absorbance bands. Thus the role of cytochromes in mitochondrial electron transport was appreciated long ago by Keilin (9), as a result of his studies of insect flight muscle using a simple spectroscope. In addition the cofactors most commonly found in biological oxidation–reduction reactions, namely the pyridine nucleotides (NAD, NADP) and the flavins are fluorescent in either their oxidized or their reduced forms. In photosynthetic systems, in addition to chlorophyll, the carotenoids, secondary 'antennae pig-ments', serve as 'molecular voltmeters' since the positions of their absorption peaks alter as a function of the electrical potential across the photosynthetic membrane (10). This property is useful for studying the mechanisms of energy transduction in photosynthesis, and stimulated an extensive search for exogenous pigments with similar properties which could be used for studies of membrane potential in a wide variety of systems (11).

In addition to exogenous pigments which register membrane potential, dyes have been employed to monitor many cellular properties; for example, cytoplasmic and organelle pH, cytoplasmic Ca^{2+} and Mg^{2+} concentration, membrane microviscosity, fate of ingested particles. More recently probes have been developed to report on processes such as apoptosis, critical to cell survival. In these cases the molecule in question is specifically designed: (i) sensitively to register the parameter of interest; and (ii) simply to reach its ultimate destination. Probes of cytoplasmic Ca^{2+} or pH exhibit characteristic fluorescence or absorbance depending on the cations in their environment. Membrane-permeant esters of the indicators diffuse to the compartment where endogenous enzymes catalyse their hydrolysis to generate the active species *in situ*. More specific delivery may be accomplished by exploiting natural endocytotic and/or biosynthetic processes. The range of possible probe experiments is so great that it is impractical to describe each in detail. The publications of Molecular Probes (their catalogue, the series of leaflets called BioProbes and their very user-friendly web site http://www.probes.com) provide the most extensive introduction to the types of probe experiments available and the literature from which those approaches were drawn. It is not always appreciated how simple many of the experiments are to perform using equipment available in most laboratories.

2.4.1 Criteria for chromophore selection

The primary decision, and often the most difficult, is the choice of a suitable chromophore which will provide information pertinent to the problem under consideration. In the case of intrinsic chromophores this must be a choice of system, namely the one in which the chromophore occurs and in which it exhibits the desired properties. For example, the carotenoid pigments which record the light-dependent membrane potential associated with photosynthesis. Such pigments occur in many photosynthetic membranes, but their use as 'molecular voltmeters' is most easily accomplished in preparations, known as chromatophores, from certain photosynthetic bacteria such as *Rhodobacter sphaeroides* or *Rhodopseudomonas capsulata*. Again, studies of oxygen consumption and delivery *in vivo* require highly vascularized tissues which contain many mitochondria, for example, brain, liver or cardiac muscle (although in the latter case the myoglobin contribution cannot be resolved from that of haemoglobin). In each case, however, the oxygen carriers tend to mask the much less abundant mitochondrial pigments. It is not always easy to resolve contributions from endogenous pigments in cell suspensions as the cell density required to give a suitable output is often too large to be compatible with provision of an adequate oxygen supply.

Whether the final decision is to use an endogenous or an added chromophore there are a number of factors that always have to be taken into account (6, 12).

1. Instrument compatibility. Only chromophores with characteristics compatible with the instruments available are suitable. The main features to consider are excitation wavelength, availability of appropriate emission filters and the sensitivity of the detector. For example, many dyes with red fluorescence are usually adequately excited by 543 nm laser light and detected via the 'rhodamine' filter sets provided with commercial fluorescence microscopes; likewise dyes with green fluorescence are suited to the 'fluorescein' set up.

2. Colour. Fluorescence instruments may be able to register signals from more than one dye in the same experiment. Thus it is possible, for fluorescence microscopy, to include dyes with green and red fluorescence and monitor their output using successively, in the same sample, the 'fluorescein' and 'rhodamine' set ups. If the images are stored separately, it will subsequently be possible to combine them, in which case regions containing both red and green fluorescence will appear orange and it is possible to tell whether the molecules segregate to the same or different regions of the image. This can be particularly useful for following the movement of different materials through intracellular compartments.

3. Signal intensity. In absorbance experiments, dyes with high extinction coefficients provide the largest signals. In fluorescence experiments, the output depends on both the efficiency of absorbance (of the excitation beam) and of the fluorescence emission (usually characterized by the quantum yield (Q): a Q value of 1 indicates that every photon absorbed is re-emitted as fluorescence, a Q value of 0.1 indicates that one in every ten photons absorbed is re-emitted as fluorescence). Good compromises can often be achieved using highly fluorescent dyes with less than optimal excitation or by using weaker emitters but with very efficient excitation.

4. Stability. Nearly all chromophores will be 'bleached' by persistent high intensity excitation. This is because the excited state of the chromophore usually has enhanced chemical reactivity compared to the ground state, so strong excitation promotes formation of non-fluorescent/non-absorbing material, especially if oxygen is present.

5. Background interference. All biological materials absorb light at one wavelength or another. Membranes contain components which absorb near-ultraviolet and scatter blue light strongly. Probes with fluorescence/absorbance in the red region of the spectrum are usually the most useful. Remember, however, that many photometers are less sensitive to red light than they are to blue light; so that the advantages gained by choosing a probe with little optical 'overlap' with native pigments are offset by the decreased sensitivity of the apparatus. The fluorescence of extrinsic probes will be severely quenched if the emission of the fluorophore overlaps with intense absorbance bands of intrinsic chromophores. Cellular autofluorescence (due, at least in part, to flavins) is green/yellow and can interfere with experiments using fluorescein-based chromophores. There is little autofluorescence in the orange/red region of the spectrum where rhodamine derivatives emit. In favourable circumstances, changes in fluorescence may be observed even when absorbance changes are masked by endogenous pigments (13). It is always good practice to view materials without your probe present to check that endogenous signals are unlikely to complicate the interpretation of your experiment.

2.4.2 Delivery of exogenous probes

Optical probes used to study membranes are usually lipophilic and can be stored as stock solutions in solvents such as ethanol or dimethyl sulphoxide. The stock solution is diluted at least 200-fold with the membrane preparation and the dye reaches its destination by diffusion. Such labelling is non-specific and in multi-membrane systems such as intact cells, the probes partition unevenly among the various compartments. This is helpful if the dye concentrates in the region of interest, as cyanine dyes do in mitochondria (7) and weak bases do in acidic endosomes (14, 15). However, if cytoplasmic pH (Section 3.3.2) or plasma membrane potential (Section 3.2.1) are the object of interest, these dyes cannot be used in a straightforward way unless measures are taken to suppress responses from other compartments. Interpretation is more complex when different systems are labelled. For example, the cyanine dye diO-C_3-(6) has been reported as an excellent mitochondrial stain, when used at 'low' concentration (less than 10^{-7} M (7)), and as a stain for endoplasmic reticulum (16) when used at 'high' concentration (greater than 10^{-6} M). The reason for this is that high concentrations of diO-C_3-(6) poison mitochondria (they inhibit complex 1 of the respiratory chain (17)), dead mitochondria release accumulated dye which is now available to accumulate in endoplasmic reticulum.

Slightly more specific labelling can be achieved by delivering an inactive (usually an ester) form of the dye which is locally converted to the active form by endogenous esterases. This is the technique used to probe cytoplasmic Ca^{2+} and H^+ levels with fluorescent analogues of EDTA (18,19). Specificity is achieved by the location of active esterases in the compartment of interest (see Section 3.3.2).

Specific labelling can be achieved by coupling the chromophore covalently to a vehicle, usually an immunoglobulin or a hormone, that binds specifically to cellular receptors. This is the technique widely used to identify specific sub-cellular compartments/locations. Antibody specificity is coupled to the sensitivity of fluorescence to permit the highlighting of particular objects for which specific, often monoclonal, antibodies are available. Polyclonal antibodies specific for determinants on the monoclonal antibody are available with covalently attached chromophores, whose spectral properties match the excitation and emission criteria available in most 'fluorescence' microscopes.

2.4.3 Toxic actions of chromophores

Chromophores which are non-toxic at low concentrations may become potent inhibitors if they are concentrated in specific compartments. Cyanine dyes severely inhibit respiration at site 1 of the mitochondrial respiratory chain, providing that the inner mitochondrial membrane potential is substantial (inside negative) (18). In living cells, the inner mitochondrial membrane potential is about 180 mV and the plasma membrane potential is about 60 mV (both inside negative), and extracellular cyanine at 10^{-7} M is at electrochemical equilibrium with mitochondrial cyanine when the latter reaches 10^{-3} M. Very low concentrations of cyanines may thus adversely affect cellular respiration and energy-dependent processes (18).

2.4.4 Complex formation

The dye chosen may interact with elements of the system with a consequential change in its properties. Thus the pK_a of a pH indicator in solution may not necessarily have the same value when the indicator is bound to cellular or other constituents (see, for example, (20, 21)); indeed the spectra of the bound dye may differ from those of the free dye, making any simple evaluation of pH problematical. It may be that the dye is useful as a qualitative indicator of the parameter under investigation but cannot be calibrated because it forms complexes with agents required in the calibration protocol. It is always good practice to use several, independent calibration procedures to check for internal consistency of the dye response. Thus when calibrating optical probes of membrane potential (22, 23) the same values should be obtained when valinomycin/K^+ or uncoupler carbonyl cyanide p-trifluoromethoxyphenylhydrazone (FCCP)/H^+ are introduced.

2.5 Characterization and calibration of optical signals

Once the choice of chromophore and system has been made it is essential to characterize their optical properties. The spectrum of the chromophore (absorbance or fluorescence) *in situ* may contain unexpected contributions which will limit the scope of the experiment. Such investigations establish the signal-to-noise ratio for the intended observations (noise being used in its most general sense).

It is unfortunately the case that the calibration of optical signals, be it with pH, transmembrane potential or cation concentration, often destroys the experimental system. For example, ionophores irreversibly modify the membrane potential during the calibration of dye indicators. Likewise, entrapped indicators of cell pH or Ca^{2+} are calibrated only after the cells have been completely permeabilized. Once a suitable calibration procedure is established, it is possible to record the desired information from the experimental system.

3 Examples

3.1 Oxidation–reduction state of tissue mitochondria

The properties of mitochondrial suspensions have long been the centre of intense study and the characteristics of the pigments of the respiratory chain have been resolved by dual-wavelength spectroscopy (4; Section 2.4.1). However, mitochondrial preparations may not reflect all the properties found *in vivo*. Direct recording of the respiratory pigments *in situ*, by absorbance and/or fluorescence spectroscopy, provides one way of resolving this dilemma. In anaesthetized animals spectroscopy of exposed tissues (24, 25) reports the oxidation–reduction state of the electron transport chain and the oxygenation of haemoglobin/myoglobin. Unfortunately the rapid fluctuations of O_2 delivery and consumption make it difficult to establish convenient steady states. An alternative procedure is to freeze the tissue rapidly and then to obtain spectra more leisurely at low temperatures.

Protocol 1

Reflectance spectra of freeze-trapped rat kidney

Equipment and reagents

- Liquid nitrogen
- Aluminium tongs for freeze-clamping ('Wollenberger' tongs (26))
- Wavelength scanning, dual-wavelength spectrophotometer
- Anaesthetized rat

Method

1 Dissect the anaesthetized rat to reveal the kidneys.

2 Chill the Wollenberger tongs in liquid nitrogen. A polystyrene ice bucket is a convenient vessel in which to keep the liquid nitrogen; it is easy to break dewar vessels with aluminium tongs.

3 Bring the cooled tongs close to the kidney.

4 Remove the kidney, place on the cooled tongs and squeeze the tongs together.

5 Keeping the tongs closed return them to the ice bucket containing liquid nitrogen.

6 Open the tongs, remove the crushed, frozen kidney with **plastic** forceps and store it in suitably labelled aluminium foil in liquid nitrogen.

7 To record reflectance spectra unwrap the foil surrounding the kidney—in liquid nitrogen using plastic forceps.

8 Chill the common terminal of the optical fibre connected to the spectrophotometer (1).

9 Under liquid nitrogen press the common terminal of the optical fibre against the smooth frozen face of the kidney.

Protocol 1 continued

10 Record the reflectance spectrum—repeat steps 8–10 for several different regions of the frozen tissue.

11 Use the white surface of the polystyrene ice bucket as a 'baseline' for computing the absolute reflectance spectra of the kidney.

12 Use different areas of the frozen kidney surface for computing difference spectra.

Figure 1 shows reflectance spectra of freeze-trapped rat kidneys. Three different kidneys were assessed: a normally perfused kidney, and kidneys rendered ischaemic by tying a ligature round their arterial supply, and freeze-trapped either 1 min or 10 min later. The main features of the visible reflectance spectra arise from oxyhaemoglobin (peaks at 542 and 577 nm), haemoglobin (peak at 555 nm) and ferrocytochromes aa_3 (maximum at 605 nm) and c (maximum at 550 nm). The excess of haemoglobin means that in absolute spectra the ferrocytochrome contributions appear as shoulders rather than discrete maxima. In normal kidney (*Figure 1,* trace a) only oxyhaemoglobin is observed indicating adequate oxygen supply and adequate mitochondrial respiration. After 1 min of ischaemia (*Figure 1,* trace b), the oxyhaemoglobin content is much reduced and the shoulders at 605 and 550 nm indicate reduction of the mitochondrial respiratory chain as the supply of oxygen to the point of consumption, fails. The general features of the spectrum after 10 min of ischaemia are similar to those

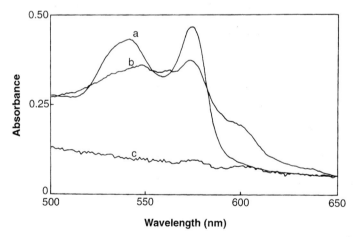

Figure 1 Reflectance spectra of freeze-trapped rat kidney. Kidneys from anaesthetized rats were freeze-trapped (26) before, 1 min and 10 min after occluding the renal blood supply. Reflectance spectra were recorded with a bifurcated light guide attached to a scanning, dual-wavelength spectrophotometer (1, 24) using 700 nm as the reference wavelength. The baseline was the spectrum of white polystyrene. Spectra labelled a and b were recorded from the normally perfused and 1 min ischaemic kidneys, respectively. Spectrum c is the difference between the 1 min and 10 min ischaemic kidneys. Reproduced with permission from C. L. Bashford and M. Stubbs (1986) *Biochemical Society Transactions,* **14,** 1213–1214. © The Biochemical Society.

found after 1 min and the *difference* between these two states (*Figure 1,* trace c) is featureless and gives an indication of the baseline for this experiment. Interestingly, oxyhaemoglobin is not absent from the ischaemic tissue suggesting that there are metabolic compartments where oxygen is delivered but not consumed (27, 28).

Similar experiments with freeze-trapped gerbil cerebral cortex (24, 29; and see *Figure 2*) indicate that only in extreme anoxia does reduction of the mitochondrial pigments occur. *Figure 2* shows the difference reflectance spectrum obtained when the reflectance spectrum from a sample obtained from an animal after breathing 5% oxygen for 1 min is subtracted from that taken from an animal which had been breathing 0% oxygen for 1 min. In both situations the haemoglobin was largely deoxygenated and the difference spectrum of the whole tissue closely resembles that of reduced *minus* oxidized mitochondria. This shows that the respiratory chain still has enough oxygen under conditions where tissue haemoglobin is almost completely deoxygenated; or put another way the large difference in affinity for oxygen exhibited by mitochondria and haemoglobin *in vitro* is still apparent *in vivo*.

The experiments illustrated in *Figures 1* and *2* employed a microprocessor-controlled, scanning, dual-wavelength spectrophotometer which was capable of storing and manipulating spectra.

3.2 Membrane potential of cells and organelles

Most biological membranes have the ability to sustain electrostatic potential differences between the aqueous phases which they separate. This potential,

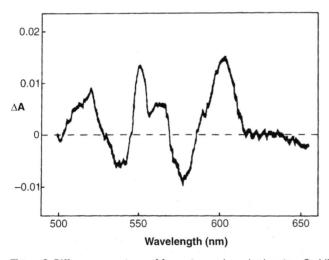

Figure 2 Difference spectrum of freeze-trapped cerebral cortex. Gerbil brains were freeze-trapped (43) 1 min after the anaesthetized animal began breathing either 5% O_2, 90% N_2, 5% CO_2 or 95% N_2, 5% CO_2. Reflectance spectra were recorded with a bifurcated light guide attached to a scanning, dual-wavelength spectrophotometer (1, 24) from the right cerebral cortex at 77 K, and the difference spectrum was computed from the spectra from each sample. Figure redrawn from data originally published in (29).

usually called the membrane potential, can play a significant part in membrane activities; it is, for example, critical for signalling in excitable cells and makes a contribution to the accumulation of nutrients driven by the electrochemical Na^+ gradient across the plasma membrane of many cells. In subcellular organelles, such as chloroplasts and mitochondria, the membrane potential is an important 'high energy' intermediate in the synthesis of ATP. In only a limited number of circumstances can the potential be recorded directly and other methods such as the use of dyes with specific sensitivity to membrane potential are available (30).

Many dyes change either their absorbance or their fluorescence in response to a change in membrane potential (11). However, it is often difficult accurately to calibrate the signal with known potentials. In one procedure, the membrane permeability to a particular cation, usually K^+ or H^+, is increased by adding a reagent called an ionophore. In the presence of sufficient ionophore, such that all other permeabilities are relatively small, the membrane potential (V) will approach that predicted by the Nernst equation:

$$V = (-RT/F)\ln([M^+]_i/[M^+]_o) \qquad 1$$

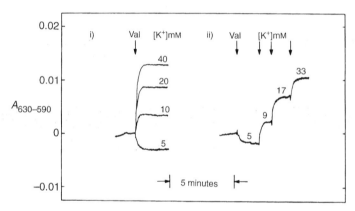

Figure 3 Measurement of Lettré cell membrane potential using oxonol-V. (i) 5×10^6 Lettré cells/ml were suspended in 5 mM HEPES, 1 mM $MgCl_2$, 2.2×10^{-6} M oxonol-V, pH adjusted to 7.4 with NaOH at 33°C and 155 mM NaCl *plus* KCl with the final K^+ concentration indicated. Valinomycin was added as indicated to give a final concentration of 1×10^{-6} g/ml. Traces from four separate experiments are superimposed such that the signal before the addition of valinomycin is identical. The extracellular K^+ concentration at which valinomycin would give no change in absorbance was estimated to be 7.1 mM by interpolation. The intracellular K^+ concentration was 73.3 mM giving a membrane potential (see *equation 1*) under these conditions of –62 mV. (ii) 5×10^6 Lettré cells/ml were suspended in 150 mM NaCl, 5 mM KCl, 5 mM HEPES, 1 mM $MgCl_2$, 2.2×10^{-6} M oxonol-V, pH adjusted to 7.4 with NaOH at 33°C. Valinomycin (1×10^{-6} g/ml) and KCl to give the concentration indicated were added as shown by the arrows. The extracellular K^+ concentration at which valinomycin would give no change in absorbance was estimated to be 7.1 mM by interpolation, the intracellular K^+ concentration was 63.3 mM giving a membrane potential (see *equation 1*) under these conditions of –60 mV. Reproduced with permission from Bashford, C. L., Alder, G. M., Gray, M. A., Micklem, K. J., Taylor, C. C., Turek, P. J. and Pasternak, C. A. (1985) *J. Cell. Physiology*, **123**, 326.

where M^+ is the relevant ion, R the gas constant, T the absolute temperature and F Faraday's constant. At 37 °C this relation simplifies to:

$$V(mV) = -61.5 \log ([M^+]_i/[M^+]_o) \qquad\qquad 2$$

which shows that, for a 10-fold change in cation gradient, you obtain a 61.5 mV change in potential. The cation gradient can be manipulated by altering the medium concentration of M^+ and hence a calibration curve of optical signal and potential can be obtained in the presence of ionophore.

3.2.1 Plasma membrane potential of animal cells

The membrane potential of whole cells is more difficult to assess than that of isolated organelles with optical indicators because cells contain many sub-cellular compartments, and it is necessary to isolate the response of the plasma membrane from that of any other cell membrane. For example, the intense staining of mitochondria by cyanine dyes (7), which are membrane-permeant cations, means that if you choose a cyanine dye, or other lipophilic cation, to measure plasma membrane potential it is important to establish conditions where the mitochondrial potential is disabled. If an anionic dye is chosen, this problem is overcome but there is, then, very little entry of dye into the cells unless the dye is itself reasonably lipophilic. Thus, for studies of whole cells, we employ an oxonol dye with a phenyl substituent. The procedures for measuring plasma membrane potential with this dye, oxonol-V (31), are illustrated in *Figure 3* and *Protocol 2*.

Protocol 2

Measurement of plasma membrane potential using oxonol-V and cells in suspension

Equipment and reagents

- Isotonic saline; e.g. 0.15 M NaCl, 0.005 M KCl, 0.005 M HEPES, 0.001 M MgCl₂ pH adjusted to 7.4 with NaOH. **Do not include** serum or other albumin-containing media because albumin/protein will sequester the oxonol and none of the dye will interact with the cells
- Pipettes

- Stock solutions of oxonol-V, valinomycin and FCCP, 0.001 M in DMSO or ethanol
- Stirrer
- 1-cm square cuvettes
- Recording spectrophotometer with a thermostatted cell holder

Method

1 Wash the cells in the saline and resuspend them at a cytocrit of about 0.5–1% v/v.

2 Transfer 1 ml of the cell suspension to a cuvette and place this in the spectrometer, select the appropriate wavelength(s) and then zero the absorbance/fluorescence reading.

3 Stir in oxonol-V to give a final concentration of $1-5 \times 10^{-6}$ M; record the signal until it reaches a steady value—this takes up to 10 min.

4 Then stir in valinomycin or FCCP to give a final concentration of $1-5 \times 10^{-6}$ M; record the change in signal.

5 Then stir in KCl (valinomycin-treated cells) or acid/base (FCCP) treated cells to double/halve the relevant cation concentration; record the change in signal.

6 Use the changes in signal induced by valinomycin/FCCP to estimate the K^+/H^+ concentration at which addition of valinomycin/FCCP would no give no change in signal (the 'null-point').

7 Determine the internal (cytosolic) K^+/H^+ concentration and calculate the membrane potential using *equation 1*.

The first protocol is to incubate the cells in media of differing K^+ content and to assess the effect of valinomycin addition (*Figure 3(i)*). The value of the K^+ null point is obtained by interpolation. This procedure works best with cells that have rather low K^+ permeability; cells more permeable to K^+ depolarize in high K^+ media and the subsequent addition of valinomycin has little effect.

In the second protocol, after valinomycin has been added to the cells and a new steady reading obtained, increase the concentration of K^+ in the medium by adding successively (*Figure 3(ii)*) small aliquots of concentrated KCl, about 3 M is the highest manageable concentration. After taking readings at a few different extracellular K^+ concentrations, obtain the value of the null point by interpolation (as in method 1). In both methods it is sufficient to interpolate by joining data points rather than by using sophisticated least-squares protocols, especially as the theoretical form of the dye absorbance versus potential plot is unknown.

In either procedure, once the 'null-point' is known it is necessary to measure cytoplasmic K^+ concentration to calculate the potential according to *equation 1*. This can be achieved by pelleting the cells through a mixture of di-*n*-butyl-phthalate (2 parts) and dinonylphthalate (1 part) with a density of 1.02; cells pellet through such an oil within 10 s in a Beckman microcentrifuge B. Cell cations can be determined (after the water content; wet – dry weight) by atomic absorption spectrometry. The spillover of medium into the cell pellet is usually less than 0.05%, but this can be verified by including an extracellular space marker, such as [14C]inulin, in the incubation medium.

There is a danger that the lipophilic valinomycin/K^+ complex may precipitate lipophilic anions such as the oxonol dyes and it is advisable to check on the calibration described above with another ionophore. The uncoupling agent FCCP is useful in this respect as it selectively increases the H^+ permeability of membranes. The experimental protocol for measuring the potential using FCCP is identical to that using valinomycin except that medium pH rather than K^+ is varied by addition of acid or base (33). Calculation of the membrane potential using the pH null-point requires a knowledge of cell pH (usually close to 7. 1) which has to be determined separately, for example, by using indicators (see below).

A more direct optical method has been introduced whereby a membrane-permeant, charged dye equilibrates across the membrane of interest. Ideally the dye will equilibrate rapidly and show minimal binding to materials on either side of the membrane (34). Using quantitative confocal fluorescence micro-scopy, the fluorescence of the dye on either side of the membrane is measured; as fluorescence intensity is linearly proportional to fluorophore concentration, the measured intensities can be substituted for the ion concentrations ($[M^+]_i$ and $[M^+]_o$) in *equation 1* and the potential obtained (34). This technique is ultimately limited by the spatial resolution of the confocal microscope—it works well for large compartments in whole cells. Recent computational approaches to image analysis have allowed the spatial resolution to be increased to the level where mitochondrial membrane potential can be monitored by such means in individual cells (35).

3.3 pH of cellular compartments

3.3.1 pH of endosomes

Most eukaryotic cells internalize components of their plasma membrane and associated exogenous material by endocytosis. The subsequent fate of the endosomes may be complex but it usually involves a step in which the vesicle milieu is acidified (36). The pH of such an endosome can be monitored directly if a pH-sensitive dye can be attached either to the relevant piece of membrane or to a suitable extracellular vehicle which is endocytosed. The dye most commonly used is fluorescein because it is easily visualized with a fluorescence microscope and because its isothiocyanate derivatives can be covalently conjugated to the amino groups in proteins (for receptor-mediated endocytosis) or to dextran. Furthermore the fluorescence of fluorescein and its conjugates is sensitive to pH in the range 4.5–7.5 (20). The endosomes can be loaded with the fluorescein derivative by incubation in a suitable physiological medium and the extracellular dye is subsequently removed by washing (*Protocol 3*). The calibration of the fluorescence signal is illustrated in *Figure 4(a)*: after recording the signal, the pH of all the cellular compartments is set equal to the bulk pH by adding the ionophore monensin; the bulk pH is then systematically reduced until the original signal is restored and the final value of bulk pH equals that originally present in the endosomes. The major difficulty of this technique arises from cellular autofluorescence which may, itself, be pH-dependent.

Protocol 3

Labelling of endosomes with fluorescein conjugates

Equipment and reagents

- Culture medium
- Fluorescein conjugate
- Fluorimeter or fluorescence microscope
- Media for mounting cells—slides and cover slips

Method

1 Incubate the cells in culture medium containing the fluorescein conjugate (present at ~1 mg/ml) at 37°C. The conjugate enters cells rapidly (within 10 min) and is subsequently transferred to secondary lysosomes (20). Endocytosis and intracellular membrane traffic are markedly reduced at lower than physiological temperatures. At 4°C only binding of ligands to cell surface receptors occurs, so receptors can be 'loaded' at low temperature and endocytosis subsequently 'triggered' by raising the temperature.

2 Remove the extracellular conjugate by washing the cells with cold (4°C), unlabelled medium. The residual fluorescence arises from within the cells.

3 Place the labelled cells in a fluorimeter/fluorescence microscope and record the fluorescence intensity. Cells attached to a solid substratum can be monitored with light guides (1), or by fitting the substrate diagonally in a fluorescence cuvette and using a conventional fluorimeter (the final optical arrangement resembles the 'front face' system used to record fluorescence from opaque or strongly absorbing samples). It is important not to move the substrate during signal calibration.

4 Calibrate the signal with extracellular pH using monensin and the protocol illustrated in *Figure 4(a)*.

A more qualitative estimate of endosome pH can be obtained by monitoring the distribution of a permeant, fluorescent amine such as 9-aminoacridine or acridine orange whose fluorescence varies with concentration. Such amines accumulate in acid compartments because the protonated form of the dye is much less membrane permeant than the free base. Indeed, if the latter equilibrates readily across membranes then the pH within the vesicle is given by the relationship (15):

$$pH_{in} = pH_{out} - \log ([AH^+]_{in}/[AH^+]_{out}) \qquad 3$$

normally $pH_{out} < pK - 1$ so that $[AH^+] \cong [AH^+] + [A]$. 9-Aminoacridine (10 mM) exhibits a yellow (1) and acridine orange a red rather than the blue fluorescence seen at low dye concentrations. Cells incubated with 0.01 mM 9-aminoacridine may have many yellow vesicles when examined by fluorescence microscopy using the fluorescein settings; this indicates that such vesicles have accumulated the dye by at least two orders of magnitude and that the internal pH must be in the range 5–6.

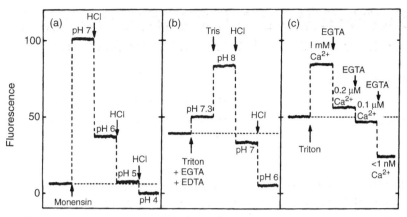

Figure 4 Calibration of intracellular probes of pH and Ca^{2+}. (a) Cells labelled with
fluorescein-dextran (*Protocol 3*; 20) in Hanks balanced salts solution buffered with 20 mM
HEPES, pH 7.0 (NaOH) were excited at 493 nm and fluorescence monitored at 520 nm. The
initial signal (6 units) was calibrated by adding as indicated: 1 μg/ml monensin (final
concentration) to equalize the pH of all compartments in the system (20) followed by HCl to
bring the solution pH to the values indicated. The original signal from the cells corresponds
to that found at pH 4.8 in the presence of monensin indicating that the fluorescein-dextran
occupied a compartment whose pH was 4.8. Records should be corrected for cellular
autofluorescence (fluorescence of unlabelled cells) and, after calibration, the supernatant
fluorescence should be measured and found to be <10% of that found in the presence of
cells. (b) Thymocytes labelled with quene-1 (19) were suspended at 5×10^6 cells/ml in a
medium containing the inorganic salts of RPMI 1640 (without phenol red) and supplemented
with 11 mM glucose and 10 mM HEPES, pH 7.3 (NaOH) at 37°C. Fluorescence at 530 nm
was excited at 390 nm; the signal (49 units) was calibrated by permeabilizing the cells with
0.05% Triton X-100 (Triton) in the presence of 0.5 mM EGTA and 0.5 mM EDTA. More than
90% of the dye is released by the Triton and it is titrated by the addition of Tris and HCl to
give the pH values indicated. The signal before permeabilization corresponds to that found
for pH 7.1 in the presence of Triton indicating that cytoplasmic pH was 7.1 when the
medium pH was 7.3. (c) Lymphocytes labelled with quin-2 (18) were suspended at 5×10^6
cells/ml in 130 mM KCl, 20 mM NaCl, 1 mM $CaCl_2$, 1 mM $MgCl_2$, 5 mM glucose, 10 mM
MOPS pH 7.4 (NaOH). Fluorescence at 500 nm was excited at 339 nm. The signal was
calibrated by permeabilizing the cells with 0.05% Triton X-100 (Triton) and subsequent
titration of the medium with EGTA, $MgCl_2$ and KOH to give 0.2 and 0.1×10^{-6} M Ca^{2+} as
indicated, with 1 mM free Mg^{2+}, pH 7.05 (18). 'Zero' Ca^{2+} (<1 nM) was achieved by adding
2 mM EGTA and Tris to give a pH > 8.3. Autofluorescence from unlabelled cells has been
subtracted from each reading and the fluorescence of unpermeabilized lymphocytes
corresponds to a Ca^{2+} level of 150 nm. All traces in this figure are redrawn from data
originally published in (18–20).

3.3.2 pH or [Ca^{2+}] of cytoplasm or isolated membrane vesicles

Trapped dyes will also record the pH or calcium ion concentration in the
cytoplasm or sealed compartments. For this purpose indicators of a type similar
to quene-1 (19) or quin-2 (18) are convenient.

Protocol 4

Measurement of cytoplasmic Ca^{2+} concentration or pH using acetoxymethylester-linked indicators

Equipment and reagents

- Isotonic saline; e.g. 0.15 M NaCl, 0.005 M KCl, 0.005 M HEPES, 0.001 M $MgCl_2$ pH adjusted to 7.4 with NaOH. It helps to include up to 0.025 M $NaHCO_3$ in the medium (replacing some of the NaCl) so that intracellular pH remains slightly alkaline during hydrolysis of the acetoxymethyl ester
- Pipettes
- Stirrer

- Stock solutions of acetoxymethylester-linked indicator, 0.01 M in DMSO, and Triton X-100, Ca^{2+}, EDTA, EGTA, acid and base, in water, for calibration titrations
- 1-cm square cuvettes
- Recording spectrophotometer with a thermostatted cell holder
- Water bath thermostatted to 37°C

Method

1. Wash the cells in the saline and resuspend them at a cytocrit of about 0.5–1% v/v.

2. Transfer 1 ml of the cell suspension to a cuvette and place this in the spectrometer, select the appropriate wavelength(s) and then zero the absorbance/fluorescence reading.

3. Stir in acetoxymethylester-linked indicator to give a final concentration of 1–5 × 10^{-6} M.

4. Follow the hydrolysis of the ester by monitoring the change in absorbance/ fluorescence.

5. When ester hydrolysis is more than 75% complete, this may take 30–90 min, pellet the cells, resuspend them in unlabelled isotonic saline and incubate them for a further 30 min at 37°C.

6. Then pellet the cells once more and resuspend them in isotonic saline for absorbance/fluorescence measurements.

7. Calibrate the optical signals with extracellular pH or Ca^{2+} after permeabilizing the cells with Triton X-100 (see *Figure 4*).

A word of caution about the cleaning of cuvettes is appropriate here; the use of chromic acid may introduce fluorescence quenchers. These can usually be removed by soaking the cuvettes in solutions of chelating agents such as EDTA.

Unfortunately some cells seem not to retain the acid form of these pH and Ca^{2+} indicating dyes very long even though the free acid is supposed to be membrane impermeant. It is very important, therefore, to check that the signal observed does indeed originate from within cells; this is best done by pelleting a sample of the suspension and verifying that the supernatant lacks dye (check the supernatant fluorescence). If leakage is particularly troublesome, reasonable data can be obtained by loading the cells for only a short time (much less than the time taken for complete hydrolysis of the ester), rapidly washing the cells and immediately making the pH or Ca^{2+} determination. Unfortunately this leads to an inevitable wastage of rather expensive starting material.

It is also possible to record pH or $[Ca^{2+}]$ directly using the appropriate indicators and quantitative confocal microscopy in a similar manner to that used to measure plasma and mitochondrial membrane potential directly (see Section 3.2.1). The great advantage of the 'direct' technique is that it removes the requirement for the destructive titration/calibration procedures used by the other methods. This opens up the possibility of following dynamic changes in pH and cell calcium directly.

3.4 Membrane cycling and recycling

3.4.1 Endocytosis and exocytosis

During endocytosis dyes, such as FM1–43 which are non-fluorescent in aqueous media but brightly fluorescent when membrane bound, accumulate in the inner leaflet of the endocytotic vesicles where there fluorescence is readily detected. This provides a useful fluorescence assay of endocytosis. Furthermore if dye is removed from the medium the cell-associated fluorescence diminishes as the endocytotic vesicles undergo further rounds of exocytosis—the dye washing away into the bath. Indeed the rate of decrease in fluorescence intensity is a good measure of the rate of exocytosis (37). Recently quantitative confocal fluorescence microscopy (38) has revealed that FM1–43 is released 'quantally' during exocytosis, confirming that this is an essentially 'all-or-none' process. Intriguingly the size of the fluorescence 'quantum' is time-independent, strongly suggesting that in these cells, hippocampal neurons, endocytotic vesicles undergo exocytosis before they have the chance to 'mix' with newly formed vesicles arising from intracellular membranes. Once again the power of quantitative image analysis coupled with the exquisite resolution and sensitivity of the confocal microscope has opened up cell biology to the quantitative approaches of physical chemistry.

In our own studies with tumour cells in culture we have found a completely different pattern of FM1–43 behaviour (39). Here the dye accumulates in vesicular bodies within the cell quite rapidly—as if endocytosis is occurring. However, once in the cells the dye remains there for extended periods (up to a week) suggesting that there is not the same tight link between endo- and exo-cytosis in this material.

3.4.2 Trafficking of intracellular proteins

With the advent of very high resolution fluorescence microscopes it is perhaps not surprising that new approaches have been developed to study the movement of materials through cells. In particular, the gene for the naturally fluorescent 'green fluorescent protein' (GFP) of the jellyfish *Aequoria victoria* has been cloned, isolated and subjected to deliberate genetic modification. GFP has been fused with a viral glycoprotein so that the synthesis and processing of the glycoprotein can be monitored in real time and space (40). These results show how newly synthesized material moves between the endoplasmic reticulum and the Golgi apparatus. More recently the fluorophore of GFP has been mutated to make it sensitive, in one study, to the putative pH values of cellular compartments (41) and, in another, to cytoplasmic Ca^{2+} levels (42). In each case, the advantages of the system are that the photosensitive material is guided directly to the arena of interest by the normal biosynthetic activities of the cell. Astonishingly it appears that adding the GFP to endogenous material does not significantly alter its pattern of processing.

4 Future prospects

The elegance and simplicity of optical experiments ensures that considerable efforts will continue to be expended to develop photometric and fluorometric assays of cellular membrane function. The power to direct sensitive chromophores precisely to interesting destinations and the ability to assess them quantitatively at the single cell level promises to open up the dynamics of membrane biology to the rigorous analyses of chemistry and physics. The new indicators of pH and cell calcium will help to unravel the role of these cations in cell stimulation and differentiation, particularly at the single cell level. Finally, existing procedures will be used, on account of the wide applicability of the techniques, on systems inaccessible to more elaborate experiments permitting a proper investigation of the full range of biological diversity.

Acknowledgements

I would like to thank Mrs B. J. Bashford for her patient preparation of the diagrams and the Royal Society, Biotechnology and Biological Sciences Research Council and the Cell Surface Research Fund for financial support. This chapter is an extensive revision of Chapter 7 in *Biological membranes: A practical approach* (ed. J. B. C. Findlay and W. H. Evans), published by IRL Press in 1987; any material reproduced is by permission of Oxford University Press.

References

1. Harris, D. A. and Bashford, C. L. (eds) (1987) *Spectrophotometry and spectrofluorimetry. A practical approach.* IRL Press, Oxford and Washington, DC.
2. Miller, J. N. (1981) *Standards in fluorescence spectrometry.* Chapman and Hall, London.
3. Bangham, A. D., De Gier, J. and Greville, G. D. (1967) *Chem. Phys. Lipids,* **1**, 225.
4. Chance, B. and Williams, G. R. (1955) *J. Biol. Chem.,* **217**, 395.
5. Chance, B., Schoener, B., Oshino, R., Itshak, F. and Nakase, Y. (1979) *J. Biol. Chem.,* **254**, 4764.
6. Harbig, K., Chance, B., Kovach, A. G. B. and Reivich, M. (1976) *J. Appl. Physiol.,* **41**, 480.
7. Johnson, L. V., Walsh, M. L., Bockus, B. J. and Chen, L. B. (1981) *J. Cell Biol.,* **88**, 526.
8. Waggoner, A. S. (1979) *Methods Enzymol.,* **55**, 689.
9. Keilin, D. (1925) *Proc. R. Soc. Lond. B,* **98**, 312.

10. Jackson, J. B. and Crofts, A. R. (1969) *FEBS Lett.,* **4**, 185.
11. Cohen, L. B., Salzberg, B. M., Davila, H. V., Ross, W. N., Landowne, D., Waggoner, A. S. and Wang, C. H. (1974) *J. Membr. Biol.,* **19**, 1.
12. Molecular Probes (1998) *Bioprobes,* **29**, 16.
13. Kauppinen, R. A. and Hassinen, I. E. (1984) *Am. J. Physiol.,* **247**, H508.
14. Reijngoud, D. J. and Tager, J. M. (1973) *Biochim. Biophvs. Acta,* **297**, 176.
15. Rottenberg, H. (1979) *Methods Enzymol.,* 55, 547.
16. Terasaki, M. (1989) *Meth. Cell Biol.,* **29**, 125.
17. Montecucco, C., Pozzan, T. and Rink, T. J. (1979) *Biochim. Biophys. Acta,* **552**, 552.
18. Tsien, R. Y., Pozzan, T. and Rink, T. J. (1982) *J. Cell Biol.,* **94**, 325.
19. Rogers, J., Hesketh, T. R., Smith, G. A. and Metcalfe, J. C. (1983) *J. Biol. Chem.,* **258**, 5994.
20. Geisow, M. J. (1984) *Exp. Cell Res.,* **150**, 29.
21. Barber, J. (1986) In *Ion interactions in energy transfer biomembranes* (ed. G. C. Papageorgiou, J. Barber and S. Papa), p. 15. Plenum Press, London.
22. Akerman, K. E. O. and Wikstrom, M. K. F. (1976) *FEBS Lett.,* **68**, 191.
23. Bashford, C. L., Alder, G. M., Gray, M. A., Micklem, K. J., Taylor, C. C., Turek, P. J. and Pasternak, C. A. (1985) *J. Cell. Physiol.,* **123**, 326.
24. Bashford, C. L., Barlow, C. H., Chance, B., Haselgrove, J. C. and Sorge, J. (1982) *Am. J. Physiol.,* **242**, C265.
25. Jobsis, F. F., Keizer, J. H., LaManna, J. C. and Rosenthal, M. (1977) *J. Appl. Physiol.,* **43**, 858.
26. Wollenberger, A., Ristau, O. and Schoffa, G. (1960) *Pflugers Arch. ges. Physiol.,* **270**, 399.
27. Epstein, F. H., Balaban, R. S. and Ross, B. D. (1982) *Am. J. Physiol,* **243**, F356.
28. Bashford, L. and Stubbs, M. (1986) *Biochem. Soc. Trans.,* **14**, 1213.
29. Bashford, C. L., Barlow, C. H., Chance, B. and Haselgrove, J. (1980) *FEBS Lett.,* **113**, 78.
30. Bashford, C. L. (1981) *Biosci. Rep.,* **1**, 183.
31. Smith, J. C., Russ, P., Cooperman, B. S. and Chance, B. (1976) *Biochemistry,* **15**, 5094.
32. Hoffman, J. F. and Laris, P. C. (1974) *J. Physiol (Lond),* **239**, 519.
33. Bashford, C. L. and Pasternak, C. A. (1984) *J. Membrane Biol.,* **79**, 275.
34. Loew, L. M. (1998) In *Cell biology: A laboratory handbook* (ed. J. E. Celis), Vol. 3, p. 375. Academic Press, San Diego.
35. Fink, C., Morgan, F. and Loew, L. M. (1998) *Biophys. J.,* **75**, 1648.
36. Geisow, M. J. and Evans, W. H. (1984) *Exp. Cell Res.,* **150**, 36.
37. Betz, W. J. and Bewick, G. S. (1992) *Science,* **255**, 200.
38. Murthy, V. N. and Stevens, C. F. (1998) *Nature,* **392**, 497.
39. Schaefer, M., Djamgoz, M. B. A. and Bashford, C. L. (1999) in preparation.
40. Presley, J. F., Cole, N. B., Schroer, T. A., Hirschberg, K., Zaal, K. J. M. and Lippincott-Schwartz, J. (1997) *Nature,* **389**, 81.
41. Miesenbock, G., De Angelis, D. A. and Rothman, J. E. (1998) *Nature,* **394**, 193.
42. Miyawaki, A., Llopis, J., Helm, R., McCaffery, J. M., Adams, J. A., Ikura, M. and Tsien, R. Y. (1997) *Nature,* **388**, 882.
43. Kerr, S. E. (1935) *J. Biol. Chem.,* **110**, 625.

Chapter 12

Use of optical spectroscopic methods to study the thermodynamic stability of proteins

Maurice R. Eftink and Haripada Maity

Department of Chemistry, 112 Coulter Hall, The University of Mississippi, MS 38677, USA

1 Introduction

The biophysical characterization of globular proteins will almost always include some type of study of the unfolding of protein to obtain thermodynamic parameters. The basic idea is that a transition between a native and unfolded state, induced by temperature, pH, or denaturant concentration, can serve as a standard reaction for obtaining a thermodynamic measure of the stability of the native state. For example, the free energy change for the unfolding reaction can be used to compare the stability of a set of mutant forms of a protein (1–4).

This type of analysis is based both on assumptions of the thermodynamic model for the unfolding process and on assumptions in the way the data are analysed; some of these assumptions and their limitations will be discussed below.

There are a variety of methods that can be used to monitor an unfolding process. A common method is differential scanning calorimetry, DSC, which measures the variation in the specific heat of a protein-containing solution as a protein is thermally unfolded (5–7). DSC is a popular method for this purpose, but optical methods can also provide suitable information for tracking the unfolding of a protein The spectroscopic signals for the native and unfolded states of a protein can give some insight regarding the structure of the states, and often can provide advantages of economy, ease of measurement and amenability to a wide range of sample concentration. The optical spectroscopic methods that have been used most often for this purpose are absorption spectroscopy, circular dichroism and fluorescence, which will be discussed in this chapter. A key to each of these methods and their use in protein unfolding studies is that the signal is a mole fraction weighted average of the signals of each thermodynamic state. That is, the observed signal, S, can be expressed as

$$S = \sum X_i S_i \qquad 1$$

where X_i is the mole fraction of species i and s_i is the intrinsic signal of species i. In order for a particular spectroscopic signal to be useful for tracking a $N \longleftrightarrow U$ transition of a protein, the signal must be sufficiently different for the N and U states.

In this chapter we will first discuss thermodynamic models, and their assumptions, for protein unfolding reactions. Then we will discuss advantages and limitations of these three individual optical methods. Since most of the fundamentals of optical spectroscopy are covered in other chapters in this volume, emphasis will be placed on thermodynamic models and practical matters related to the application of the spectroscopic methods.

2 Basic thermodynamic principles

Globular proteins are almost always considered to exist, under some conditions, as a native state, N, that has a defined three-dimensional structure and generally has some type of biological activity. If the N state is subjected to high (or low) temperature, extreme pH, high concentrations of chemical denaturants (e.g. urea or guanidine-HCl (Gdn-HCl)), or pressure, the organized native structure can be lost in a more or less cooperative manner to form an unfolded state, U. A thorough discussion of an unfolded state is beyond the scope of this chapter (see, for example, the article by Dill and Shortle (8)), but it is considered to have a much more random and fluctuating structure, with amino acid side chains much more exposed to solvent than they are in the native state. We use the term 'unfolded' in this chapter, as opposed to 'denatured', to stress that the unfolded state can refold to the native state if the perturbing condition is removed. We refer to 'denaturation' as a process that yields an altered structure that cannot readily refold to the native state.

The simplest model for a protein unfolding reaction is a two-state model, $N \longleftrightarrow U$. An important issue is whether or not the unfolding process really follows this model or involves one or more equilibrium folding intermediates (i.e. $N \longleftrightarrow I \longleftrightarrow U$). If one assumes an incorrect model, the resulting thermodynamic parameters will have little meaning (5, 9). However, it is very difficult to determine experimentally that there is one or more intermediate states, and the standard approach is to assume the process to be two-state, unless the data clearly indicate otherwise.

2.1 The two-state model

For a two-state transition between a native state, N, and an unfolded state, U, the equilibrium unfolding constant, K_{un}, partition coefficient, Q, mole fraction of each state, X_i, and standard free energy change for the unfolding reaction are defined as follows.

$$N \rightleftharpoons U \qquad\qquad 2$$

$$K_{un} = \frac{[U]}{[N]} \qquad\qquad 3$$

$$Q = 1 + K_{un} \qquad\qquad 4$$

$$X_N = 1/Q \,; \; X_U = K_{un}/Q = 1 - X_N \qquad\qquad 5$$

$$\Delta G^\circ_{un} = -RT \ln K_{un} \qquad\qquad 6$$

The value of K_{un} (or ΔG°_{un}) will depend on the extent of progress along the perturbation axis (temperature, T, pressure, P, pH, etc.). The following sections give various functions that are generally accepted as describing the dependence of ΔG°_{un} on these perturbation axes. One of the equations below, when combined with those above and *equation 1* can describe spectral data as a function of temperature or chemical denaturant, in terms of the two-state model.

2.2 Thermal unfolding

The accepted relationship for the temperature induced unfolding of a protein is

$$\Delta G^\circ_{un}\,(T) = \Delta H^\circ_{un} - T\Delta S^\circ_{un} \qquad\qquad 7a$$

$$\Delta G^\circ_{un}\,(T) = \Delta H^\circ_{o,un} + \Delta C_p\,(T - T_o) - T[\Delta S^\circ_{o,un} + \Delta C_p \ln (T\,/\,T_o)] \qquad\qquad 7b$$

where ΔH°_{un} and ΔS°_{un} are the enthalpy and entropy changes for unfolding. Both ΔH°_{un} and ΔS°_{un} may be temperature dependent, when the heat capacity change, ΔC_p, has a non-zero value. In this case, *equation 7b* (the Gibbs–Helmholtz equation) applies, where the $\Delta H^\circ_{o,un}$ and $\Delta S^\circ_{o,un}$ are values at some reference temperature, T_o (e.g. 0 or 20 °C). If the reference temperature is selected to be the high temperature unfolding temperature, T_G, then *equation 7b* can be recast as

$$\Delta G^\circ_{un}\,(T) = \Delta H^\circ_{G,un}\left(1 - \frac{T}{T_G}\right) - \Delta C_p\left(T_G - T + T \ln \left(\frac{T}{T_G}\right)\right) \qquad\qquad 8$$

In this case, $\Delta H^\circ_{G,un}$ (the value of ΔH°_{un} at $T = T_G$), ΔC_p and T_G would be fitting parameters, whereas in *equation 7b* the fitting parameters would be $\Delta H^\circ_{o,un}$, $\Delta S^\circ_{o,un}$ and ΔC_p.

The heat capacity change for unfolding of proteins has been found to be positive and to be related to the increase in solvent exposure of apolar side chains (1–3). In other words, a positive ΔC_p is a result of the hydrophobic effect and a consequence is that the $\Delta G^\circ_{un}(T)$ for unfolding of a protein will have a non-linear dependence on temperature, reaching a maximum at some temperature and showing both high temperature and low temperature induced unfolding.

2.3 Denaturant induced unfolding

The following equation applies to this process

$$\Delta G^\circ_{un}([d]) = \Delta G^\circ_{o,un} - m\,[d] \qquad\qquad 9$$

where $\Delta G^\circ_{o,un}([d])$ is the free energy change for unfolding at the reference condition in the absence of denaturant and m is a denaturant susceptibility parameter $(= -\partial\Delta G^\circ_{un}/\partial[d])$. This equation, referred to as the linear extrapolation model, LEM, is the most widely used relationship for describing denaturant induced unfolding (10,11). Admittedly, this is an empirical relationship, but it appears to adequately describe the pattern for denaturant induced unfolding of a number of proteins. There are other models (and equations) that have been suggested to describe denaturant induced unfolding, including a binding model (12, 13) and a solvent partitioning model (12), and a combination of the two (14), but the advantage and popularity of the LEM model is that it has only two fitting parameters, $\Delta G^{\circ}_{0,un}$ and m, for describing data. The $\Delta G^{\circ}_{0,un}$ is the parameter of prime interest, since it is a direct measure of the stability of a protein in the absence of denaturant, and this parameter applies to the ambient solvent conditions, which can be moderate temperature and pH (e.g. 20°C and pH 7). By contrast, it is more difficult to determine $\Delta G^{\circ}_{0,un}$ at 20°C from thermal unfolding studies, since it requires precise determination of $\Delta H^{\circ}_{0,un}$, $\Delta S^{\circ}_{0,un}$ and ΔC_p and use of these parameters to extrapolate back from the unfolding temperature (which is usually above 50°C) to 20°C.

Whereas the $\Delta G^{\circ}_{0,un}$ is a measure of thermodynamic stability of a protein, the m value also provides structural insights. m values have been suggested to correlate with the change in solvent-accessible apolar surface area upon unfolding of the protein (12, 15). That is, a relatively large m value (which corresponds to a high susceptibility of the unfolding reaction to denaturant and a steep plot of X_U or X_N versus [d]) has been taken to indicate that there is a large change in the exposure of apolar side chains on unfolding. This might apply to a protein that has an extensive core of apolar side chains that are exposed upon denaturation. Conversely, a small m value (and a less steep dependence of X_U on [d]) would indicate that there is a lesser change in the exposure of apolar side chains, which might apply to a protein that has few buried apolar side chains in the N state (or a case in which the apolar side chains are already on the surface in the N state, so that there is little change in their exposure on unfolding).

2.4 Acid induced unfolding

A relationship for acid induced unfolding is given by *equation 10*. This is the simplest relationship of this type (16), and assumes that there are n equivalent acid dissociating groups on a protein that all have the same $pK_{a,U}$ in the unfolded state, and that they are all perturbed to have a $pK_{a,N}$ in the N state that is at least 2 pH units lower than $pK_{a,U}$. If the pK_a shift is not at least 2 pH units, *equation 11* should be used. If there are more than one type of perturbed amino acid residue and/or if the residues are perturbed by less than 2 pH units, then the equation must be expanded further to include several more fitting parameters, which usually makes fitting data untractable.

$$\Delta G^{\circ}_{un}(pH) = \Delta G^{\circ}_{0,un} - RT \ln \left[(1 + [H^+]/K_{a,U})^n \right] \qquad 10$$

$$\Delta G^{\circ}_{un}(pH) = \Delta G^{\circ}_{0,un} - RT \ln \left[\frac{(1 + [H^+]/K_{a,U})^n}{(1 + [H^+]/K_{a,N})^n} \right] \qquad 11$$

Equation 10 includes as fitting parameters the free energy of unfolding at neutral pH, $\Delta G^{\circ}_{0,un}$, the number of such assumed residues, n, and their $pK_{a,U}$ in the unfolded state. Presumably n should be an integer and $pK_{a,U}$ should be approximately equal to the values for such amino acids as glutamate, aspartate (for which $pK_{a,U}$ should be about 4–4.3) or histidine (for which $pK_{a,U}$ should be around 6.5).

2.5 Pressure induced unfolding

The relationship for pressure, P, induced unfolding of proteins is given by *equation 12*, where $\Delta G^\circ_{o,un}$ is again the value of the free energy change at 1 atmosphere pressure and $\Delta V_{un} = V_U - V_N$ is the difference in volume of the unfolded and native states.

$$\Delta G^\circ_{un}(P) = \Delta G^\circ_{o,un} - \Delta V_{un}(P^\circ - P) \qquad 12$$

Pressure induced unfolding studies are more rarely performed since a specialized high pressure cell is required.

2.6 Simulations

Shown in *Figure 1* are simulated plots of X_N versus perturbant axis for various types of perturbations. Shown in *Figure 2* are corresponding simulated plots of ΔG°_{un} versus perturbant axis, all for the two-state unfolding reaction and for thermodynamic parameters given in the legend to *Figure 1*. The symbols represent simulated data points for which X_N is 0.05 to 0.95, the range over which this value can be determined with reasonable accuracy. The plots illustrate the length of the extrapolation needed to determine $\Delta G^\circ_{o,un}$ from typical unfolding data.

2.7 Unfolding of oligomeric proteins

There have been several studies of the stability of homodimeric or other oligomeric proteins (30–35). These proteins are interesting as models for understand-

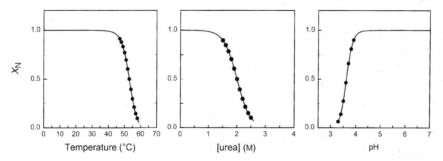

Figure 1 Simulated plots of the mole fraction of the native state, X_N, versus perturbation axis for a two-state transition described by the following thermodynamic parameters. The circles indicate the anticipated range of data that is usable for calculating the unfolding equilibrium constant (or ΔG°_{un}). The curves were simulated for a two-state model using the following parameters and *equations 3–9* and *11*. $\Delta H^\circ_{o,un} = -20\,000$ cal/mol at $T_o = 0\,°C$, $\Delta S^\circ_{o,un} = -90$ cal/mol·K, $\Delta C_p = 200$ cal/mol·K, $m = 4.0$ kcal/mol·M, $pK_{a,U} = 4.5$, $pK_{a,N} = 2.5$, and $n = 4$.

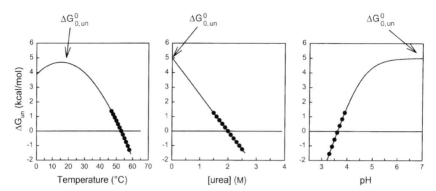

Figure 2 Simulated plots of ΔG°_{un} versus perturbation axis for the simulations in *Figure 1*. The circles are the anticipated range of data over which reliable values of ΔG°_{un} can be determined.

ing intermolecular protein–protein interactions. For a dimeric protein there is a general question of whether the protein unfolds in a two-state manner, D $\longleftrightarrow$ U, or whether there is an intermediate state, which might be either an altered dimeric state, D', or a folded (or partially folded) monomer species, M. Models for these two situations are as follows:

$$D \rightleftharpoons D' \rightleftharpoons 2U$$
$$D \rightleftharpoons 2M \rightleftharpoons 2U$$

12a

The latter situation is particularly interesting because the intermediate state, M, may or may not have a structured conformation similar to that of the monomeric units of the dimer.

As an example, consider the lambda phage protein *Cro*, which is a homodimeric protein having subunit molecular weight of 7351 (36). This protein, like many other dimeric DNA binding proteins, has two helix-turn-helix reading heads and the two major domains of the protein are interlinked by a beta-sheet formed by their C-terminal segments. Shown in *Figure 3* are urea induced unfolding data determined by monitoring a CD signal. A cooperative transition is observed that can be well fitted by a two-state model, N $\longleftrightarrow$ U. However, the data can also be fitted by a D $\longleftrightarrow$ 2 U model. If the latter model is correct, then one would expect to see a protein concentration dependence of the unfolding data. In fact, this is observed, as shown in *Figure 3*. For a D $\longleftrightarrow$ 2U. model, the relationships between the observed CD signal, S_{exp}, the mole fractions of dimer, X_D, and unfolded monomer, X_U, and the unfolding equilibrium constant ($K_{un} = [X_U]^2/[X_D]$) will be given by:

$$S_{exp} = \sum X_i S_i$$
$$X_U = \frac{K_{un}^2 + 8K_{un}[P]_0^{1/2} - K_{un}}{4[P]_0} ; X_D = 1 - X_U$$

13

where $[P]_0$ is the total protein concentration (expressed as monomeric form), S_i is the relative CD signal of species i and where K_{un} will depend on the per-

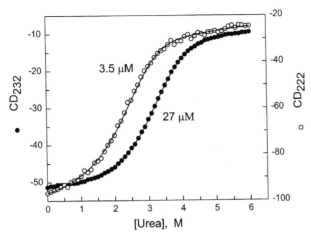

Figure 3 Urea induced unfolding of *Cro* repressor as a function of total protein concentration, $[P]_o$, and as measured by changes in CD signal (at 222 nm for the less concentrated protein sample and at 232 nm for the more concentrated sample). The solid curves are a global fit of the D $\rightleftharpoons$ 2U model with the parameters given in the text.

turbant as given by one of the above equations. That is, the transition should depend not only on the value of K_{un} (or $\Delta G^\circ_{o,un}$) and m (in the case of urea or Gdn-HCl induced unfolding), but also on the total subunit concentration, $[P]_o$.

The data in *Figure 3* are for total monomer concentrations of $[P]_o = 3.5$ μM and 27 μM. As protein concentration is lowered, the dimer becomes destabilized. The solid lines through the two data sets are global fits of *equation 13* to the multiple data sets with a single value of $\Delta G^\circ_{o,un} = 11.2$ kcal/mol and $m = 1.6$ kcal/mol·M.

The following paragraphs give a detailed experimental description of the unfolding study with *Cro*.

Protocol 1

Observation of the unfolding of a protein

Equipment and supplies

- A fresh, filtered solution of 20 mM NaH$_2$PO$_4$, pH 7.0

- Since urea slowly decomposes to form cyanate, stock solutions should be prepared fresh every day or two

- Access to a CD instrument, such as an Aviv 62DS spectropolarimeter, equipped with a thermoelectric (or thermojacketed, with a circulating water bath) cell holder to maintain constant temperature

- Quartz cuvettes

- A stock solution of the protein dialysed against (or, if lyophilized, dissolved in) the same buffer. The concentration will depend upon the optical techniques being used. For CD studies, a 50–200 μM protein stock solution should allow a minimal volume to be added to the denaturant solution to give a working concentration in the 5–20 μM range

- Pipettes

Method

1 Set up a series of solutions (5 ml) of buffer containing urea concentrations between 0–8 M urea.

2 Take 2 ml of each of these solutions and add to each an aliquot of 100–200 μl of the stock solution of protein, thus forming a series of solutions having the same protein concentration and varying denaturant concentration.

3 Incubate the solutions overnight.

4 Calculate the concentration of urea in each solution by taking a small sample of the different urea solutions, measuring the difference between the refractive index of the urea solution and the buffer, ΔN, and using the following equation (10):

$$[urea] = 117.66\Delta N + 29.753\Delta N^2 + 185.56\Delta N^3$$

(It is usually best to measure the refractive index of the solutions to which protein has not been added and then make a small correction of dilution. However, in cases where the protein concentration is low, it is acceptable to make the refractive index measurement with that solution.)

5 Measure the CD signal at a chosen wavelength for each sample in the series. The far UV region of 220–230 nm is sensitive to secondary structure changes. Lower wavelengths can be used, but one must make certain that light absorbance by the denaturant does not become a problem.

6 Plot the data as described in this chapter.

See Pace and coworkers for a thorough description of this general protocol (10).

Protocol 2

Observation of the unfolding of a protein, e.g. lambda *Cro*

Equipment and supplies

- A fresh, filtered solution of 20 mM NaH_2PO_4, pH 7.0 containing the desired final concentration of the protein (e.g. prepared by addition of an aliquot from a concentrated protein stock solution). Prepare approximately 2.0 ml of this zero-denaturant solution

- Access to a CD instrument, such as an Aviv 62DS spectropolarimeter, equipped with a thermoelectric (or thermojacketed, with a circulating water bath) cell holder to maintain constant temperature and a magnetic stirrer and flea for stirring

- A solution of 8 M urea in 20 mM NaH_2PO_4, pH 7.0 also containing the same final protein concentration (from addition of an aliquot of concentrated stock). This is the concentrated denaturant solution. Depending on the midpoint urea concentration for 50% unfolding, one will need from 2 ml (for a low midpoint) to 6 ml of this solution (for a high midpoint)

- Quartz cuvettes

- Pipettes

Protocol 2 continued

Method

1 Incubate both the zero denaturant and the concentrated denaturant solutions containing protein at the desired temperature to achieve thermal equilibration.

2 Place 1.8 ml of the zero denaturant solution into the cuvette and then into the spectropolarimeter, the cell block of which is controlled to the desired temperature.

3 Measure the CD signal of this first solution as in *Protocol 1*.

4 Remove an aliquot (e.g. 50–200 μl) of the solution from the cuvette and then add the same volume of the concentrated denaturant solution to the cuvette.

5 Mix and allow 1–5 min for chemical equilibration. (If after this incubation period the CD signal is still changing with time, then a longer incubation is needed.)

6 Measure the CD signal of the solution.

7 Repeat steps 4–6 until the concentration of urea in the cuvette has reached a point where the signal no longer changes.

8 Plot the data as described in this chapter.

If a computer interfaced syringe pump is available, this titration procedure can be automated, as described below and in (19). This reference also discusses an acquisition routine for determining when the signal has equilibrated.

See *Figure 3* for an example of data.

A second approach, which works well for rapidly unfolding/folding proteins (and/or for cases in which a researcher has a computer interfaced syringe pump), is to make measurements on a sample as denaturant is incrementally (or continuously) added.

When performing denaturant titrations using fluorescence, either *Protocol 1* or *2* can be followed. When monitoring the fluorescence of tryptophan residues in a protein, excitation and emission wavelengths of 290 nm and 350 nm, respectively, are typical. It is important to also measure the 'fluorescence' signal from the denaturant solution and to subtract this 'blank' signal from the fluorescence signal from the protein solutions.

In acquiring the data in *Figure 3* we prepared two protein samples, each with the same protein concentration (by dilution from a more concentrated protein stock solution), but one with and one without the highest concentration of denaturant (e.g. around 7–8 M urea). We placed 1.8 ml of the 'without' sample and a magnetic stirring flea into a standard 1 cm quartz cuvette. This was then placed in an Aviv 62DS spectropolarimeter, equipped with a thermoelectric cell holder (with temperature set at 20 °C) and a magnetic stirrer. Approximately three times the volume of the 'with denaturant' sample was loaded into a syringe pump, with the delivery tip directed into the sample cuvette. We then recorded the CD signal at 222 nm, with averaging of the raw ellipticity signal over about

15 s and digital storage. An aliquot of the 'with denaturant' sample was then delivered (following the removal of an equal volume of sample from the cuvette by the second channel of the syringe pump, in order to maintain a constant total volume in the cuvette; we find that this strategy of ensuring a constant solution height in the cuvette works best for good mixing of the solutions by the stirring flea, and the removal of a volume is also needed to avoid overfilling the cuvette in cases where large total aliquot volumes are required). Following a waiting period for mixing and chemical equilibration (e.g. 1–5 min), the signal is again recorded. (See (19) for more details and for a description of a program for controlling the data acquisiting and for determining when the signal has equilibrated.)

The raw signal versus denaturant concentration data was then analyzed by fitting a combination of equations 1, 9, and 13 to the data using a nonlinear least-squares program (19). (In this particular case, the total protein concentration is also needed for equations 13, since this is a dimeric protein; for the unfolding of a monomeric protein, the fitting equation for a simpler N $\rightleftharpoons$ U model would be a combination of equations 1, 5, 6, and 9.) In this fitting procedure, the baseline slopes for the native signal and unfolded signal regions were assumed to be linear, with these slopes being allowed to vary as fitting parameters. The two other fitting parameters were the $\Delta G^\circ_{o,un}$ (or K_{un}) and m in equation 9.

3 Practical considerations and deviations from the two-state model

3.1 Existence of an equilibrium intermediate(s)

What if there is one or more unfolding intermediate (i.e. the transition is not two-state)? As mentioned above it is often very difficult to make this determination. Elsewhere (9, 17) we have shown extensive simulations to illustrate the difficulty in distinguishing between two-state and three-state transitions when performing denaturant induced unfolding studies.

3.2 Kinetic considerations

What if the transition is reversible, but equilibrium is not reached during the course of the experimental measurement? For example, if one measures a spectroscopic signal to track protein unfolding as a function of temperature, what will be the effect of a relatively slow approach to equilibrium as temperature is scanned? Lepcock and coworkers (18) have presented interesting simulations pertinent to this question for differential scanning calorimetry data. Such simulations are relevant to spectral thermal scanning studies as well. In general, if the unfolding transition is slow compared to the thermal scan rate, the appearance of the transition will be skewed. In general, the kinetics of the conformational transition will be slower at the lower temperature side of the transition curve and the kinetics will speed up at higher temperature. As a result, what would otherwise be a nearly symmetrical plot of X_N versus T (i.e. if the scan rate were infinitely slow) will be skewed to appear to be sharper than it should be. If this effect is not realized, a researcher fitting raw data would find apparent $\Delta H^\circ_{G,un}$ and T_G values that are larger in magnitude than they should be.

Likewise, for other types of perturbations, if the kinetics of the unfolding reaction are relatively slow and if a scanning or real-time titration procedure is employed, then it is important to demonstrate that the scan or titration does not occur too rapidly so as to cause a skewing of the apparent transition. A way to experimentally demonstrate that a scan or titration is being performed slowly enough to allow equilibration is to perform experiments at two or more scan rates and then choose a slow enough scan rate that gives unchanging values for the apparent mid-point (and steepness) of the transition. Alternatively, we have a data acquisition routine that enables us to acquire spectral data for a protein, add titrant (or change temperature) and then automatically pause for various intervals until the spectroscopic signal is no longer changing with time. This procedure is described in Ramsay et al. (19) for denaturant induced unfolding reactions, using CD and fluorescence to track the unfolding of a protein.

3.3 Irreversibility

What if the unfolding process is irreversible? A common problem and concern in protein unfolding studies involves irreversibility of the process. This may be observed as an aggregation under denaturing conditions and/or an inability to recover biological activity or characteristic spectroscopic or other physical properties of the native state when the perturbing conditions are removed. For example, a common way to observe such irreversibility is to notice a hysteresis in plots of a signal versus temperature during upward, followed by downward scans.

Such irreversibility may be modelled as

$$N \rightleftharpoons U \xrightarrow{k_d} D$$

where D is a permanently denatured form of the protein and k_d is the apparent rate constant for the irreversible process. If, for example, k_d has a steep dependence on temperature, acid or base, or if k_D depends on protein concentration, then the irreversibility of the process will increase as the perturbing condition is made more extreme. That is, it still may be possible to characterize the $N \rightleftharpoons U$ process by avoiding extreme perturbing conditions. If the $U \rightarrow D$ reaction cannot be avoided, the result will be a skewing of the shape of the transition, making the recovery of valid thermodynamic data unreliable (18).

Another model for an irreversible process is one in which there is an equilibrium intermediate, $N \rightleftharpoons I \rightleftharpoons U$, and the I species reacts in an irreversible manner. There has been some discussion lately of this mechanism with the idea that an intermediate may have a greater exposure of apolar side chains and that this might lead to an enhanced tendency of the intermediate species to self-aggregate (20).

3.4 **Baseline considerations**

When spectroscopic signals are used to track the unfolding of a protein, there are always minor problems associated with baseline trends. Whether measuring absorbance, CD or fluorescence signals (especially the latter), it is usually found that the intrinsic spectroscopic signal of the N or U state of a protein will depend on the perturbing condition. That is, the absorbance, CD or fluorescence intensity of a protein will usually depend on temperature, chemical denaturant concentration, pH or pressure, so that there will be non-zero baseline slopes. The following paragraphs will comment on effects of temperature and chemical denaturant, these being the most commonly used perturbants, on these spectroscopic signals.

The absorbance of any solution will depend slightly on temperature, due to thermal expansion (and hence dilution) of a chromophore. Also, changes in refractive index with temperature can lead to apparent absorbance changes. With pure samples of chemical denaturants, such as urea and Gdn-HCl, there is usually a very small dependence of baseline absorbances on denaturant concentration, provided that the excitation wavelength is one where the denaturant is transparent.

The intrinsic temperature dependence of CD signals of a native and unfolded protein is difficult to assess, but the trends appear to be small. For the proteins studied in our laboratory we find that the far-UV CD signal of an unfolded protein decreases (becomes more negative) with increasing temperature and becomes more positive with increasing concentration of urea or Gdn-HCl (21). We note that other laboratories seem to find similar patterns (22–23).

The fluorescence intensity of tryptophan (or N-acetyl-L-tryptophanamide), the principle fluorophore in proteins, increases slightly with increasing urea concentration (24, 25). The trend is a bit larger with Gdn-HCl. Our experience is that such baseline trends with these chemical denaturants are larger for the U form of a protein than for the N form. The largest fluorescence baseline trends occur with temperature. While these baseline trends can be significant, we find that this does not cause a serious problem in using fluorescence to study thermal unfolding of proteins, if there is a large enough difference in the fluorescence signal of the N and U states. The temperature dependence of the fluorescence intensity of either the N or U states is frequently treated as a linear slope. Invariably, the slope of the N state will be less than the slope of a U state and these slopes will always be negative (i.e. fluorescence decreases with increasing temperature due to thermally activated quenching processes, see Chapter 2). Although a linear assumption is almost always made by researchers, we actually know from model system studies (e.g. the fluorescence of tryptophan or indole as a function of temperature) that plots of fluorescence intensity versus temperature are not linear. Instead, the fluorescence quantum yield, Φ_F (which is directly proportional to fluorescence intensity, F), is more accurately described by the relationship (26, 27)

$$\Phi_F = \frac{k_f}{k_f + k_{nfo} + \Sigma A_{nfi} \exp\left(-E_{a,nfi}/RT\right)} \qquad 14$$

where k_f is the rate constant for radiative decay (which is assumed to be independent of temperature), k_{nfo} is the rate constant for temperature-independent non-radiative processes (e.g. intersystem crossing to the triplet state), and A_{nfi} and $E_{a,nfi}$ are the Arrhenius factor and activation energy for quenching of fluorescence. If the E_a value is relatively small, it will be difficult to tell that a plot of Φ_F (or F) versus temperature is not linear. Fluorophores that are buried in the apolar interior of a protein tend to have smaller E_a values for thermal quenching. Likewise, solvent exposed fluorophores, as in an unfolded state of a protein, will tend to have a larger temperature dependence (28).

A conventional way to analyse protein unfolding data is the graphical approach in which linear baselines are drawn through the regions where the spectral signal is dominated by the N or U states. This, of course, can introduce human biases, particularly when the baseline region is limited. After drawing the two baselines, the mole fraction of the N and U states are calculated from the extent of the displacement of a data point (in the transition region) from these two baselines. The preferred method is to avoid the straight edge and to use nonlinear least-squares computational methods to fit equations to data, including slopes for the baseline regions as fitting parameters (11, 29).

3.5 Interfering substances

The presence of an interfering substance can lead to spurious changes in optical spectroscopic signals as well as perturbations of the chemical equilibrium. The latter situation, such as the case of metal ions that preferentially bind to either the native or unfolded state, are particular chemical problems associated with each system and there is not much that can be said other than 'researcher be aware'. The controlled variation in the concentration of these chemically perturbing species can reveal the problem and can sometimes lead to understanding of coupled equilibria.

Cases where an interfering substance contributes to the apparent spectroscopic signal, without directly interacting with the protein, can also be a serious problem. If, for example, there is a contaminant that has a constant absorbance, ellipticity or fluorescence signal (as a function of the perturbant axis of temperature, denaturant concentration, etc.), then the contaminant will not be much of a problem, other than the fact that it diminishes the relative amplitude of the signal from the protein. If a contaminating substance undergoes a change in signal as a function of the perturbant axis, this can lead to difficult problems, such as giving the appearance that a protein unfolding reaction is multi-state, when it is actually two-state. It is, of course, the responsibility of a researcher to identify, eliminate or minimize such contaminating signals, but sometimes the presence of a contaminating signal can be surprising. For example, a contaminating fluorescence signal can alter apparent absorbance or ellipticity measurements, since photomultipliers measure photons whether they are transmitted directly or luminesced (placement of the detection photomultiplier further from the cuvette minimizes this measurement of fluorescent photons). Also, a strong absorbance signal from a contaminant (the signal of which changes along the perturbant axis) can distort the apparent fluorescence or ellipticity signal of a protein. The distortion of the fluorescence signal by an absorbing contaminant is a consequence of the inner filter effect (see Chapter 2). The distortion of an ellipticity signal is a consequence of the fact the CD instrument usually measures a difference in absorbance divided by a total absorbance. For example, when using CD to perform a pH induced unfolding of a protein, we were almost mislead on one occasion by the contribution from the chelating agent, EDTA. This chelator does not have an intrinsic CD signal itself, so the standard procedure of substracting a blank did not correct for the problem. The absorbance of EDTA in the 220–230 nm range varies greatly with pH, as its carboxylate groups undergo proton dissociation, thus causing a perturbation of the relative ellipticity of the protein sample in an indirect manner. Scattered light, due either to contaminants (which usually should be removable by micropore filtration or centrifugation) or the self-association of the protein under study, can also diminish a CD signal, or perturb a fluorescence signal as Rayleigh scattering.

3.6 Global analysis

A very useful strategy when performing spectroscopic studies of the unfolding a proteins is to do a global analysis over multiple data sets (see Chapter 8) (19, 29). For example, shown in *Figure 4* are data for the Gdn-HCl induced unfolding of *Staphylococcal* nuclease A, in which changes in fluorescence intensity and circular dichroism ellipticity were measured simultaneously on the same sample using a multi-dimensional spectrophotometer. The solid line is a global nonlinear least-squares fit of a two-state unfolding model to both data sets.

4 Advantages of different spectroscopic signals

4.1 Absorbance

When using absorbance spectroscopy (difference spectroscopy) to monitor conformational transitions in proteins, the measurements usually focus on absorb-

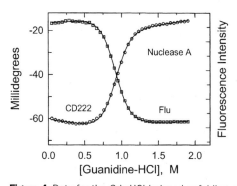

Figure 4 Data for the Gdn-HCl induced unfolding of *Staphylococcal* nuclease A, monitored by changes in CD signal at 222 nm and by the fluorescence intensity at 340 nm (excitation at 295 nm). The solid curves are a global nonlinear least-squares fit of a two-state model to the data with fitting parameters $\Delta G^\circ_{o,un}$ = 5.32 kcal/mol and m = 5.83 kcal/mol·M.

Table 1 Spectral properties of some intrinsic and extrinsic protein probes[a]

Chromophore	Absorbance		Fluorescence	
	λ_{max} (nm)	ε ($\times 10^{-3} M^{-1} cm^{-1}$)	λ_{max} (nm)	Φ_F
Tryptophan	280	5.6	355	0.13
Tyrosine	275	1.4	304	0.14
Phenylalanine	258	0.2	282	0.021
FAD	450	11	530	0.03
Pyridoxamine phosphate	325	8.3	390	0.14
Dansyl chloride	330	3.4	510	~0.1
1,5-IEDANS	360	6.8	480	~0.5
PRODAN	364 (342)[b]	14.5	531 (401)[b]	–
Fluorescein derivatives	495	42–85	516–525	0.3–0.5
Rhodamine derivatives	560	12	580	~0.7
Pyrene derivatives	342	40	383	0.25
1,8-ANS	355	~6	515	0.004
1,8-ANS-apoMb	374	6.8	454	0.98
Coumarin derivatives	~390	26–30	~460	~0.7

See (38) for original references for values.

[a] Values are for neutral aqueous solution at ~20°C, unless stated otherwise. The values for the dansyl chloride, IAEDANS, fluorescein, rhodamine, pyrene and coumarin derivatives are approximate values for protein adducts in aqueous solution.

[b] Values are for hexane as solvent.

ance changes in the aromatic region of the spectrum. In Table 1 is shown general information of the spectral properties of the three aromatic amino acids, tryptophan, tyrosine and phenylalanine. Since tryptophan and tyrosine have a much large molar extinction coefficient than does phenylalanine, the discussion will be restricted to the former pair.

Both the absorbance spectra of the indole ring of tryptophan and the phenol ring of tyrosine show a sensitivity to solvent polarity. With increasing solvent polarity there is usually a red shift in the absorbance of indole and phenol. Consequently, one would expect that the unfolding of protein, and increase in solvent exposure of tryptophan and tyrosine residues, would usually lead to a blue shift in absorbance. For tryptophan, it is typical to monitor the absorbance at 291–294 nm when studying conformational changes in proteins (there is also a significant difference spectrum around 284 nm, but this can be obscured by tyrosine in proteins). That is, one will usually see a decrease in the absorbance at 293 nm upon unfolding of a tryptophan-containing protein with denaturants, pH or temperature. Likewise, for tyrosine there is a peak in the difference spectrum at 285–288 nm, which can be used to study protein unfolding in tyrosine-containing proteins. It should be noted in particular that tryptophan's absorbance is also sensitive to the local electrostatic field, so that changes in indole-charge interactions can cause either red or blue shifts upon protein unfolding.

Absorbance measurements will require 0.01 to 0.1 mM protein solutions to achieve reasonable signal-to-noise in 1 cm pathlength cells. Thermal scans are fairly easy to perform, with thermoelectric cell holders or thermal-jacketed cells and a circulating bath. For denaturant induced unfolding studies, the common procedure is to prepare a series of solutions having the same protein concentration and varying denaturant concentrations. When either urea or Gdn-HCl is used as denaturant, the absorbance signals show a significant and fairly linear baseline slope. These slopes are probably due to changes in refractive index of the solution caused by the addition of denaturant. In thermal scans, the baseline slopes are smaller and the temperature dependence of the absorbance is usually not linear, making analysis of such data a bit more precarious.

The advantages of absorbance measurements are the ready availability, ease of use and low cost of the instrumentation. The biggest disadvantage is that it is less sensitive than the other two methods described below.

4.2 Circular dichroism

CD is a very commonly used method for studying protein conformational changes. The CD spectrum of a protein includes the far-UV region from 180–250 nm, which is dominated by the absorbance of peptide bonds. The CD signal in this region senses the secondary structure (primarily α-helix) of a protein. The CD spectrum in the aromatic UV region from 250–300 nm senses the chirality around the aromatic side chains. Hence there is usually a structured aromatic CD spectrum (or signature) for the native state of a protein and an absence of spectral features for the unfolded state; changes in the aromatic CD region are thus thought to reflect changes in tertiary structure of a protein.

CD is a moderately sensitive method in the far-UV spectral region, requiring in the range of 0.01 mM solutions. Modern CD instruments can be purchased with thermoelectric cell holders for thermal scans and with automated titrator syringe pumps for chemical denaturant titrations. Also, the instruments will be capable of data averaging and on-line data acquisition.

When performing CD measurements it is necessary to pay attention to the buffer and salts being used, particularly if one wishes to make measurements below 200 nm, since various buffers and salts can absorb a significant amount of light in the far-UV. Schmid (24) has provided a very useful table of the absorbances of common buffer components. Also, chemical denaturants will absorb light in the far-UV below 210 nm, which must be considered when designing unfolding studies.

The baseline slopes of CD signals versus temperature or chemical denaturant concentration are usually assumed to be linear. In the far-UV region we find the baseline slopes for the native state of a protein are close to zero for both temperature and chemical denaturant. For unfolded proteins we usually find the baseline slopes to the negative for temperature and positive for urea and Gdn-HCl.

Aromatic-UV CD signals can be used for studying unfolding transitions, but the CD signals are much smaller, thus having lower signal-to-noise as compared to far-UV CD or fluorescence data.

The advantage of CD measurements over the other optical methods is that the far-UV signals observe changes over the entire protein (i.e. its secondary structure). Also, the nature of the signal change can be related to structural changes (e.g. loss of ellipticity at 222 nm can be attributed to loss of α-helix). A disadvantage is that CD instruments are moderately expensive and less available than standard absorbance and fluorescence instruments. However, CD instruments are marketed primarily for the biochemistry–protein–peptide research community and an excellent selection of accessories for protein unfolding studies is available.

4.3 Fluorescence

When using steady-state fluorescence to study protein unfolding, the most commonly used fluorophores are the intrinsic tryptophan and tyrosine residues. Since tryptophan has a larger molar extinction coefficient, a redder absorbance, and is found in most proteins, this fluorophore is the one most frequently employed in fluorescence studies. The sensitivity of fluorescence detection is such that concentrations as low as 10^{-8} M can be studied with commercial fluorimeters, and, by selecting smaller cell pathlengths or longer excitation wavelengths, concentrations as high as 10^{-4} M can be studied. A very important property of tryptophan fluorescence is that it is very dependent on the environment of the indole side chain, making tryptophan fluorescence responsive to the structure of a protein.

The emission maximum of tryptophan in proteins can range from 308 nm for a residue buried in a completely apolar environment (as in the single tryptophan variant of azurin (37)) to 350 nm for a fully solvent-exposed residue (38–40). Due to their apolar character, tryptophan residues are usually either fully or partially buried in the three-dimensional structures of proteins and are expected to have emission maxima from 320 to 340 nm. Unfolding of a protein, and concomitant increase in the solvent exposure of tryptophan residues will invariably lead to a red-shift in the fluorescence of a tryptophan-containing protein.

The fluorescence intensity, or quantum yield Φ_F, of a tryptophan residue in a protein also varies over a wide range. The Φ_F of single-tryptophan-containing proteins can range from 0.01 to 0.4 in globular proteins (40). The wide range appears to be due to a combination of factors, including the closeness of the indole side chain to intramolecular quenching groups, such as peptide bonds and certain amino acid side chains (41–43). There does not appear to be such a wide dispersion of Φ_F values for unfolded proteins. The consequence of this variation in Φ_F, along with the above mentioned red-shift, is that the fluorescence intensity of tryptophan-containing proteins will almost always change at either the red or blue edge of the emission spectrum when a protein unfolds.

The fluorescence intensity (either at a single wavelength, or integrated over the entire emission envelope) is a signal that follows *equation 1* and can be used to extract thermodynamic information. The apparent emission maximum of a protein sample does not follow *equation 1*, so one must exercise caution in its use. As we have shown by simulations elsewhere (25), if there is a significant increase or decrease in Φ_F of a protein upon unfolding (as well as a red shift), the measurement of the apparent emission maximum will not give a true reflection of the change in population of native and unfolded states, so that any recovered thermodynamic parameters will be under- or over-estimates of the actual values. While this problem is probably recognized by most people using fluorescence, there is still a strong desire to be able to use emission maxima in studying protein unfolding, since the maxima are not dependent on protein concentration (except possibly for associating systems) or lamp intensity and thus would make it more convenient in the measurement of a series of solutions. A way to use emission maxima in thermodynamic studies is to do curve fitting of the composite spectra. That is, it should be possible to fit the fluorescence spectra at intermediate points along a transition (e.g. when $X_N = 0.1$ to 0.9) as a linear combination of the basis spectra of the native and unfolded species times their relative population.

Another steady-state fluorescence signal that can be measured fairly routinely is the fluorescence anisotropy, r. The anisotropy value for the fluorescence of a tryptophan residue will depend on its rotational freedom (as expressed by its rotational correlation time, ϕ) and its fluorescence decay time, τ, with immobilized and short-lived excited states having largest r values. Since there is a range of fluorescence lifetime and rotational correlation times for tryptophan residues in native proteins, and since these ϕ and τ values will usually have a narrower range for the unfolded state, it follows that r values will usually change upon unfolding a protein. However, as we have discussed elsewhere (25), r values do not follow *equation 1*, but instead follow

$$r = \frac{\Sigma\, \Phi_i X_i r_i}{\Sigma\, \Phi_i X_i} \qquad\qquad 15$$

That is, the observed r value depends not only on the r_i value and mole fraction, X_i, for the N and U states, but r also depends on the quantum yield of each state, Φ_i. If one of the states has a larger quantum yield than the other, the result will be a skewing of the plot of r versus perturbation axis toward the more dominant fluorescing state. In principle, if the Φ_F (or relative intensity) of the native and unfolded states are known, one can still use r values to obtain thermodynamic information. We have shown elsewhere that plots of fluorescence anisotropy versus denaturant concentration for the unfolding of a protein can give the same thermodynamic parameters as fluorescence versus intensity data, if the above equation is used (17). Also, it must also be realized that anisotropy measurements have more noise than simple intensity measurements, so that it is preferable to just make intensity measurements, both because of the better quality of the latter data and the more straightforward analysis.

If a protein contains tyrosine but not tryptophan residues, the fluorescence of tyrosine can also be used for unfolding studies. The emission maximum of tyrosine is very insensitive to environment (and protein conformational state) and its Φ_F is also relatively insensitive to environment. These factors, coupled with the lower extinction coefficient of tyrosines, make it less valuable as a reporter group. The advantage of tyrosine fluorescence is that there are usually several such residues in a protein, which partially overcomes the lower extinction coefficient. Besides tyrosine and tryptophan, extrinsic fluorescence probes can also be used. For example, dansyl groups can be covalently attached to cysteine or lysine residues. (See (44) for a detailed description of several extrinsic probes.)

In cases where there is not a significant change in fluorescence intensity upon unfolding of a protein, it may be possible to enhance the signal change by adding a solute quencher or by changing the solvent to D_2O. A solute quencher, such as acrylamide or iodide, will usually be able to quench the fluorescence of exposed fluorophores in the unfolded state of a protein to a much greater extent that for the same fluorophores in a native protein. Alternatively, heavy water is known to increase the fluorescence intensity of some fluorophores and may have different degrees of enhancement for buried and exposed fluorophores.

Fluorescence instrumentation is relatively widely available and can be purchased with various configurations and accessories over a wide range of prices. As with CD instruments, some fluorescence instruments are designed to allow automated thermal scans and/or titrations, usually with digital acquisition.

To summarize, the advantages of fluorescence methods for studying protein unfolding reactions are the wide concentration range that can be measured and the sensitivity of the signal to the microenvironment of the fluorophore. Also, fluorescence signals of the native and unfolded state can provide some low resolution structural information about these states (at least with respect to the microenvironment of the fluorophores).

5 Concluding remarks

Optical spectroscopic techniques provide several advantages in studies of the kinetics and thermodynamics of protein unfolding. This chapter has attempted to summarize many of these advantages, as they apply to equilibrium studies, and to also highlight the assumptions involved and to point out some practical considerations.

Acknowledgements

We acknowledge the contributions of recent graduate students and post-doctoral associates to our work in this area. In particular, the contributions of Glen Ramsay and Roxana Ionescu are noted. Some of the unpublished work presented in this chapter was supported by NSF grant MCB 9808635.

References

1. Schellman, J. A. (1987(*Annu. Rev. Biophys. Biophys. Chem.* **16**, 115–137; Becktel, W. J. and Schellman, J. A. (1987) *Biopolymers* **26**, 1859–1877.

2. Privalov, P. (1989) *Annu. Rev. Biophys. Biophys. Chem.* **18**, 47–69.

3. Robertson, A. D. and Murphy, K. P. (1997) *Chemical Physics* **97**, 1251–1267.

4. Shortle, D. A., Meeker, A. K. and Freire, E. (1988) *Biochemistry* **27**, 4761–4768.

5. Lumry, R., Biltonen, R. and Brandts, J. F. (1966) *Biopolymers* **4**, 917–944.

6. Sturtevant, J. M. (1987) *Annu. Rev. Phys. Chem.* **38**, 463–488.

7. Grikko, Y. V., Privalov, P. L., Sturtevant, J. M. and Venyaminov, S. Y. (1988) *Proc. Natl Acad. Sci. USA* **85**, 3343–3347.

8. Dill, K. A. and Shortle, D. (1991) *Annu. Rev. Biochem.* **60**, 795–825.

9. Eftink, M. R. and Ionescu, R. (1997) *Biophys. Chem.* **64**, 175–197.

10. Pace, C. N. (1986) *Methods Enzymol.* **131**, 266–280; Pace, C. N., Shirley, B. A. and Thomson, J. A. (1989) In *Protein structure and function: A practical approach* (ed. T. E. Creighton) pp. 311–330. IRL Press, Oxford.

11. Santoro, M. M. and Bolen, D. W. (1988) *Biochemistry* **27**, 8063–8068; Min, Y. and Bolen, D. W. (1995) *Biochemistry* **34**, 3771–3781.

12. Tanford, C. (1970) *Adv. Protein Chem.* **23**, 1–95; Aune, K. and Tanford, C. (1969) *Biochemistry* **8**, 4586–4590.

13. Makhatdze, G. I. and Privalov, P. L. (1992) *J. Mol. Biol.* **226**, 491–505.

14. Staniforth, R. A., Burston, S. G., Smith, C. J., Jackson, G. S., Badcoe, I. G., Atkinson, T., Holbrook, J. J. and Clarke, A. R. (1993) *Biochemistry* **32**, 3842–3851.

15. Myers, J. K., Pace, C. N. and Scholtz, J. M. (1995) *Protein Sci.* **4**, 2138–2148.

16. Barrick, D. and Baldwin, R. L. (1993) *Biochemistry* **32**, 3790–3796.

17. Eftink, M. R. (1998) *Biochemistry (Moscow)* **63**, 276–284.

18. Lepcock, J. R., Ritchie, K. P., Kolios, M. C., Rodahl, A. M., Heinz, K. A. and Kruuv, J. (1992) *Biochemistry* **31**, 12706–12712.

19. Ramsay, G. D., Ionescu, R. and Eftink, M. R. (1995) *Biophys. J.* **69**, 801–707.

20. Safar, J., Roller, P. P., Gajdusek, D. C. and Gibbs, C. J., Jr (1993) *J. Biol. Chem.* **268**, 20276–20284.

21. Inoescu, R. M. and Eftink, M. R. (1997) *Biochemistry* **36**, 1129–1140.

22. Kuhlman, B. and Raleigh, D. P. (1998) *Protein Sci.* **7**, 2405–2412.

23. DeKoster, G. T. and Robertson, A. D. (1995) *J. Mol. Biol.* **249**, 529–534.

24. Schmid, F. X. (1989) In *Protein structure. A practical approach* (ed. T. E. Creighton), pp. 251–285. IRL Press, Oxford.

25. Eftink, M. R. (1994) *Biophys. J.* **66**, 482–501.

26. Kirby, E. P. and Steiner, R. F. (1970) *J. Phys. Chem.* **74**, 4480–4490.

27. Robbins, R. J., Fleming, G. R., Beddard, G. S., Robinson, G. W., Thistlethwaite, P. J. and Woolfe, G. J. (1980) *J. Am. Chem. Soc.* **102**, 6271–6279.

28. Burstein, E. A., Vedenkina, N. S. and Ivkova, M. N. (1973) *Photochem. Photobiol.* **18**, 263–279.

29. Ramsay, G. D. and Eftink, M. R. (1994) *Methods Enzymol.* **240**, 615–645.

30. Jaenicke, R. (1991) *Biochemistry* **30**, 3147–3161.

31. Gittleman, M. S. and Matthews, C. R. (1990) *Biochemistry* **29**, 7011–7020.

32. Eftink, M. R., Helton, K. J., Beavers, A. and Ramsay, G. D. (1994) *Biochemistry* **33**, 10220–10228.

33. Bowie, J. U. and Sauer, R. T. (1989) *Biochemistry* **28**, 7139–7143.

34. Pakula, A. A. and Sauer, R. T. (1989) *Proteins: Struct. Funct. genet.* **5**, 202–210.

35. Herold, M. and Kirschner, K. (1990) *Biochemistry* **29**, 1907–1913.

36. Bolotina, I. A., Kurochkin, A. V. and Kircpichnikov, M. P. (1983) *FEBS Letters* **155**, 291–294.

37. Hutnik, C. M. and Szabo, A. G. (1989) *Biochemistry* **28**, 3923–3934.

38. Eftink, M. R. (1991) *Methods Biochem. Anal,.* **35**, 127–205.

39. Lakowicz, J. R. (1983) *Principles of fluorescence spectroscopy*. Plenum Press, New York.

40. Beechem, J. M. and Brand, L. (1985) *Annu. Rev. Biochem.* **54**, 43–71.

41. Busheuva, T. L., Busel, E. P. and Burstein, E. A. (1975) *Stud. Biophys.* **52**, 41–52.

42. Chen, Y. and Barkley, M. D. (1998) *Biochemistry* **37**, 9976–9982.

43. Ricci, R. W. and Nesta, J. M. (1976) *J. Phys. Chem.* **80**, 974–980.

44. Haugland, R. P. (1996) *Handbook of fluorescent probes and research chemicals* (6th edn). Molecular Probes, Eugene, OR 97402.

Chapter 13

The use of spectroscopic techniques in the study of DNA stability

John Santalucia, Jr
Department of Chemistry, Wayne State University, Detroit, MI 48202, USA

1 Introduction

Accurate determination of nucleic acid thermodynamics has become increasingly important in understanding biological function as well as applications in biotechnology and pharmaceuticals. Knowledge of the thermodynamics of DNA hybridization and secondary structure formation is necessary for understanding DNA replication fidelity (1), mismatch repair efficiency (2) and the mechanism of DNA triplet repeat diseases (3). In addition, RNA folding thermodynamics are an important aspect of understanding ribozyme catalysis, as well as understanding the regulation of protein expression, mRNA stability and the mechanism of protein synthesis by the ribosome (4). With the genome sequencing era upon us (5), it will increasingly become important to predict the folding and hybridization thermodynamics of DNA and RNA, so that accurate diagnostic tests for genetic and infectious diseases can be developed. Thus, there is a need to develop a database of accurate thermodynamic parameters for different nucleic acid folding motifs (4).

This chapter describes practical aspects of the application of UV absorbance temperature profiles to determine the thermodynamics of nucleic acid structural transitions. Protocols and practical advice are presented for issues not normally addressed in the primary literature but that are crucial for the determination of reliable thermodynamics, such as sequence design, sample preparation, choice of buffer, protocols for determining strand concentrations and mixing strands, design of microvolume cuvettes and cell holder, instrumental requirements, data analysis methods, and sources of error. References to the primary literature and reviews are also provided where appropriate. *Sections of this chapter have been adapted from previous reviews and are reprinted with permission from the Annual Review of Biochemistry, Volume 62 © 1993, by Annual Reviews www.AnnualReviews.org (6) and with permission from Biopolymers © 1997, by John Wiley & Sons, Inc. (4).*

2 Overview of UV melting

The temperature-induced transition between native and random coil states of a nucleic acid can be conveniently monitored by ultraviolet (UV) absorbance (see (4, 9, 10) for reviews). The reason for this is that stacked bases have a smaller absorption per base than unstacked bases; this is called hypochromicity (11, 12) which is defined as:

$$\%\text{hypochromicity} = 100 \, \frac{(A_{\text{denatured}} - A_{\text{native}})}{A_{\text{native}}} \qquad 1$$

where $A_{\text{denatured}}$ and A_{native} are the absorbances at high and low temperature, respectively. The absorbance versus temperature profile is commonly referred to as a UV absorbance melting curve (*Figure 1*). As the temperature increases, the ratio of molecules in the single-stranded versus native states increases, resulting in an increase in the UV absorbance. The melting temperature, T_M, is defined as the temperature at which half of the strands are in the native state and half are in the 'random coil' state. Whereas many methods such as circular dichroism and NMR can be used to monitor thermal denaturation, UV absorbance is the most sensitive due to the high molar absorption of the bases (6).

The simplest way to derive thermodynamic parameters from UV melting data is to apply a van't Hoff analysis of the data by assuming a two-state model (i.e. native and denatured states) and that the difference in heat capacities of the native and denatured states, $\Delta C_p{}^\circ$, is zero (13–16) (more complex models are described in Section 5). At each temperature the absorbance can be used to calculate the fraction of strands in the native and denatured states, thereby allowing the calculation of an equilibrium constant (10). Thus, the absorbance versus temperature profile is used to determine the temperature dependence of

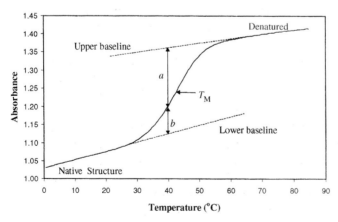

Figure 1 Typical experimental UV melting profile. At a given temperature, the fraction of strands in the duplex state, α, is given by the ratio $a/(a + b)$, where a and b are the respective vertical distances from the upper and lower baselines to the experimental melting curve. The temperature at which α is equal to 0.5 is defined as the melting temperature, T_M. The data are for CGTGTCTCC•GGAGTCACG at 2.62×10^{-4} M total strand concentration dissolved in 1.0 M NaCl, pH 7 solution (2).

the equilibrium constant, allowing the calculation of ΔH° and ΔS° for the transition (10) from a van't Hoff plot (i.e. $\ln K$ versus $1/T$ plot). ΔG°_T is then calculated using *equation 2* (discussed further in Section 5).

$$\Delta G^\circ_T = -RT \ln K = \Delta H^\circ - T\Delta S^\circ \qquad\qquad 2$$

In practice, however, to obtain accurate thermodynamics from the shape of a melting curve it is necessary to subtract upper and lower temperature baselines because the extinction coefficients of the denatured and folded states are temperature dependent (9, 14, 16). The details of one commonly used program for fitting UV melting curves were published recently (17).

For sequences that form self-complementary duplexes, the melting temperature, T_M, is calculated using *equation 3*.

$$T_M = \frac{\Delta H^\circ}{\Delta S^\circ + R \ln C_T} \qquad\qquad 3$$

where R is the gas constant (1.987 cal/K mol) and C_T is the total oligonucleotide strand concentration. For non-self-complementary molecules, C_T in *equation 2.3* is replaced by $C_T/4$ if the strands are in equal concentration, or by $(C_A - C_B/2)$ if the strands are in different concentration, where C_A and C_B are the concentrations of the more concentrated and less concentrated strands, respectively.

Observing the transition with a different technique (e.g. calorimetry, NMR, or circular dichroism) can test the two-state assumption. The observation of isosbestic or isodichroic points in the absorbance or circular dichroism spectra, respectively, is diagnostic of a two-state transition (6). Alternatively, the melting profile can be monitored at a different wavelength, so that contributions from different nucleotides are emphasized. Deriving thermodynamics from the concentration dependence of the T_M (see below) can also test the two-state assumption. If all methods give the same results, the two-state approximation is validated since the methods have different sensitivities to each species.

The dependence of the T_M on the oligonucleotide strand concentration reveals the molecularity of the transition (6). For example, formation of a hairpin structure is a unimolecular process and therefore does not depend on concentration, whereas formation of a duplex is bimolecular and is concentration dependent for oligonucleotides. Polynucleotides show little concentration dependence of the T_M because double-strand initiation, which is the event dependent on concentration, is only a small fraction of the total free energy involved in the transition. Higher order complexes such as triplexes and quadruplexes show stronger concentration dependencies (10, 18), but are usually non-two-state. The concentration dependence of the T_M provides an alternative van't Hoff method for calculating folding thermodynamics. For bimolecular reactions, thermodynamic parameters can be derived by rearranging *equation 3* to give *equation 4* and plotting reciprocal melting temperature (in Kelvins) versus logarithm of C_T (19) (*Figure 2*).

$$\frac{1}{T_M} = \frac{R}{\Delta H^\circ} \ln C_T + \frac{\Delta S^\circ}{\Delta H^\circ} \qquad\qquad 4$$

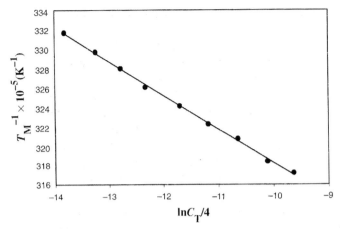

Figure 2 Typical $1/T_M$ versus $\ln C_T/4$ plot. The data are for the duplex CGTGTCTCC•GGAGTCACG at strand concentrations from 2.62×10^{-4} M to 4.00×10^{-6} M in 1.0 M NaCl, pH 7 solution (2). The thermodynamic parameters derived from the plot are: $\Delta G^\circ_{37} = -6.90 \pm 0.21$ kcal/mol, $\Delta H^\circ = -56.8 \pm 0.9$ kcal/mol and $\Delta S^\circ = -160.9 \pm 2.1$ e.u. The thermodynamic parameters derived from the average of the fits are: $\Delta G^\circ_{37}, = -6.89 \pm 0.10$ kcal/mol, $\Delta H^\circ = -63.7 \pm 2.1$ kcal/mol and $\Delta S^\circ = -183.2 \pm 6.4$ e.u.

The concentration should be varied more than a factor of fifty for reliable measurements of ΔH° and ΔS°. The general equations for two-state melting of complexes of any molecularity are provided in an elegant paper by Marky and Breslauer (18). If the molecularity is known from UV absorbance mixing curves, where absorbance is measured while varying the stoichiometry of the reacting strands (20), or from another technique such as NMR, size exclusion chromatography or non-denaturing gel electrophoresis, then *equation 4* can be used to test the two-state approximation. For example, if significant concentrations of intermediates are present, then the ΔH° from *equation 2* (from fitting the shape of the curves) will be expected to be smaller than the ΔH° from *equation 4* (21). On the other hand, if the transition is known to be two state, differences in ΔH° from *equation 2* and *equation 4* can reveal the molecularity. For example, a triplex would give a ΔH° value from *equation 2* which is 50% larger than that from *equation 4*, and would indicate that equations appropriate for triplexes not duplexes need to be used (18, 21).

It is important to note that the ΔH° and ΔS° determined by van't Hoff analysis are determined at the T_M. If large temperature extrapolation is required (greater than $\sim 20\,^\circ$C away from T_M), then heat capacity effects should be accounted for. Previous data have indicated that ΔC°_p is usually small for nucleic acids (14, 15). Based on these data, an approximate estimate of ΔC°_p is in the range of 0–120 entropy units (cal K^{-1} mol) per base pair and ΔC°_p is assumed to be temperature independent. Given ΔC°_p, the ΔH°_T, ΔS°_T, and ΔG°_T can be calculated more accurately than *equation 2* with the *equations 5* (14).

$$\Delta H^\circ_T = \Delta H^\circ_{Tm} + \Delta C^\circ_p \times (T - T_m) \qquad \text{5a}$$
$$\Delta S^\circ_T = \Delta S^\circ_{Tm} + \Delta C^\circ_p \times \ln (T/T_m) \qquad \text{5b}$$
$$\Delta G^\circ_T = \Delta H^\circ_{Tm} + \Delta C^\circ_p \times (T - T_m) - T\Delta S^\circ_{Tm} - T\Delta C^\circ_p \times \ln (T/T_m) \qquad \text{5c}$$

Due to enthalpy–entropy compensation, $\Delta G°_T$ is relatively insensitive to even large values of $\Delta C_p°$ (22).

In large molecules, like tRNA or group I introns, melting is not two state so more complicated models are required to derive thermodynamic parameters (6). UV melting, however, can reveal information on the stepwise mechanism of thermal denaturation. For example, UV melting of *E. coli* formylmethionine tRNA shows several transitions. First, the tertiary interactions are broken and then individual helices (secondary structure) are melted (23, 24). The transitions are assigned with the help of other techniques such as circular dichroism, temperature-jump kinetics, NMR or chemical modification. Since the hypochromicity of A•U and G•C base pairs are maximum at 260 and 280 nm, respectively (25), the ratio of A•U and G•C base pairs broken during each transition can be determined by measuring the hypochromicities at both wavelengths (10). For example, the melting curve of the *Tetrahymena thermophila* large subunit group I intron involves three major transitions (26). The first occurs at 36°C and is attributed to partial unfolding of the tertiary structure, the second transition occurs at 68°C with the 260 nm hypochromicity greater than the 280 nm hypochromicity suggesting the transition involves an A•U rich region. The third transition occurs at 72°C with the 280 nm hypochromicity greater than the 260 nm hypochromicity, suggesting that the transition involves G•C rich regions (26, 27). These results are consistent with the general notion that tertiary structure is less stable than secondary structure in RNA. This is one reason for the success of secondary structure predictions for RNA (28).

2.1 Strengths and weaknesses of UV melting and calorimetry

Microcalorimetry, including differential scanning calorimetry (DSC) and isothermal titration calorimetry (ITC) (29), provides complementary information to that provided by UV melting. UV detection offers the advantages of high sensitivity so that small sample sizes are required (typically ~3 A_{260} units are required for a full set of curves), and the standard errors in $\Delta H°$ and $\Delta S°$ are typically about 5–8% (4). Due to compensating errors in $\Delta H°$ and $\Delta S°$ (4, 14, 30, 31), a van't Hoff analysis of UV melting curves provides very precise measurements of $\Delta G°_{37}$ (standard error ±2–5%) and T_M (±0.5–1.0 °C). Microcalorimetry offers the advantages that transition enthalpy changes are directly measured, are model independent, and recent improvements in instrument design allow $\Delta H°$ for nucleic acid samples to be determined with standard errors of 2–5% (32). Calorimetric methods, however, require substantially larger sample sizes (typically 30–50 A_{260} units for a full set of measurements). The enthalpy–entropy compensation of errors is not as evident in DSC studies (33), and errors in the $\Delta G°_{37}$ and T_M are often much larger than those from UV melting studies. Thus, DSC measurements are usually accompanied by UV melting experiments to determine accurate $\Delta G°_{37}$ and T_M values (34). While DSC provides model-independent thermodynamic parameters, often the calorimetric $\Delta H°$ is used in conjunction with the T_M determined from UV melting and the oligonucleotide concentration to obtain the entropy for the reaction from *equation 3*. This amounts to applying the two-state approximation to the data. Further, while it is true that DSC provides a model independent measurement of the heat released in the transition between two temperatures even for non-two-state reactions, it does not reveal the nature of the states involved in the transition (10).

3 **Sample**

3.1 **Sequence design**

Design of optimal nucleic acid sequences to address specific scientific questions or to determine specific incremental contributions to overall thermodynamics is something of an art. There are, however, a number of simple guidelines that we routinely use that minimize the number of samples that give artefactual data. In general, a little time spent on careful experimental design results in a large saving in time and confusing results in the long run. The two most basic principles are (i) careful design of individual sequences that do not have the potential to form undesired structures and (ii) to make multiple measurements for each unknown so that sequences with anomalous thermodynamics are readily identified. On-line servers for secondary structure prediction of single-stranded RNA and DNA are very useful for deducing potential alternative structures (see http://www.ibc.wustl/~zuker and http://JSL1.chem.wayne.edu/). Algorithms for the prediction of duplexes and higher molecularity complexes are currently under development in my laboratory. No matter how carefully one designs sequences, however, there are always occasional examples where new motifs with surprising stability are discovered, for example, the exceptional stability of tandem GA mismatches in DNA and RNA was unanticipated (35, 36). Sequences with runs of three or more guanines should be avoided since they form G-quartet structures and consistently yield lower than expected $\Delta H°$ parameters (34). Sequences should also be designed to have T_M values close to physiological temperature (37°C) to minimize temperature extrapolation errors (particularly if $\Delta C_p°$ is large) (13). Sequences should also be designed to have T_M values between 20 to 75°C to allow upper and lower baselines to be adequately defined.

For hairpins, sodium concentration below 0.1 M is recommended to minimize bimolecular self-complementary internal loop formation (37). The predicted thermodynamics from RNA-MFOLD (28, 38) or DNA-MFOLD (39) programs are accurate enough to deduce if the putative internal loop will form in high population (to use MFOLD the two strands should be connected by the dummy sequence LLL (M. Zuker and J. SantaLucia, unpublished). Simulating the predicted populations of the hairpin and duplex forms by solving the simultaneous equilibria is also recommended (31, 36).

For bimolecular structures (duplexes), it is generally recommended that self-complementary sequences be avoided, since they have high propensity to form competing hairpin structures. Even non-self-complementary duplexes can sometimes form competing structures, such as 'slipped' self-complementary duplexes or hairpins (31, 36, 40). To design non-self-complementary single strands that are unlikely to form hairpins, it is recommended that they be folded with MFOLD and that a 'dot plot' is made to search for the possibility of self-complementary duplex formation by each strand (36). On the other hand, self-complementary duplexes offer the advantages that only one sequence needs to be synthesized and a given motif (e.g. mismatch or dangling-end) can be included twice in the duplex, thus cutting in half the experimental error for the contribution of a motif. To minimize hairpin formation for self-complementary duplexes, it is recommended that measurements be carried out in high salt conditions (e.g. 1 M NaCl) and to design the middle of the duplex to be G–C rich and the ends of the duplex to be A–T rich (41). The very ends of duplexes, however, should be G–C whenever possible to minimize 'end-fraying' artefacts (34, 41). In studying the thermodynamic contributions of molecules with multiple motifs that require multiple linear regression analysis, the rank of the occurrence matrix should be tested to ensure that it is possible to solve for all parameters uniquely; this issue has been thoroughly addressed in the literature (42, 43) for Watson-Crick pairs and by my laboratory for mismatches (2, 31). It is also important to realize that at salt concentrations below 1 M, melting is length dependent and the corrections provided in (44) are recommended.

The design of more complex nucleic acid structures requires particular care. A general principle that has been applied toward the synthesis of multi-branched loops (45) and novel structures (46) is to maximize 'orthogonality' of the base pairing potential of different stems. The idea is to design the different stems with the lowest potential for forming alternative stem structures. If a complex structure requires a hairpin, the exceptionally stable UUCG (for RNA) or GNRA (for DNA or RNA) or GNA (for DNA) sequences are recommended (47, 48). Often, it is advisable to study the thermodynamics of complex structures, by reducing the molecularity of the system by connecting two strands by a stable tetraloop. This strategy has been effectively applied toward the study of coaxial stacking interactions (49, 50). Another concern for large nucleic acids or systems with tightly bound ligands is that they often exhibit slow kinetics. Thus, slow heating rates are recommended for these systems.

3.2 Redundant design of motifs

Since UV detection requires so little material, it is usually possible to synthesize multiple molecules with a given structural motif. Essentially, this allows for multiple measurements for each unknown (e.g. thermodynamic loop contribution), so that molecules with anomalous thermodynamics due to unforeseen formation of alternative structures are readily identified, particularly by using resampling statistical analysis (31). This strategy also has the advantage that statistical errors can be reduced by taking advantage of the $1/(n - v)^{1/2}$ dependence of the error propagation, where n is the number of measurements and v is the number of parameters determined from the data (51). For example, we recently combined measurements from our laboratory and others in the literature (108 sequences) to determine the 10 Watson–Crick nearest-neighbour propagation $\Delta G°_{37}$ parameters and two initiation parameters (31). The average standard error in the $\Delta G°_{37}$ of each of the 108 sequences is estimated as 0.4 kcal/mol; thus, the average standard error in the 12 unknown parameters is estimated as 0.4 kcal/mol/$(108 - 12)^{1/2}$ = 0.04 kcal/mol.

In our actual study (31), we rigorously propagated individual errors of experimental measurements to the derived parameters using the variance–covariance matrix during singular value decomposition (52) and the results are in good agreement with the average 0.04 kcal/mol error calculated above. This example dramatically illustrates the value of performing a large number of measurements to reduce statistical uncertainties in derived parameters. It should be noted, however, that the propagated errors reflect precision of the determined parameters and not the accuracy, because of the possible presence of systematic errors in measurements. Another important point is that the derived parameters are also inaccurate because the imposed *model itself* is limited (e.g. due to next-nearest-neighbour interactions). The issue of error propagation during regression analysis from the measured thermodynamics to the calculated parameters is discussed in detail in our previous work (4, 31, 53). We have also extended this type of analysis (i.e. multiple measurements for each unknown) to the determination of mismatch contributions (31) and this methodology is readily extended to the determination of other motifs.

3.3 Sample preparation

Accurate determination of nucleic acid thermodynamics requires that samples are pure (typically >95% purity), free of dust and degassed. Both automated chemical synthesis as well as *in vitro* enzymatic methods of DNA and RNA synthesis are now routine (54–57). Sample purification usually consists of HPLC (58, 59), denaturing gel electrophoresis or thin layer chromatography (60). Just because a sample is purified by these methods does not mean, however, that the sample is free of carried over salts and other contaminants which can affect nucleic acid melting temperature and thermodynamics. Reliable methods for sample desalting include: SEP-PAK purification (Waters Inc.) (*Protocol 1*), G25 size exclusion chromatography and continuous flow dialysis (we recommend the apparatus from BRL Inc.). Ethanol precipitation is also a good method for removal of large amounts of salt, but generally samples still require further desalting by the above methods to be ready for UV melting experiments. Trace amounts of divalent metals can sometimes lead to RNA hydrolysis or dramatically affect nucleic acid structural transitions (26, 61). The best procedure for removal of trace divalent metals, which can cause RNA hydrolysis, is to add a few grains of Chelex-100 to the RNA or DNA sample, and then remove them by decanting (using a thin gel-loading plastic pipette tip) (62). Alternatively, the sample can be dialysed against 10 mM EDTA, which does not efficiently pass through 1000 Dalton cut-off dialysis membrane.

Protocol 1

SEP-PAK desalting (7)

Equipment and reagents

- Solution A: 10 mM ammonium bicarbonate (or ammonium acetate) adjusted to pH 4.5 with 1 M HCl. pH 4.5 is critical for efficient sample loading of most sequences. It is important that the buffer used is volatile so that it does not carry over into your final sample preparation. Occasionally, we find that adjusting the pH to 7 helps sequences that are rich in A, C or U (T) to stick to the column.

- Solution B: 30% acetonitrile (HPLC grade)/70 % deionized water with no added buffer.
- Nucleic acid sample
- SEP-PAK column
- 'Speed-Vac' evaporator or equivalent
- 1.5 ml centrifuge tubes
- 0.2 micron Millipore filters
- Quartz cuvettes

Method

1 (Optional) Filter Solutions A and B through a 0.2 micron Millipore filter. This will remove dust and sterilize the buffers. Do not autoclave the buffers because acetonitrile and ammonium bicarbonate are volatile.

2 Add 5 ml of Solution A to the dried nucleic acid sample.

3 Attach the SEP-PAK column (C-18 reverse phase column) to a 10 ml syringe barrel and clamp the syringe barrel to a ring stand. Equilibrate the SEP-PAK column: (i) rinse with 10 ml of 100% HPLC grade acetonitrile and (ii) rinse with 5 ml of Solution A. Air bubbles that can impede the solvent flow are easily removed by using a glass Pasteur pipette to 'swish' solvent in the syringe barrel. Gravity is sufficient force for all steps if air bubbles are removed. If desired, the flow rate can be increased by repeatedly pushing the syringe plunger ~0.5 cm into the barrel and then removing the plunger. It is important to *never let the SEP-PAK go dry*.

4 Set up a series of 1.5 ml centrifuge tubes to collect fractions (including the sample load and wash). Directly load the sample onto the column (without letting the SEP-PAK go dry). Wash with 5 ml Solution A (this desalts the oligomer and also removes other small molecule impurities). Wash with 5 ml Solution B (30% acetonitrile).

5 Quantify number of A_{260} units (see *Protocol 2*) of all fractions including the sample load and wash. Combine only the fractions eluted with Solution B that contain large amounts of oligonucleotide. Do not combine the load and wash fractions since they are not desalted.

6 Dry the sample either by lyophilization or by centrifugal evaporation (i.e. use the 'speed-vac' apparatus).

Protocol 2

Measurement of optical density units

The concept of absorbance units at 260 nm, A_{260} (also known as optical density units, ODU) often causes confusion for students. By definition, one A_{260} unit is the amount of sample that gives a UV absorbance at λ = 260 nm of 1.0 if the sample is dissolved in 1 ml volume and measured in a 1-cm pathlength cuvette. Since aborbance is proportional to concentration (Beer's Law) and the volume is 1 ml, the number of A_{260} units of a sample is proportional to the number of moles of strands. Knowledge of the number of A_{260} units is important for planning UV melting experiments to optimize sensitivity.

Method

1   'Zero' the spectrophotometer by placing deionized water or buffer in a cuvette and use the UV spectrophotometer's calibration routine.

2   Dissolve the DNA (or RNA) in 1 ml of deionized water, put into a 1-cm pathlength microvolume cuvette, and measure the absorbance at 260 nm. If the absorbance is less than 2, then the absorbance reading is the number of A_{260} units. If the sample is dissolved in a volume other than 1 ml, then use the following equation:

$$\text{Total } A_{260} \text{ units} = A \times V/l$$

where A is the measured absorbance, V is the sample volume in ml and l is the pathlength in cm.

3   If the absorbance is greater than 2, then the sample must be diluted (usually 10-fold is sufficient) or the absorbance measured in a short pathlength cuvette. Take 0.1 ml of the original sample and add 0.9 ml of water and put in the 1 cm cuvette. Remeasure the absorbance. Multiply the absorbance reading by the dilution factor, D (a factor of 10 in this example), according to the following equation:

$$\text{Total } A_{260} \text{ units} = A \times D/l$$

4   To determine the number of moles of sample, we simply rearrange Beer's Law and use the molar extinction coefficient at 260 nm, ε, calculated from the nearest-neighbour model (10, 63):

$$\text{Moles} = (\text{Total } A_{260} \text{ units})/\varepsilon$$

Note that if the oligonucleotide folds at room temperature, the extinction coefficient must be corrected for hypochromicity or the absorbance measured at ~85°C.

5   Aliquot the volume of solution that contains the desired number of absorbance units (typically 3.0 A_{260} units) of oligonucleotide for the melting experiment, and evaporate to dryness. See *Protocol 4* for a description of a procedure to accurately mix non-self-complementary strands in equal concentration.

3.4 Choice of buffer

Choice of buffer is an important yet often overlooked aspect of performing reliable UV thermal denaturation studies. Among the most important considerations are the buffer pK_a, buffer capacity, compatibility with divalent metals and temperature dependence of pK_a. *Table 1* lists the properties of the most commonly used buffers for UV melting experiments. A list of other non-complexing buffers over the entire pH range has recently been described (64). Despite the wide use in the literature of Tris as a buffer, it is *not recommended* because of its large temperature dependent pK_a (consider that between 20 to 80°C the pK_a of Tris changes by 1.7 units!). For routine measurements near pH 7 in the absence of Mg^{2+}, phosphate is the buffer of choice. Phosphate buffer is not appropriate, however, for solutions that include Mg^{2+}, since it forms tight Bjerrum ion pairs (which decreases the available concentration of divalent metal) often resulting in precipitation of magnesium phosphate (particularly at pH > 7). Cacodylate is an excellent buffer between pH 5–7 and is compatible with Mg^{2+} at pH > 6 (below pH 6, however, we have occasionally observed unusual complexation of magnesium cacodylate with RNA; S. Varma and J. SantaLucia, unpublished results). Cacodylate also offers the advantage that it is antibacterial, thereby facilitating long-term storage.

Choice of a buffer with an appropriate pK_a to give sufficient buffer capacity at the desired pH is important to prevent changes in pH as a result of a nucleic acid structural transition. Since nucleic acid concentrations for UV melting studies are routinely 10^{-4} to 10^{-6} M and most nucleic acid structural transitions do not involve the absorption or release of very many equivalents of acid (C^+–GC triple helices are a notable exception), a buffer capacity of greater than 1 mM is usually sufficient to prevent large fluctuations in pH during a melting curve. For example, we and others have observed that the adenine N1 of A^+–C

Table 1 Commonly used buffers for UV thermal denaturation studies[a]

Buffer	pK_a (@ 25°C, $I = 0.1$ M)	$\Delta pK_a/\Delta T$ (°C^{-1})	Compatible with divalent metals	References
Sodium cacodylate	6.27	−0.0015	Y	b, c
Sodium phosphate	7.20	−0.0028	N	c
MES	6.26	−0.011	Y	a, c
Tris[d]	8.06	−0.028	N	c
PIPES	6.80	−0.0085	Y	c
HEPES	7.55	−0.014	Y	c
Acetic acid	4.76	+0.0002	?	c

[a] See reference (64) for a complete list of non-complexing buffers.

[b] (67).

[c] (66).

[d] Tris is *not recommended* for thermal denaturation studies due to the large temperature dependence of its pK_a.

mismatches has a pK_a of ~6 and we thus performed UV melting studies at pH 5 (where adenine N1 is ~90% protonated) and pH 7 (where adenine N1 is ~90% deprotonated) (65). In that study 20 mM cacodylate (pK_a of cacodylate ≈ 6.0 in 1 M NaCl according to the Davies equation, see (66)) was used to buffer the solution, because at pH 5 ~90% of the buffer capacity is available to prevent raising the pH. At pH 7, on the other hand, ~90% of the buffer capacity is available to prevent lowering the pH. Another consideration in studies that plan to study structural transitions at widely varying ionic strengths is that the pK_a values of all buffers are ionic strength dependent. The empirical extension of the Debye–Huckel equation by Davies is sufficient for most applications (66).

Protocol 3

Sample degassing

Equipment

- Speed-Vac evaporator or equivalent
- Quartz microcuvettes

Method

1   The major dissolved gas in aqueous solutions is O_2. To remove dissolved oxygen, we recommend *gently* bubbling argon through the melting buffer. The buffer can then be added to the dried nucleic acid sample without introducing gas by gentle pipetting. Samples also spontaneously degas during the annealing step of the UV melting experiment (*Protocol 6*).

2   An alternative procedure is to fill the quartz microcuvette with the desired sample and to place the cuvette in a centrifuge tube. The sample is then briefly (approximately 1 min) centrifuged at low speed under vacuum (a 'speed-vac' works well for this). The only drawback to this procedure is that occasionally it is difficult to remove the cuvette from the centrifuge tube.

Nucleic acid stability at different pH values is also a consideration. At pH < 2, DNA is susceptible to depurination (usually not a problem for RNA), while for RNA at pH > 9 strand cleavage by hydrolysis is an issue. RNA strand cleavage is often catalysed by the presence of traces of divalent metals even at neutral pH, and thus 0.1 mM disodium EDTA is added to RNA solutions for thermal denaturation studies. On the other hand, many interesting RNA structural transitions occur in the presence of Mg^{2+} (26, 61), and in these cases it is recommended that the RNA solutions be prepared freshly and that temperatures greater than 80 °C be avoided if possible. Also at pH < 4 cytosine and adenine bases are protonated, usually resulting in severe destabilization of double helical structures and thus favouring the single-stranded state. Above pH 12, DNA damage can be a concern particularly at temperatures above 80 °C. Thus, the practical range of pH accessible to UV thermal denaturation studies of nucleic acids is 4–9.

Protocol 4

Mixing equal concentrations of non-self-complementary strands[a] (8)

Equipment and reagents

- Desalted strand A
- Desalted strand B
- Melting buffer

- 1.5 ml centrifuge tubes
- Quartz cuvettes (0.1 cm pathlength)
- Speed-Vac evaporator or equivalent

Method

1 Aliquot ~2.5 ODU of desalted strands A and B in separate tubes. Evaporate both samples to dryness and dissolve each sample in 100 μl melting buffer (not water).

2 Determine the concentration of each strand by placing 60 μl of each sample in separate 0.1 cm pathlength microcuvettes and measuring their absorbances at 260 nm and at high temperature (we usually use 90 °C). The concentrations, C_A and C_B, are then determined from their respective absorbances, single-stranded extinction coefficients (calculated from the nearest neighbour model (10, 63)), and using Beer's Law. It is important to measure the absorbances at high temperature since strands often form self-structure at low temperatures resulting in a hypochromic absorbance.

3 Use the following equations to calculate the volumes of strands A and B to be mixed:

$$V_A = 160 \times C_B \div (C_A + C_B)$$
$$V_B = 160 \times C_A \div (C_A + C_B)$$

4 Mix together volumes V_A and V_B of strands A and B, respectively (note $V_A + V_B = 160$ μl). The resulting solution has equal concentrations of the two strands within about 10%. It is important to use gentle pipetting back and forth to achieve good mixing.

5 To calculate the average ε for the mixed sample (which will be used later to calculate the total strand concentration, C_T), use the equation:

$$\varepsilon_{mixed} = (\varepsilon_1 + \varepsilon_2)/2$$

[a]This procedure yields 160 μl of DNA with strands in equal concentration that can be used for the UV melting dilution series (Protocol 6) experiment. If a different volume is desired, then simply substitute the desired volume for 160 in all of the equations below.

Protocol 5

Sample dilution scheme (8)

The following protocol provides 10 concentrations over a 100-fold range and utilizes custom manufactured microcuvettes[a] (see text).

Equipment

- Quartz microcuvettes (0.1 cm pathlength)

Method

1 Start with 4 A_{260} units of dry sample. Dissolve in 160 μl of the desired melting buffer. If a non-self-complementary duplex is melted, then use the solution resulting from *Protocol 4*. We usually perform the melting experiment at 280 nm for Dilution series 1 and at 260 nm for Dilution series 2 so that a larger concentration range can be studied.

2 Dilution series 1:

   (i) Take 60 μl (out of the 160 μl total) and place in cell 1 (0.1 cm). Check that the absorbance of this sample is below 1.6 at the desired wavelength, dilute with melting buffer if necessary.

   (ii) To the 100 μl remaining solution, add 70 μl of melting buffer, take 60 μl (out of the 170 μl total) and place in cell 2 (0.1 cm).

   (iii) To the 110 μl remaining solution, add 77 μl of melting buffer, take 60 μl (out of the 187 μl total) and place in cell 3 (0.1 cm).

   (iv) To the 127 μl remaining solution, add 89 μl of melting buffer, take 120 μl (out of the 216 μl total) and place in cell 4 (0.2 cm).

   (v) To the 96 μl remaining solution, add 67 μl of melting buffer, take 120 μl (out of the 163 μl total) and place in cell 5 (0.2 cm). 43 μl is leftover for the next dilution series.

3 Dilution series 2: Combine the contents of cells 2, 3 and 4 (300 μl total) with the left over solution from Dilution series 1 (43 μl) to give a total of 343 μl. If you suspect sample degradation from the previous melt (as might occur for RNA in the presence of divalent metals), then prepare a fresh sample for the second dilution series. If sample evaporation occurred in one of the cuvettes in Dilution series 1 (as evidenced by a large air bubble in the cuvette), then do not combine this sample since the salt concentration in that sample will be more concentrated than desired.

   (i) To the combined volume of 343 μl add 600 μl of melting buffer. Take 300 μl (out of the 943 μl total) and place in cell 1 (0.5 cm). Check that the absorbance of this sample is below 1.6 at the desired wavelength, dilute with melting buffer if necessary.

   (ii) To the 643 μl remaining solution, add 400 μl of melting buffer, take 300 μl (out of the 1043 μl total) and place in cell 2 (0.5 cm).

   (iii) To the 743 μl remaining solution, add 470 μl of melting buffer, take 600 μl (out of the 1213 μl total) and place in cell 3 (1.0 cm).

(iv) To the 613 μl remaining solution, add 390 μl of melting buffer, take 600 μl (out of the 1003 μl total) and place in cell 4 (1.0 cm).

(v) To the 403 μl remaining solution, add 230 μl of melting buffer, take 600 μl (out of the 633 μl total) and place in cell 5 (1.0 cm).

[a]Important notes on filling cuvettes: Before filling cuvettes, visually inspect them for cleanliness (see *Protocol 8*). *When filling cuvettes, be sure to leave a small air space* between the top of the sample and the stopper, to allow for the ~4% volume expansion that occurs for water between 0 to 95°C. If you do not do this, then either the cap will pop off during the melt or the cuvette will crack. It is also important to use gentle pipetting back and forth to achieve good mixing for each of the dilutions. Also be sure to insert a 1 cm × 1 cm strip of Teflon tape between the stopper and the cuvette to provide a tight seal. *Warning:* The stopper must be inserted with sufficient force so that it will not pop off during the melting curve. On the other hand, if the stopper is forced in too tightly you will crack the cuvette. There is no need to run a buffer control for each melting curve, since subtracting the buffer data from the sample data adds noise to the data; thus, all five cuvettes are filled with sample.

4 Instrumentation

4.1 Microvolume cuvettes and aluminium cuvette adapters

Standard 1 cm × 1 cm cuvettes are not appropriate for UV melting studies of nucleic acids because they require too much sample and they are difficult to heat evenly. A much preferred option is to have microvolume cuvettes custom manu-factured (both Precision Glass Inc. and Helma Cells produce high quality cuvettes). We use microvolume cuvettes with the following pathlengths and volumes: 0.1 cm path, 60 μl; 0.2 cm path, 120 μl; 0.5 cm path, 300 μl; 0.8 cm path, 480 μl; 1.0 cm path, 600 μl. With this set of cuvettes, we routinely investigate an 80–100-fold range in oligonucleotide concentration (*Protocol 5*). If millimolar concentrations are desired (e.g. from an NMR sample), then the 0.1 cm path cells can have a 0.09 cm thick quartz window inserted to give a net path of 0.01 cm (10). In designing the cuvettes it is important to minimize the sample volume, minimize the amount of quartz (since quartz is a thermal insulator), and provide a design that lasts at least 200 melting curves before stress leads to cracking. The cuvettes also need to have a tightly fitted Teflon stopper or screw cap, or be covered with Dow silicone oil (Corning, 200-fluid 20 centipoise viscosity) (do not overfill to prevent spillage during the volume expansion that occurs with temp-erature increases). Detailed drawings of our microvolume cuvettes can be ob-tained from the following web address: http://JSL1.chem.wayne.edu or upon request from the author.

In order to fit the microvolume cuvettes into the standard 1.25 cm × 1.25 cm sample chamber, it is necessary to have custom adapters machined from aluminium which properly position the microcuvette in the light beam and provide optimal thermal contact with the thermoelectric controller. Since the microcuvettes are different sizes, separate aluminium adapters are required for the different pathlength cuvettes. An important consideration in the design of the aluminium adapter is the exact position of the light beam and that the adapter be snug enough to provide good thermal contact and yet have some tolerance to allow for volume expansion during the melting curve. The adapter should also allow for the slight variance in stopper heights. Detailed drawings of the aluminium adapters can also be obtained from the web address given above.

4.2 Spectrophotometer

There are several instrumental requirements for precision UV melting curve analysis. Accuracy and stability of the absorbance reading at least in the forth decimal place is required. This is routinely achieved with double-beam instruments (e.g. our Aviv 14DS is stable in the fifth decimal place) as well as well-designed single-beam instruments such as the Gilford-250/260 or Beckman-DU instruments. The instrument also needs to be interfaced to a temperature controller (see below) and the absorbance and temperature readings need to be continuously output to a computer file for later data processing. It is important that the raw absorbance data are output, since many instruments apply 'box-car' or other averaging methods which make melting curves look smoother, but distort the shape of the melting curve and decrease the temperature resolution so that subtle transitions are missed; both of these effects negatively affect the quality of the derived thermodynamic parameters. The sample chamber should be connected to a compressed nitrogen source with flow controller. This allows the sample compartment to be purged with nitrogen so that water condensation is prevented at low temperatures.

The decisive factors in choice of commercial instruments for UV melting curves are the method of temperature control, method of temperature measurement and the number of cuvettes in the cell compartment. Peltier effect heating/cooling is the preferred method since circulating water baths with jacketed cuvettes usually do not change temperature quickly enough to be convenient (cooling is particularly slow, often taking several hours). All instruments with Peltier-effect heaters are not equivalent; many of the commercially available spectrophotometers can cool only to 10°C and heat to ~80°C. The Peltier-effect heaters in the Aviv 14DS allow routine heating/cooling in the entire 0–100°C range. The Aviv temperature controller can actually go to significantly higher or lower temperatures than this range if appropriate solvent mixtures are used to prevent sample freezing and/or boiling. An important maintenance tip is to periodically (every 3 months or so) blow out the sample compartment and thermoelectric controller with compressed nitrogen. This removes dust which can significantly decrease the performance of the thermoelectric controller. A thermoelectric controller with multiple cuvettes is essential if many melting curves need to be recorded. The Aviv 14DS is equipped with a five-cuvette thermoelectric controller, that allows us to routinely record 15 melting curves per day. This makes studying the concentration dependence of multi-molecular complexes routine. A design feature of the Aviv thermoelectric controller is that the samples are rotated in a circle rather than the linear configuration used in most commercially available instruments. This ensures uniform heating of the samples. The temperature controller also needs to be flexible and programmable so that melting curves can be recorded at different heating rates in both the forward and reverse directions (e.g. to check for hysteresis, which is an indicator of non-equilibrium heating rate).

The method of temperature measurement is absolutely critical to obtaining accurate thermal denaturation profiles. In principle, the best method would be to directly measure the temperature of each sample during the melting experiment with a separate temperature probe inserted into each sample. In practice, this is not easy to engineer because each of the cuvettes is sealed with a Teflon stopper and there is very little room to fit a temperature probe. Thus, most thermo-electric devices have a solid state temperature transducer (e.g. from Analog Devices Inc.) mounted somewhere in the sample changer (in the Aviv 14DS the temperature transducer is mounted in the centre spindle which receives the cuvette holder. The reading from the transducer is calibrated periodically by performing a mock UV thermal denaturation curve and measuring the voltage produced by a type K thermocouple (or using a 100 ohm porcelainized platinum resistance temperature detector (68)) inserted into a buffer filled cuvette, to simulate as much as possible the conditions of an actual melting experiment. This method of calibration provides temperature precision of ~0.1°C and accuracy of ~0.3°C, which is sufficient for most applications. Studies of polymer melting transitions require more rigorous temperature measurement and the method described by Blake and co-workers is recommended (69, 70).

Protocol 6

Typical melting protocol

Equipment

- UV spectrophotometer with thermostated cell holding
- Quartz cuvettes/microcuvettes

Method

1 Turn on the instrument at least half an hour before making measurements to allow the D_2 lamp to warm-up to minimize instrument drift. Purge the sample compartment with nitrogen at least 20 min before a melting curve is recorded. Place buffer in one cuvette and 'zero' the absorbance reading.

2 Load filled microcuvettes (*Protocols 4* and *5*) into the aluminium cell holders and place into the thermoelectric controller.

3 Set the wavelength to 260 nm and raise the temperature to 90°C for 5 min. This 'annealing' helps equilibrate the structure and to degas the sample. Record the absorbances in a laboratory notebook for future calculation of strand concentrations. If optimum hypochromicity is desired, then record a high-temperature wavelength scan while at high temperature.

4 Ramp the temperature down to 0°C (we actually go to −3°C for solutions in 1 M NaCl) over ~10 min. If the wavelength for optimum hypochromicity needs to be

determined, then record a low-temperature wavelength scan once the sample is equilibrated at 0 °C.

5   Input the high and low temperature absorbance versus wavelength data into a spreadsheet (e.g. Microsoft Excel) and calculate the hypochromicity at each wavelength using the *equation 1*. The optimum wavelength has the largest hypochromicity. It is also important to consider that the spectrometer signal-to-noise ratio is optimum at an absorbance of 0.454 and that the signal-to-noise ratio decreases dramatically below 0.2 or above 1.8.

6   Once the samples are equilibrated at 0 °C, begin ramping up the temperature at a rate of 0.25, 0.5, 0.8 or 1.0 °C/min. The absorbance reading is usually averaged over a 5 s period and sample changing requires ~ 1 s. Thus each cell has the absorbance read every 30 s (corresponding to 0.5 °C for a 1 °C/min heating rate). Repeating the melting curve at slower heating rates is a good way to check that the heating rate is slow enough for equilibrium to be achieved.

7   When the melting curve is complete, note the high temperature absorbances of all cells in your laboratory notebook. Compare these results with those measured in step 3. If any of the absorbances are significantly higher, this indicates that the Teflon cap probably popped off during the experiment resulting in evaporation; data for such cells should be rejected.

8   Ramp the temperature down to 25 °C. Remove samples, clean and dry the cuvettes (see *Protocol 8*). Load a new set of samples and go to step 1 or shut down the spectrophotometer according to manufacturer recommendations.

9   Occasionally (about 1 in every 100 melting curves), sample degassing can occur during the melting curve (or a dust particle might move) resulting in a spurious absorbance reading at one temperature. Visually inspect the melting curve data file and manually remove any data points that are anomalous. Save the data file on a computer disk (give it a systematic name and record the name in your notebook). The data are now ready for fitting and analysis (see Section 5).

Protocol 7

Sample recovery

UV melting studies typically use only 3 A_{260} units for a full set of curves and thus samples are routinely discarded or archived at –70 °C in case future analysis is required. If a sample is precious, then it may be desirable to recover the sample:

1   Generally, it is recommended that the sample purity be checked by analytical HPLC or gel electrophoresis to ensure that the sample has not degraded.

2   Typically, sample desalting by the methods above (see text and *Protocol 1*) is sufficient for further experiments to be performed.

3   Treatment with Chelex-100 is crucial if the sample was previously dissolved in a buffer containing multivalent metals.

Protocol 7 continued

4   In the event that Dow silicone oil (Corning, 200-fluid 20 centipoise viscosity) was used to prevent sample evaporation, sample recovery can be readily obtained by pipetting out the entire sample (including oil) onto a Teflon dish. The oil will stick to the Teflon as the bead of sample is rolled around the dish (10).

Protocol 8

Cleaning of quartz cuvettes

Equipment and reagents

- Plastic gel loading pipette tip
- Plastic scissor clamps
- 50% nitric acid
- Methanol (HPLC or spectroscopic grade)
- Compressed nitrogen

Safety precaution: Wear safety goggles and latex gloves, since nitric acid is caustic and mutagenic.

Method

1   After a melting curve is complete, remove the sample using a plastic gel loading pipette tip.

2   Clamp the cuvette with locking plastic scissor clamps (the scissor clamp should grip the frosted sides of the cuvette to prevent scratching the optical surfaces). Remove traces of your sample by rinsing the cuvettes with distilled water in a squirt bottle. The water is most easily removed from the cuvette with a syringe that has the needle capped with a short length of narrow gauge plastic tubing (to prevent the syringe needle from scratching the quartz). We do not recommend shaking the cuvette upside down to remove the liquid contents since this often results in the cuvette slipping out of the clamps and breaking the cuvette.

3   Fill the cuvettes with 50% nitric acid (if the nitric acid is brown it has degraded and a fresh solution should be made). Place the filled cuvette into a 50% nitric acid bath for ~10 min. Cover the acid bath with a watchglass to prevent dust from getting in.

4   Remove the sample from the nitric acid using the plastic scissor clamps and re-move the nitric acid using a pipette (not the syringe, since the nitric acid will corrode the metal syringe needle).

5   Rinse several times with water and *only then* rinse several times with methanol (HPLC or spectroscopic grade). *Never allow methanol to come into contact with the nitric acid*. Dry the cuvette with compressed nitrogen (be sure to hold the cuvette firmly as the nitrogen pressure can cause it to be dropped).

6   Store the cuvettes on a benchtop covered with dust-free cloth. Cover the cuvettes with an inverted beaker to prevent dust accumulation. Cuvettes stored this way are ready to use for the next melting experiment without further cleaning (unless you do not trust the cleaning technique of your labmates).

5 Data analysis

5.1 Curve fitting to calculate thermodynamic parameters

The methods for fitting UV melting curves and determining thermodynamics have been extensively reviewed (4, 9, 10, 14, 16–18). The optimal choice of fitting method depends on a number of factors including molecularity of the transition, melting temperature of the transition (i.e. whether sufficient upper or lower baseline is available) and which approximations are introduced (e.g. two-state model with ΔC_p° equal to zero). This chapter presents the methods for fitting unimolecular and bimolecular transitions since they form the majority of measurements made on nucleic acids. For higher molecularities the reader is referred to the primary literature (9, 18, 40). It is appropriate to reiterate here that careful design of oligonucleotide sequences (see Sections 3.1 and 3.2) which minimize the potential for the formation of undesired structures significantly simplifies data analysis and interpretation.

5.1.1 Unimolecular transitions

Unimolecular structural transitions involve the equilibrium between random coil and ordered states such as single-stranded stacking (71) and/or hairpin formation (48).

$$A_{RC} \longleftrightarrow A_F \qquad\qquad 6$$

where A_{RC} and A_F are the random coil and folded states of sequence A. Before analysing the melting data, it is essential that the concentration independence of the transition be experimentally verified. This is accomplished simply by recording the UV melting profile over at least a factor of 10 in oligonucleotide concentration and plotting the normalized absorbance curves (see Section 5.2). Note that it is important that the *entire shape of the curve* is concentration independent, not just the T_M. Observation of small concentration dependence (particularly at high oligonucleotide concentrations) usually suggests the presence of end-to-end aggregation, non-specific aggregation, or the presence of bimolecular internal loop formation. These artefacts can be minimized by recording the melting curves at low salt concentration and low oligonucleotide concentration.

To derive the enthalpy and entropy changes for a random-coil to hairpin transition it is necessary to determine the temperature dependence of the equilibrium constant by analysing the shape of the melting curve. The work of Turner and co-workers indicates that to derive accurate thermodynamics from the shape of a melting curve it is necessary to subtract the effects of sloped upper and lower baselines (14, 16) (*Figure 1*). Sloped baselines arise because the extinction coefficients of both the folded and denatured states are temperature dependent, although the physical origin of the temperature dependence is largely unknown. The simplest interpretation of sloping baselines is that they indicate that some non-two-state behaviour is occurring, such as helix fraying, single-strand stacking, or changes in hydration with temperature (71). The best procedure is to assume linear baselines with the following temperature dependent extinction coefficients:

$$\varepsilon_F(T) = m_F \times T + b_F \qquad\qquad 7$$

$$\varepsilon_{RC}(T) = m_{RC} \times T + b_{RC} \qquad\qquad 8$$

where $\varepsilon_F(T)$ and $\varepsilon_{RC}(T)$ are the temperature dependent extinction coefficients of the folded and 'random coil' states and m and b are the slopes and intercepts.

For a hairpin transition assuming a two-state model, the fraction of strands in the folded state, α, is given by:

$$\alpha = [A_F]/C_T \qquad\qquad 9$$

$$\alpha = \frac{a}{a + b} \qquad\qquad 10$$

where a and b are the respective vertical distances from the upper and lower baselines to the experimental melting curve (*Figure 1*). The temperature at which α is equal to 0.5 is defined as the melting temperature, T_M. The fraction of strands in the random coil state is given by $1 - \alpha$. Thus the equilibrium constant at a particular temperature is given by:

$$K = \frac{[A_F]}{[A_{RC}]} = \frac{\alpha}{1 - \alpha} \qquad\qquad 11$$

By using sloping baselines, one is effectively removing from the shape analysis those molecules in the sample with intermediate states. Hence, only those molecules in the sample that actually melt two-state are used to calculate the thermodynamics. Whilst technically this is somewhat alarming, there is general agreement that the use of sloping baselines provides the most reliable thermodynamic data (10, 14, 16). Assuming a two-state model with ΔC_p° equal to zero (i.e. the van't Hoff approximation), the ΔH° and ΔS° can be derived from the slope and intercept of a ln$K$ versus $1/T$ plot by rearranging *equation 2*:

$$\ln K = \frac{-\Delta H^{\circ}}{RT} + \frac{\Delta S^{\circ}}{R} \qquad\qquad 12$$

In practice, only the data points for $0.15 \leq \alpha \leq 0.85$ are used due to experimental uncertainty in the K values outside of this range (10).

An alternative and preferred method to determine thermodynamics is to directly fit the entire absorbance versus temperature curve (9, 10, 14, 17). The total absorbance at temperature T, $A(T)$, is given by the sum of Beer's law contributions of each component in the solution:

$$A(T) = \varepsilon_{RC}(T) \times l \times [A_{RC}] + \varepsilon_F(T) \times l \times [A_F] \qquad\qquad 13$$

where l is the pathlength. Substituting the linear approximations for the baselines (i.e. *equations 7* and *8*) and *equation 9* gives:

$$A(T) = C_T \times l \times [(m_{RC} \times T + b_{RC}) \times \alpha(T) + (m_F \times T + b_F) \times (1 - \alpha(T))] \qquad\qquad 14$$

According to *equations 11* and *12*, $\alpha(T)$ is determined by ΔH° and ΔS°. Thus the experimental absorbance versus temperature curve is fit using multiple non-linear regression analysis, which amounts to a six parameter fit of the parameters: $\Delta H°$, $\Delta S°$, m_{RC}, b_{RC}, m_F and b_F (14). In practice, the data are truncated to include the data within ~30 °C of the T_M so that the baseline slopes in the transition region are accurately accounted for. Consult (17) for details on a widely used program for curve fitting. It should be noted that this method does not work well in cases where the upper or lower baselines are not well defined due to a high or low T_M value. In cases where the upper or lower baselines are not well defined, it is recommended that thermodynamics be derived using the absorbance derivative method described in (72) and (18). It should also be noted that observation of a large lower baseline slope often suggests the presence of structural intermediates which compromise the validity of the van't Hoff method (37). As stated in Section 2, the van't Hoff method provides two-state thermodynamic parameters at the T_M. In principle, $\Delta C_p°$ could be added as a seventh unknown in the curve fitting so that more accurate temperature extrapolations are possible, though the noise level in a typical absorbance experiment may not justify such a treatment (32, 73).

5.1.2 Bimolecular transitions

The van't Hoff curve fitting methods presented above for hairpin transitions are general, but for higher molecularity the equilibrium expressions are different (*equations 6, 9* and *11*) (9, 10, 18). For self-complementary duplexes the relevant equilibrium equations are:

$$2 A_{RC} \longleftrightarrow A_2 \qquad\qquad 15$$

$$\alpha = \frac{2[A_2]}{C_T} \qquad\qquad 16$$

$$K = \frac{[A_2]}{[A_{RC}]^2} = \frac{\alpha}{2 \times (1 - \alpha)^2 C_T} \qquad\qquad 17$$

For non-self-complementary duplexes the relevant equilibrium equations are:

$$A_{RC} + B_{RC} \longleftrightarrow AB \qquad\qquad 18$$

$$\alpha = \frac{2[AB]}{C_T} \qquad\qquad 19$$

$$K = \frac{[AB]}{[A_{RC}][B_{RC}]} = \frac{2\alpha}{(1 - \alpha)^2 C_T} \qquad\qquad 20$$

Note the fourfold difference in *equations 17* and *20* which is ultimately reflected in the different concentration dependence of the T_M for self-complementary versus non-self-complementary sequences (*equation 3*).

As described in Section 2, the concentration dependence of the T_M provides an alternative van't Hoff method for determining thermodynamics. In this method, the transition $\Delta H°$ and $\Delta S°$ are obtained by *equation 4* from the slope and intercept of a $1/T_M$ against $\ln C_T/4$ plot. Since the T_M is relatively insensitive to non-two-state behaviour and to the choice of baselines, $1/T_M$ versus $\log C_T$ plots provide very reliable $\Delta H°$, $\Delta S°$, and $\Delta G°_{37}$ parameters (16), particularly if the T_M is measured over a large concentration range. It is worth noting that the T_M *cannot* be accurately determined from an inflection point of the melting curve determined from the maximum of the temperature derivative of the absorbance (9, 10). It is necessary to subtract the upper and lower temperature baselines before performing the differentiation, since they affect the position of the apparent T_M (10, 14). The preferred method for calculating the T_M is to use *equation 3* using the $\Delta H°$ and $\Delta S°$ determined from the fit of the shape of the curve with sloped baselines, as described above (14, 16).

Since curve fitting and $1/T_M$ against $\ln C_T/4$ plot methods depend differently on the two-state approximation, agreement of $\Delta H°$ obtained from both methods provides a test of the validity of the two-state approximation (6, 10). Agreement within 10–15% of $\Delta H°$ parameters from a $1/T_M$ versus $\ln C_T$ plot and from fits of the shapes of melting curves is generally regarded as indicative of a two-state transition (4), though caution is recommended. This criterion suggests the standard error in the van't Hoff $\Delta H°$ parameters from optical melting is about 5–8%. In most cases that have been studied by both optical melting and calorimetry, the $\Delta H°$ values are in agreement if the two ways of analysing the optical data give the same value within 10% (15, 16, 34). Agreement between enthalpy changes determined by different methods is a necessary, but not a sufficient criterion to definitively establish two-state behaviour (4, 18). For example, the self-complementary DNA sequence (CGTT$\underline{G}$CG$\underline{T}$AACG)$_2$ yielded $\Delta H°$ parameters from a $1/T_M$ versus $\ln C_T$ plot and from fits of the shapes of optical melting curves that agree within 10% for the single strand to duplex transition. The temperature dependence of NMR spectra and comparison of that sequence with thermodynamics for other sequences, however, revealed that it actually melts through a hairpin intermediate (31).

An alternative criterion for two-state thermodynamics is to compare the enthalpy changes from optical melting and calorimetry (15, 16, 34). For transitions with large $\Delta C_p°$, it appears that van't Hoff analyses and calorimetric data provide systematically different $\Delta H°$ parameters and the possible origins of these differences have been recently reviewed (32, 73). Even agreement between van't Hoff analysis of optical melting and calorimetry does not guarantee a two-state transition. The $\Delta H°$ values for the DNA duplex, (GCGTACGCATGCG•CGCATGTGTACGC), are in agreement, but the transition is not two-state as evidenced by the melting of the individual single strands (31, 40). Known exceptions of this type are rare, however.

For two-state transitions, the $\Delta H°$ parameters obtained by both methods are equally reliable, so it is best to report the average of parameters from the $1/T_M$ versus $\ln C_T$ plot and from the fits of the shapes of curves (41, 74). Obtaining $\Delta H°$ from melting curves measured at different wavelengths can be used to check the two-state approximation (6). Importantly, agreement between enthalpy changes determined by different methods is a necessary, but not a sufficient criterion to establish two-state behaviour (16, 27, 38). Additional methods for validation of the two-state model are described in Section 2.

A van't Hoff analysis of UV melting data cannot be used to reliably measure the thermodynamics of molecules with non-two-state transitions. In the case where the transition is clearly non-two-state (particularly if end-fraying or internal loop structures are significantly populated), a statistical mechanical approach may be able to yield reliable thermodynamics in some circumstances (75, 76). One drawback of the statistical mechanical approach is that the data are fitted with more parameters than justified by the signal-to-noise ratio of the data, and thus loop parameters are introduced in an *ad hoc* fashion. Recently, we derived a method for treating UV melting curves by a three-state model (31) in which a self-complementary sequence was allowed to form a hairpin intermediate, utilizing a coupled equilibrium method described previously for an RNA with duplex intermediates (36). Such an approach is easily generalized to non-self-complementary duplexes, with each strand able to fold into hairpin intermediates. A better approach, however, is to carefully design oligonucleotides so that the possibility for alternative structure formation is minimized. Performing the control melting experiments of the individual single strands reveals if hairpin states are likely to be significantly populated. It is worth noting that actual single-strand melts *should not* be directly subtracted from duplex melting curves, since in duplex melting the competition with the duplex state significantly reduces the population of hairpin conformation (36). Instead, a fully coupled three-state model is appropriate (31).

5.2 Presentation of normalized absorbance curves

Often it is useful to present several UV absorbance curves on the same figure to illustrate differences in T_M or hypochromicity. To display the curves on the same scale, it is necessary to normalize the curves either to high temperature or to both high and low temperature. To normalize different melting curves to the high temperature absorbance, A_{HT}, *equation 21* is applied to each curve and then the data are plotted on the same graph.

$$A_{Norm} = \frac{A}{A_{HT}} \qquad\qquad 21$$

where A_{Norm} is the normalized absorbance at high temperature and A is the raw absorbance data. *Equation 21* preserves the shape of the melting curve without distorting the hypochromicity. Alternatively, the data can be 'double normalized' to both high and low temperatures by applying *equation 22* to each curve and then plotting the data on the same graph.

$$A_{Norm} = \frac{A/A_{LT} - 1}{A_{HT}/A_{LT} - 1} \qquad\qquad 22$$

where A_{LT} is the low temperature absorbance. *Equation 22* makes all curves start at 0 and finish at 1 and preserves the shape of the melting curve, but information about hypochromicity is lost. *Equation 22* is useful for comparing the melting curves of different sequences (at the same concentration) or comparing the melting curves of one sequence measured at different wavelengths.

5.3 Error analysis

Careful analysis of errors is an important aspect of any experimental study. The main sources of errors in optical melting are:

(1) signal-to-noise ratio of the data;

(2) random errors due to fluctuations in sample preparation (e.g. oligonucleotide concentration, purity, volume, salt concentration, pH, errors in mixing etc.);

(3) systematic errors due to incorrect instrument calibration;

(4) systematic errors introduced as a result of poor oligonucleotide design (e.g. a sequence that can form intermediates);

(5) systematic errors due to incorrect assignment of baselines; and

(6) systematic errors due to imposing the two-state approximation and from assuming that ΔC_p° is zero.

An important distinction is the difference between precision, which reflects the experimental reproducibility of the data, and accuracy, which reflects how well the experimental measurement agrees with the real value if a perfect measurement were made (51). The first two sources of error are easy to quantify by simply reproducing one's data or analysing the sampling error in a $1/T_M$ versus $\ln C_T$ plot or the sampling error in the fitted data. The theory for determining sampling errors in $\Delta G°_{37}$, $\Delta H°$, and $\Delta S°$ from the linear regression of the $1/T_M$ versus $\ln C_T$ plot using standard statistical analysis (77) has been previously described (30). The best method to quantify systematic errors (sources 3–6 above) is to compare thermodynamic measurements on the same oligonucleotides that were independently determined from different laboratories utilizing different instrumentation and techniques. For DNA duplexes, an estimate for systematic error sources 3 and 5 can be derived from results on three sequences (CGATATCG, GAAGCTTC and GGAATTCC) that have been independently measured by two groups (34, 41, 74). The average deviations for $\Delta G°_{37}$, $\Delta H°$, $\Delta S°$ and T_M for these sequences are 3%, 6%, 6% and 1.0 °C, respectively.

It is important to understand how errors in $\Delta H°$ and $\Delta S°$ propagate to give the error in $\Delta G°_{37}$. Experimental $\Delta H°$ and $\Delta S°$ parameters are not independently determined, but instead are highly correlated, with a typical $R^2 > 0.99$ (14, 22, 31). This enthalpy–entropy compensation results in errors in $\Delta G°_{37}$ that are much smaller than would be expected if $\Delta H°$ and $\Delta S°$ were uncorrelated. Equation 4.8 of Bevington (51) provides the equation for error propagation for a general function $x = f(u,v)$:

$$(\sigma_x)^2 = (\sigma_u)^2 \left(\frac{\partial x}{\partial u}\right)^2 + (\sigma_v)^2 \left(\frac{\partial x}{\partial v}\right)^2 + 2(\sigma_{uv})^2\left(\frac{\partial x}{\partial u}\right)\left(\frac{\partial x}{\partial v}\right) \qquad 23$$

Performing the appropriate differentiation of *equation 2* and substitution into *equation 23* gives the equation for the propagation of error from $\Delta H°$ and $\Delta S°$, $\sigma_{\Delta H°}$ and $\sigma_{\Delta S°}$, to give the error in $\Delta G°_{37}$, $\sigma_{\Delta G°37}$ (30):

$$(\sigma_{\Delta G°37})^2 = (\sigma_{\Delta H°})^2 + T^2(\sigma_{\Delta S°})^2 - 2T\,(\sigma_{\Delta H°\Delta S°})^2 \qquad 24$$

where $\sigma_{\Delta H^\circ \Delta S^\circ}$ is the covariance between ΔH° and ΔS°, and T is 310.15 K. An alternative equation expresses the covariance in terms of the correlation coefficient of a plot of ΔH° versus ΔS°, $R_{\Delta H^\circ \Delta S^\circ}$ (78):

$$(\sigma_{\Delta G^\circ 37})^2 = (\sigma_{\Delta H^\circ})^2 + T^2(\sigma_{\Delta S^\circ})^2 - 2T\,(R_{\Delta H^\circ \Delta S^\circ})\,\sigma_{\Delta H^\circ}\,\sigma_{\Delta S^\circ} \qquad 25$$

Performing the appropriate differentiation of *equation 3* and substitution into *equation 23* gives the equation for the propagation of $\sigma_{\Delta H^\circ}$ and $\sigma_{\Delta S^\circ}$, to give the error in T_M, σ_{Tm} (31):

$$\sigma_{Tm}{}^2 = \left(\frac{\sigma_{\Delta H^\circ} T_M}{\Delta H^\circ}\right)^2 + \left(\frac{\sigma_{\Delta S^\circ} T_M}{\Delta H^\circ}\right)^2 - \frac{2T_M{}^3\,R_{\Delta H^\circ \Delta S^\circ}\,\sigma_{\Delta H^\circ}\,\sigma_{\Delta S^\circ}}{(\Delta H^\circ)^2} \qquad 26$$

This equation assumes that there is negligible error in the $\ln C_T$ term. This is reasonable because a 10% error in C_T propagates according to *equation 23* to a 1% error in $\ln C_T$ at oligonucleotide concentrations in the range of 10^{-5} M (7). The enthalpy–entropy compensation effect is evident in the high quality of predictions made for ΔG°_{37} and T_M (31, 41, 53).

Another way to evaluate error propagation with minimal assumptions about the experimental errors or how these errors propagate to the derived NN (nearest neighbour) parameters is to use a resampling analysis of the data (79). Consider the 'unified data set' of 108 sequences with only Watson–Crick base pairs (31). First, SVD analysis is used to determine the linear least-squares fit of the data to obtain the 10 NN parameters and the two initiation parameters as described (31, 41, 53). Since the unified data set contains 108 equations with 12 unknowns, the problem is overdetermined. A total of 30 resampling trails were used. For each trial, a different set of 68 randomly selected sequences was used in the SVD analysis to calculate the 12 unknowns. It is important to check the column rank of the stacking matrix for each trial. The results from the 30 trials were averaged and the standard deviations calculated for each NN parameter. This resampling analysis was performed for ΔG°_{37}, ΔH° and ΔS°. We also performed controls in which certain classes of sequences (e.g. all sequences from a particular lab or all sequences greater than 10 bp, etc.) were *systematically* omitted from the SVD analysis and found that in all cases the parameters agreed within the propagated experimental error of the NN from the total dataset.

References

1. Petruska, J., Goodman, M. F., Boosalis, M. S., Sowers, L. C., Cheong, C. and Tinoco, I., Jr (1988). *Proc. Natl. Acad. Sci. USA.*, **85**, 6252–6256.
2. Peyret, N., Seneviratne, P. A., Allawi, H. T. and SantaLucia, J., Jr (1999). *Biochemistry*, **38**, 3468–3477.
3. Gacy, A. M. and McMurray. (1998). *Biochemistry*, **37**, 9420–9434.
4. SantaLucia, J., Jr and Turner, D. H. (1997). *Biopolymers*, **44**, 309–319.
5. Koonin, S. E. (1998). *Science*, **279**, 36–37.
6. Jaeger, J. A., SantaLucia, J., Jr., and Tinoco, I., Jr. (1993). *Annu. Rev. Biochem.*, **62**, 255–287.
7. SantaLucia, J., Jr (1991) in *Department of Chemistry, Ph.D. Thesis*. University of Rochester, Rochester, NY.

8. Allawi, H. T. (1998) in *Department of Chemistry*, *Ph.D. Thesis*. Wayne State University, Detroit, MI.

9. Breslauer, K. J. (1995). *Methods Enzymol.*, **259**, 221–241.

10. Puglisi, J. and Tinoco, I., Jr (1989). *Methods Enzymol.*, **180**, 304–325.

11. Doty, P., Boedtker, H., Fresco, J. R., Haselkorn, R. and Litt, M. (1959). *Proc. Natl. Acad. Sci. USA*, **45**, 482–499.

12. Tinoco, I., Jr (1960). *J. Am. Chem. Soc.*, **82**, 4785–4790.

13. Freier, S. M., Kierzek, R., Jaeger, J. A., Sugimoto, N., Caruthers, M. H., Neilson, T. and Turner, D. H. (1986). *Proc. Natl. Acad. Sci. USA*, **83**, 9373–9377.

14. Petersheim, M. and Turner, D. H. (1983). *Biochemistry*, **22**, 256–263.

15. Rentzeperis, D., Ho, J. and Marky, L. A. (1993). *Biochemistry*, **32**, 2564–2572.

16. Albergo, D., Marky, L., Breslauer, K. and Turner, D. (1981). *Biochemistry*, **20**, 1409–1413.

17. McDowell, J. A. and Turner, D. H. (1996). *Biochemistry*, **35**, 14077–14089.

18. Marky, L. A. and Breslauer, K. J. (1987). *Biopolymers*, **26**, 1601–1620.

19. Borer, P. N., Dengler, B., Tinoco, I., Jr and Uhlenbeck, O. C. (1974). *J. Mol. Biol.*, **86**, 843–853.

20. Stevens, C. and Felsenfeld, G. (1964). *Biopolymers*, **2**, 293–314.

21. SantaLucia, J., Jr, Kierzek, R. and Turner, D. H. (1990). *Biochemistry*, **29**, 8813–8819.

22. Krug, R. R., Hunter, W. G. and Grieger, R. A. (1976). *J. Phys. Chem.*, **80**, 2335–2341.

23. Cole, P. E. and Crothers, D. M. (1972). *Biochemistry*, **11**, 4368–4374.

24. Crothers, D. M., Cole, P. E., Hilbers, C. W. and Schulman, R. G. (1974). *J. Mol. Biol.*, **87**, 63–88.

25. Felsenfeld, G. and Hirschman, S. Z. (1965). *J. Mol. Biol.*, **13**, 407–427.

26. Jaeger, J. A., Zuker, M. and Turner, D. H. (1990). *Biochemistry*, **29**, 10147–10158.

27. Banerjee, A. R., Jaeger, J. A. and Turner, D. H. (1993). *Biochemistry*, **32**, 152–163.

28. Mathews, D. H., Andre, T. C., Kim, J., Turner, D. H. and Zuker, M. (1997). In *Molecular modeling of nucleic acids* (ed. N. B. Leontis and J. SantaLucia, Jr), Vol. 682, pp. 246–257. A.C.S. , Washington, DC.

29. Breslauer, K., Freire, E. and Straume, M. (1992). *Methods Enzymol.*, **211**, 533–567.

30. SantaLucia, J., Jr, Kierzek, R. and Turner, D. (1991). *J. Am. Chem. Soc.*, **113**, 4313–4322.

31. Allawi, H. T. and SantaLucia, J., Jr (1997). *Biochemistry*, **36**, 10581–10594.

32. Liu, Y. and Sturtevant, J. M. (1997). *Biophys. Chem.*, **64**, 121–126.

33. Law, S. M., Eritja, R., Goodman, M. F. and Breslauer, K. J. (1996). *Biochemistry*, **35**, 12329–12337.

34. Breslauer, K. J., Frank, R., Blocker, H. and Marky, L. A. (1986). *Proc. Natl. Acad. Sci. USA*, **83**, 3746–3750.

35. Li, Y., Zon, G. and Wilson, W. D. (1991). *Biochemistry*, **30**, 7566–7572.

36. Longfellow, C. E., Kierzek, R. and Turner, D. H. (1990). *Biochemistry*, **29**, 278–285.

37. SantaLucia, J., Jr, Kierzek, R. and Turner, D. H. (1992). *Science*, **256**, 217–219.

38. Zuker, M. (1989). *Science*, **244**, 48–52 .

39. SantaLucia, J., Jr and Zuker, M. (unpublished experiments).

40. Plum, G. E., Grollman, A. P., Johnson, F. and Breslauer, K. J. (1995). *Biochemistry*, **34**, 16148–16160.

41. SantaLucia, J., Jr, Allawi, H. and Seneviratne, P. A. (1996). *Biochemistry*, **35**, 3555–3562.

42. Gray, D. M. (1997). *Biopolymers*, **42**, 795–810.

43. Goldstein, R. F. and Benight, A. S. (1992). *Biopolymers*, **32**, 1679–1693.

44. SantaLucia, J., Jr (1998). *Proc. Natl. Acad. Sci. USA*, **95**, 1460–1465.

45. Ladbury, J. E., Sturtevant, J. M. and Leontis, N. B. (1994). *Biochemistry*, **33**, 6828–6833.

46. Seeman, N. C. (1990). *J. Biomol. Struct. Dyn.*, **8**, 573–581.

47. Hirao, I., Nishimura, Y., Tagawa, Y., Watanabe, K. and Miura, K. (1992). *Nucleic Acid Research,* **20**, 3891–3896.

48. Antao, V. P., Lai, S. Y. and Tinoco, I., Jr (1991). *Nucleic. Acids Res.,* **19**, 5901–5905.

49. Walter, A. E., Turner, D. H., Kim, J., Lyttle, M. H., Muller, P., Mathews, D. H. and Zuker, M. (1994). *Proc. Natl. Acad. Sci. USA,* **91**, 9218–9222.

50. Peyret, N. and SantaLucia, J., Jr (1999). *(In preparation).*

51. Bevington, P. R. (1969). *Data reduction and error analysis for the physical sciences.* McGraw-Hill, New York. pp. 58–60

52. Press, W. H., Flannery, B. P., Teukolsky, S. A. and Vetterling, W. T. (1989) pp 52–64, 498–520. Cambridge University Press, New York.

53. Xia, T., SantaLucia, J., Jr, Burkard, M. E., Kierzek, R., Schroeder, S. J., Jiao, X., Cox, C. and Turner, D. H. (1998). *Biochemistry,* **37**, 14719–14735.

54. Milligan, J. F., Groebe, D. R., Witherell, G. W. and Uhlenbeck, O. C. (1987). *Nucleic Acids Res.,* **15**, 8783–8798.

55. Capaldi, D. and Reese, C. (1994). *Nucleic Acids Res.,* **22**, 2209–2216.

56. Brown, T. and Brown, D. J. S. (1991). In *Oligonucleotides and analogues* (ed. F. Eckstein), pp. 1–24. IRL Press, Oxford.

57. Zimmer, D. and Crothers, D. (1995). *Proc. Natl. Acad. Sci. USA,* **92**, 3091–3095.

58. Arghavani, M. B. and Romano, L. J. (1995). *Anal. Biochem.,* **231**, 210–209.

59. McLaughlin, L. W. (1989). *Chem. Rev.,* **89**, 309–319.

60. Chou, S.-H., Flynn, P. and Reid, B. (1989). *Biochemistry,* **28**, 2422–2435.

61. Zarrinkar, P. P. and Williamson, J. R. (1994). *Science,* **265**, 918–924.

62. Primrose, W. U. (1993). In *NMR of macromolecules: A practical approach* (ed. G. C. K. Roberts), pp. 7–34. IRL Press, Oxford.

63. Richards, E. G. (1975). In *Handbook of biochemistry and molecular biology: Nucleic acids* (ed. G. D. Fasman), Vol. I, p. 597. CRC Press, Cleveland, OH.

64. Yu, Q., Kandegedara, A., Xu, Y. and Rorabacher, D. B. (1997). *Anal. Biochem.,* **253**, 50–56.

65. Allawi, H. T. and SantaLucia, J., Jr (1998). *Biochemistry,* **37**, 9435–9444.

66. Perrin, D. D. and Dempsey, B. (1979). *Buffers for pH and metal ion control.* Halsted Press, New York. pp. 6–20

67. Lewis, E. A., Hansen, L. D., Baca, E. J. and Temer, D. J. (1976). *J. Chem. Soc. Perkin II,* 125–128.

68. Delcourt, S. G. and Blake, R. D. (1991). *J. Biol. Chem.,* **266**, 15160–15169.

69. Blake, R. D., Vosman, F. and Tarr, C. (1981). In *Biomolecular stereodynamics* (ed. R. H. Sarma), Vol. I, pp. 439–458. Adenine Press, New York.

70. Yen, S.-W. W. and Blake, R. D. (1980). *Biopolymers,* **19**, 681–700.

71. Freier, S. M., Hill, K. O., Dewey, T. G., Marky, L. A., Breslauer, K. J. and Turner, D. H. (1981). *Biochemistry,* **20**, 1419–1426.

72. Gralla, J. and Crothers, D. M. (1973). *J. Mol. Biol.,* **73**, 497.

73. Chaires, J. B. (1997). *Biophys. Chem.,* **64**, 15–23.

74. Sugimoto, N., Nakano, S., Yoneyama, M. and Honda, K. (1996). *Nucleic Acids Res.,* **24**, 4501–4505.

75. Wartell, R. M. and Benight, A. S. (1985). *Phys. Rep.,* **126**, 67–107.

76. Vologodskii, A. V., Amirikyan, B. R., Lyubchenko, Y. L. and Frank-Kamenetskii, M. D. (1984). *J. Biomol. Struct. Dyn.,* **2**, 131–148.

77. Meyer, S. L. (1975). *Data analysis for scientists and engineers.* Wiley, New York. Chapter 19

78. Snedecor, G. W. and Cochran, W. G. (1971). *Statistical methods.* The Iowa State University Press, Ames, IA. p. 190

79. Efron, B. and Tibshirani, R. (1993). *An introduction to the bootstrap.* Chapman & Hall, London.

List of suppliers

This core list of suppliers appears in all books in the Practical Approach series.

Aladdin Biofluorescence Center, School of Pharmacy, University of Wisconsin-Madison, 425 N. Charter St., Madison, WI 53706 USA
(gvidugir@facstaff.wisc.edu).
URL: http://www.src.wisc.edu/highlights/time_flour/default.html.
Ambion, Inc., 2130 Woodward Street, 200, Austin, TX 78744-1832, USA
Amresco, 30175 Solon Industrial Parkway, Solon, OH 44139, USA
Anderman and Co. Ltd, 145 London Road, Kingston-upon-Thames, Surrey, KT2 6NH
Tel: 0181 541 0035   Fax: 0181 541 0623
Applied Photophysics Ltd, 203/205 Kingston Road, Leatherhead, Surrey, KT22 7PB, UK (http://www.apltd.co.uk/).
Applied Scientific, 154 W Harris Avenue, South San Francisco, CA 94080, USA
Aviv Instruments Inc., Lakewood, New Jersey, USA (http://www.avivinst.com/).

Beckman Coulter Inc, 4300 N Harbor Boulevard, PO Box 3100, Fullerton, CA 92834-3100, USA
Tel: 001 714 871 4848   Fax: 001 714 773 8283   Web site: www.beckman.com
Beckman Coulter (U.K.) Limited, Oakley Court, Kingsmead Business Park, London Road, High Wycombe, Buckinghamshire, HP11 1JU
Tel: 01494 441181   Fax: 01494 447558   Web site: www.beckman.com
Becton Dickinson and Co., 21 Between Towns Road, Cowley, Oxford, OX4 3LY
Tel: 01865 748844   Fax: 01865 781627   Web site: www.bd.com
Becton Dickinson and Co., 1 Becton Drive, Franklin Lakes, NJ 07417-1883, USA
Tel: 001 201 847 6800   Web site: www.bd.com
Bel-Art Products
BioDiscovery, 11150 W Olympic Blvd. Ste. 805E, Los Angeles, CA 90064, USA
Bio 101 Inc., c/o Anachem Ltd, Anachem House, 20 Charles Street, Luton, Bedfordshire, LU2 0EB
Tel: 01582 456666   Fax: 01582 391768   Web site: www.anachem.co.uk
Bio 101 Inc., PO Box 2284, La Jolla, CA 92038-2284, USA
Tel: (+1) 760 598 7299   Fax: (+)1 760 598 0116   Web site: www.bio101.com

Biologic-Science Instruments SA, Claix, France (http://www.bio-logic.fr/).

Bio-Rad Laboratories Ltd., Bio-Rad House, Maylands Avenue, Hemel Hempstead, Hertfordshire, HP2 7TD

Tel: 0181 328 2000   Fax: 0181 328 2550   Web site: www.bio-rad.com

Bio-Rad Laboratories Ltd., Division Headquarters, 1000 Alfred Noble Drive, Hercules, CA 94547, USA

Tel: (+)1 510 724 7000   Fax: (+)1 510 741 5817   Web site: www.bio-rad.com

Calbiochem-Novabiochem Corporation, 10394 Pacific Center Court, San Diego, CA 92121 USA. Mailing Address: P. O. Box 12087, La Jolla, CA 92039–2087

Tel: 800 854 3417 (Calbiochem)/800 228 9622 (Novabiochem)/619 450 9600

Fax: 800 776 0999/619 450 9600

E-mail: orders@calbiochem.com   technical@calbiochem.com

Web Site: http://www.calbiochem.com

Cartesian Technologies Inc., 17781 Sky Park Circle, Irvine, CA 92614, USA

Center for Fluorescence Spectroscopy, University of Maryland School of Medicine, 108 North Greene Street, Baltimore, MD 21201 USA.

E-mail: cfs@sg.ab.umd.edu.   URL: http://charlie.ab.umd.edu/cfs/info.html.

Cortex Technology, Hadsund, Denmark.

CP Instrument Company Ltd, PO Box 22, Bishop Stortford, Hertfordshire, CM23 3DX

Tel: 01279 757711   Fax: 01279 755785   Web site: www.cpinstrument.co.uk

Current Designs Inc., 3527 Hamilton Street, Philadelphia, PA 19104–2420, USA.

Dupont (UK) Ltd, Industrial Products Division, Wedgwood Way, Stevenage, Herts, SG1 4QN

Tel: 01438 734000   Fax: 01438 734382   Web site: www.dupont.com

Dupont Co, (Biotechnology Systems Division), PO Box 80024, Wilmington, DE 19880-002, USA

Tel: (+)1 302 774 1000   Fax: (+)1 302 774 7321   Web site: www.dupont.com

Eastman Chemical Company, 100 North Eastman Road, PO Box 511, Kingsport, TN 37662-5075, USA

Tel: (+)1 423 229 2000   Web site: www.eastman.com

Edinburgh Instruments Ltd, Riccarton, Currie, Edinburgh EH14 4AP, UK.

E-mail: sales@edinst.com.   URL: http://www.edinst.com.

Fisher Scientific UK Ltd, Bishop Meadow Road, Loughborough, Leicestershire, LE11 5RG

Tel: 01509 231166   Fax: 01509 231893   Web site: www.fisher.co.uk

Fisher Scientific, Fisher Research, 2761 Walnut Avenue, Tustin, CA 92780, USA

E-mail: lfd@uiuc.edu.   URL: http://www.physics.uiuc.edu/groups/fluorescence/.

Tel: (+)1 714 669 4600   Fax: (+)1 714 669 1613   Web site: www.fishersci.com

Fluka, P.O. Box 2060, Milwaukee, WI 53201, USA

Tel: (+)1 414 273 5013   Fax: (+)1 414 2734979   Web site: www.sigma-aldrich.com

Fluka Chemical Company Ltd, PO Box 260, CH-9471, Buchs, Switzerland
Tel: (+) 41 81 745 2828   Fax: (+) 41 81 756 5449   Web site: www.sigma-aldrich.com

GeneMachines, PO Box 2048, Menlo Park, CA 94026, USA
General Scanning Inc., 500 Arsenal Street, Watertown, MA 02472, USA
Genetic Microsystems Inc., 34 Commerce Way, Woburn, MA 01801, USA
Genetix Ltd., Unit 1, 9 Airfield Road, Christchurch, Dorset BH23 3TG
Genomic Solutions Inc., 4355 Varsity Drive Ste. E, Ann Arbor, MI 48108, USA
Genome Systems Inc., 4633 World Parkway Circle, St. Louis, MO 63134, USA
Genosys Biotechnologies Inc., Lake Front Circle, Suite 185, The Woodlands, TX 77380, USA
Globals Unlimited, Laboratory for Fluorescence Dynamics, University of Illinois at Urbana-Champaign, Department of Physics, 1110 West Green Street, Urbana, IL 61801–3080 USA.

Hewlett-Packard, 3000 Hanover Street, Palo Alto, CA 94304–1185, USA
Web site: www.Hewlett-Packard.com
Hi-Tech Scientific Ltd, Brunel Rd, Salisbury, Wiltshire, SP2 7PU, U.K., (http://www.hi-techsci.co.uk/).
Hybaid Ltd, Action Court, Ashford Road, Ashford, Middlesex, TW15 1XB
Tel: 01784 425000   Fax: 01784 248085   Web site: www.hybaid.com
Hybaid US, 8 East Forge Parkway, Franklin, MA 02038, USA
Tel: (+)1 508 541 6918   Fax: (+)1 508 541 3041   Web site: www.hybaid.com
HyClone Laboratories, 1725 South HyClone Road, Logan, UT 84321, USA
Tel: (+)1 435 753 4584   Fax: (+)1 435 753 4589   Web site: www.hyclone.com

Imaging Research Inc., Brock Univesity, 500 Glenridge Avenue, St. Catherines, Ontario L2S 3A1, Canada
Instruments S.A., Inc. (SPEX), 3880 Park Avenue, Edison, NJ 08820 USA.
URL: http://www.instrumentssa.com/.(See their web site for international offices.)
Intelligent Automation Systems, 149 Sidney Street, Cambridge, MA 02139, USA
IVEE Development AB, Forsta Langgata 26, SE-413 28 Goteborg, Sweden
Invitrogen BV, PO Box 2312, 9704 CH Groningen, The Netherlands
Tel: (+)800 5345 5345   Fax: (+)800 7890 7890   Web site: www.invitrogen.com
Invitrogen Corporation, 1600 Faraday Avenue, Carlsbad, CA 92008, USA
Tel: (+)1 760 603 7200   Fax: (+)1 760 603 7201   Web site: www.invitrogen.com

JASCO Incorporated, 8649 Commerce Drive, Easton, Maryland 21601–9903, USA
Tel: (+)1 410 822 1220   Fax: (+)1 410 822 7526   E-mail: inbox@jascoinc.com
Web site: www.jascoinc.com
JASCO Europe s.r.l Via Confalinieri 25, 22060 Cremella (Co), Italy
Tel: 039 956439   Fax: 039 958642   E-mail: jascoeur@tin.it

Kartell, Via delle Industrie 1, 20082 Noviglio, Milano, Italy
Tel (+)39 02900121   Web site: www.kartell.it

Laboratoire pour l'Utilisation du Rayonnement Electromagnetic (LURE), Bât 209D Centre Universitaire Paris-Sud, B.P. 34 - 91898 Orsay Cedex, France. URL: http://www.lure.u-psud.fr/www/bienvenue.htm.

Life Technologies Ltd, PO Box 35, Free Fountain Drive, Incsinnan Business Park, Paisley, PA4 9RF

Tel: 0800 269210   Fax: 0800 838380   Web site: www.lifetech.com

Life Technologies Inc, 9800 Medical Center Drive, Rockville, MD 20850, USA

Tel: (+)1 301 610 8000   Web site: www.lifetech.com

Merck Sharp & Dohme Research Laboratories, Neuroscience Research Centre, Terlings Park, Harlow, Essex CM20 2QR

Web site: www.msd-nrc.co.uk

Microflex, PO Box 1865, San Francisco, CA 94083-1865, USA

Millipore (UK) Ltd, The Boulevard, Blackmoor Lane, Watford, Hertfordshire, WD1 8YW

Tel: 01923 816375   Fax: 01923 818297

Web site: www.millipore.com/local/UK.htm

Millipore Corporation, 80 Ashby Road, Bedford, MA 01730, USA

Tel: (+)1 800 645 5476   Fax: (+)1 800 645 5439   Web site: www.millipore.com

Molecular Devices, Unit 6, Raleigh Court, Rutherford Way, Crawley, RH10 2PD, UK; 1311 Orleans Drive, Sunnyvale, CA 94089, USA

Molecular Dynamics, 928 East Arques Avenue, Sunnyvale, CA 94086, USA

Molecular Probes, Inc., 4849 Pitchford Avenue, PO Box 22010, Eugene OR 97402-9165, USA

MSD Sharp and Dohme GmbH, Lindenplatz 1, D-85540, Haar, Germany

Web site: www.msd-deutschland.com

National Synchrotron Light Source, Brookhaven National Laboratory, P.O. Box 5000 Upton NY 11973–5000 USA. URL: http://www.nsls.bnl.gov/.

New England Biolabs, 32 Tozer Road, Beverley, MA 01915-5510, USA

Tel: 001 978 927 5054

Nikon Corporation, Fuji Building, 2-3, 3-chome, Marunouchi, Chiyoda-ku, Tokyo 100, Japan

Tel: (+) 813 3214 5311   Fax: (+) 813 3201 5856

Web site: www.nikon.co.jp/main/index_e.htm

Nikon Inc, 1300 Walt Whitman Road, Melville, NY 11747-3064, USA

Tel: (+)1 516 547 4200   Fax: (+)1 516 547 0299   Web site: www.nikonusa.com

Nycomed Amersham plc, Amersham Place, Little Chalfont, Buckinghamshire, HP7 9NA

Tel: 01494 544000   Fax: 01494 542266   Web site: www.amersham.co.uk

Nycomed Amersham, 101 Carnegie Center, Princeton, NJ 08540, USA

Tel: (+)1 609 514 6000   Web site: www.amersham.co.uk

Olis (On-line Instrument Services), Inc., 130 Conway Drive, Suites A and B, Bogart, GA 30622–1724, USA (http://www.olisweb.com/).

Omega Optical Inc., P.O. Box 573, Brattleboro, VT, 05302 – 0573, USA.
Email: catalogsales@omegafilters.com.

Perkin-Elmer Corporation, 761 Main Avenue, Norwalk, CT 06859 USA.
URL: http://www.perkin-elmer.com/.
Perkin Elmer Ltd, Post Office Lane, Beaconsfield, Buckinghamshire, HP9 1QA
Tel: 01494 676161   Web site: www.perkin-elmer.com
Pharmacia Biotech (Biochrom) Ltd, Unit 22, Cambridge Science Park, Milton Rd,
Cambridge, Cambs, CB4 0FJ
Tel: 01223 423723   Fax: 01223 420164   Web site: www.biochrom.co.uk
Pharmacia and Upjohn Ltd, Davy Avenue, Knowlhill, Milton Keynes, Bucking-
hamshire, MK5 8PH
Tel: 01908 661101   Fax: 01908 690091   Web site: www.eu.pnu.com
Photon Technology International, Inc. (PTI), 1 Deerpark Drive, Suite F,
Monmouth Junction, NJ 08852 USA.
E-mail: dsmith@pti-nj.com.
URL: http://www.eurotek.com.pl/ptiwebsite/welcome.html (see their web site for
international offices).
Promega Corporation, 2800 Woods Hollow Road, Madison, WI 53711-5399, USA
Tel: (+)1 608 274 4330   Fax: (+)1 608 277 2516   Web site: www.promega.com
Promega UK Ltd, Delta House, Chilworth Research Centre, Southampton, SO16 7NS
Tel: 0800 378994   Fax: 0800 181037   Web site: www.promega.com

Qiagen UK Ltd, Boundary Court, Gatwick Road, Crawley, West Sussex, RH10 2AX
Tel: 01293 422911   Fax: 01293 422922   Web site: www.qiagen.com
Qiagen Inc, 28159 Avenue Stanford, Valencia, CA 91355, USA
Tel: (+)1 800 426 8157   Fax: (+)1 800 718 2056   Web site: www.qiagen.com
Quantum Northwest, 9723 W. Sunset Highway, Spokane, WA 99224–9426 USA.
E-mail: qnw@qnw.com. URL: www.qnw.com.

Research Genetics Inc., 2130 Memorial Pkwy SW, Huntsville, AL 35801, USA
Research Instrumentation Shop, University of Pennsylvania School of Medicine,
79E John Morgan Building, 3620 Hamilton Walk, Philadelphia, PA 19104–6059, USA
Roche Diagnostics Ltd, Bell Lane, Lewes, East Sussex, BN7 1LG
Tel: 01273 484644   Fax: 01273 480266   Web site: www.roche.com
Roche Diagnostics Corporation, 9115 Hague Road, PO Box 50457, Indianapolis, IN 46256,
USA
Tel: 001 317 845 2358   Fax: 001 317 576 2126   Web site: www.roche.com
Roche Diagnostics GmbH, Sandhoferstrasse 116, 68305 Mannheim, Germany
Tel: 0049 621 759 4747   Fax: 0049 621 759 4002   Web site: www.roche.com

Schleicher and Schuell Inc, Keene, NH 03431A, USA
Tel: 001 603 357 2398
Shandon Scientific Ltd, 93-96 Chadwick Road, Astmoor, Runcorn, Cheshire, WA7
1PR

Tel: 01928 566611 Web site: www.shandon.com

Shimadzu, 1, Nishinokyo, Kuwabaracho, Nakagyou-ku, Kyoto 604 8511, Japan

Tel: 81 (75) 823 1111   Fax: 81 (75) 823 1361   Web site: www.shimadzu.co.jp

Sigma-Aldrich Company Ltd, Fancy Road, Poole, Dorset, BH12 4QH

Tel: 01202 733114   Fax: 01202 715460   Web site: www.sigma-aldrich.com

Sigma Chemical Company, PO Box 14508, St Louis, MO 63178, USA

Tel: (+)1 314 771 5765   Fax: (+)1 314 771 5757   Web site: www.sigma-aldrich.com

Spectronic Instruments, Inc., 820 Linden Avenue, Rochester, NY 14625 USA.

E-Mail: info@spectronic.com.   URL: http://www.spectronic.com/ (see their web site for international offices).

Stanford Research Systems, 1290-D Reamwood Avenue, Sunnyvale, California 94089 USA.

Email: info@srsys.com.   URL: www.srsys.com/.

Stratagene Europe, Gebouw California, Hogehilweg 15, 1101 CB Amsterdam Zuidoost, The Netherlands

Tel: 00 800 9100 9100   Web site: www.stratagene.com

Stratagene Inc, 11011 North Torrey Pines Road, La Jolla, CA 92037, USA

Tel: (+)1 858 535 5400   Web site: www.stratagene.com

Starna

Tel: (+)44 (0) 181 501 5550   Web site: www.starna.com

Synchrotron Radiation Source, Warrington, Cheshire, England.

Email: srs-ulo@dl.ac.uk.   URL: http://www.dl.ac.uk/SRS/.

Synteni (Incyte Pharmaceuticals), 6519 Dumbarton Circle, Fremont, CA 94555, USA

Tel-Test, Inc., 1511 County Road 129 Po Box 1421, Friendswood, TX 77546, USA

United States Biochemical, PO Box 22400, Cleveland, OH 44122, USA

Tel: 001 216 464 9277

Varian (Head Office), Varian Australia Pty Ltd, 679 Springvale Road, Mulgrave, Victoria 3170

Tel: (+)61 1300 658 274   Fax: (+)61 1300 658 274   Web site: www.varianinc.com

Vysis Inc., 3100 Woodcreek Drive, Downers Grove, IL 60515, USA

Whatman International Ltd., Maidstone, Kent, UK

Index